T0257812

Encyclopedia of Insecticides: Pest Management

Volume V

Encyclopedia of Insecticides: Pest Management
Volume V

Edited by **Nancy Cahoy**

New York

Published by Callisto Reference,
106 Park Avenue, Suite 200,
New York, NY 10016, USA
www.callistoreference.com

Encyclopedia of Insecticides: Pest Management
Volume V
Edited by Nancy Cahoy

International Standard Book Number: 978-1-63239-266-4 (Hardback)

Printed in the United States of America.

Contents

Preface

This book provides comprehensive information regarding insecticides and their applications in the field of pest management. The first section on pest management describes bioactive natural products from sapindacea, management of potato pests, flower thrips, mango mealy bug, pear psylla, grapes pests, small fruit production, boll weevil and tsetse fly using insecticides. Second section on toxicological profile of insecticides provides information on insecticide resistance in natural population of malaria vector, role of Anopheles gambiae P450 cytochrome, genetic toxicological profile of carbofuran and pirimicarb carbamic insecticides, etc. The subject matter in this book will attract the reader's concern to support rational decisions regarding the use of pesticides.

After months of intensive research and writing, this book is the end result of all who devoted their time and efforts in the initiation and progress of this book. It will surely be a source of reference in enhancing the required knowledge of the new developments in the area. During the course of developing this book, certain measures such as accuracy, authenticity and research focused analytical studies were given preference in order to produce a comprehensive book in the area of study.

This book would not have been possible without the efforts of the authors and the publisher. I extend my sincere thanks to them. Secondly, I express my gratitude to my family and well-wishers. And most importantly, I thank my students for constantly expressing their willingness and curiosity in enhancing their knowledge in the field, which encourages me to take up further research projects for the advancement of the area.

Editor

Part 1

Pest Management

Pest Management Strategies for Potato Insect Pests in the Pacific Northwest of the United States

Silvia I. Rondon
Oregon State University, Hermiston Agricultural Research and
Extension Center, Hermiston, OR,
USA

1. Introduction

This publication addresses pest management guidelines for insects that attack potatoes in the Pacific Northwest of the United States (Fig. 1). Non-chemical control options are strongly encouraged; however, proper use of insecticides has proven effective when used as an additional tool in integrated pest management programs. In general terms, Integrated Pest Management (IPM) is defined as a comprehensive approach of pest control (the term pest includes insects, weeds, diseases) that when combined, reduces the number of pest densities to a level tolerable by the crop. Traditional management practices include the use of host-plant resistance, cultural, mechanical, biological and chemical means of control. Overall, scientific research-based knowledge is required to implement a successful and functional management program.

Fig. 1. The Pacific Northwest includes Washington and Oregon in the U.S. and the Canadian province of British Columbia. Geographical limits are indicated: California (south), the Pacific Ocean (west), and Idaho (east).

Insects, both as pests and beneficials, have been studied by entomologists for a very long time. Of the millions of insects on the planet, only a fraction causes problems for crops and humans. Specific to this discussion, fewer than a dozen species cause most of the insect damage to potato production in the Pacific Northwest. Many of these pests infest at a specific stage of crop growth (e.g., seed corn maggot damage is more severe at planting), while others are permanent residents before planting, at planting and during crop development (e.g., wireworms) (Fig. 2). A number of pests are important to all western U.S. regions such as aphids, especially green peach aphid and potato aphid, both important vectors of leafroll virus, and more recently PVY. Other pests are important in specific regions such as beet leafhoppers in northeastern Oregon and southeastern Washington. This chapter will briefly provide a general profile of the potato crop in the region and will also discuss the most important insect pests affecting potato production. A general description of the pest, biology, ecology, monitoring and control will be provided. No specific recommendations are made regarding insecticide compounds. To obtain more information refer to the most current version of the Pacific Northwest Insect Control Handbook located at http://uspest.org/pnw/insects or consult your local extension office.

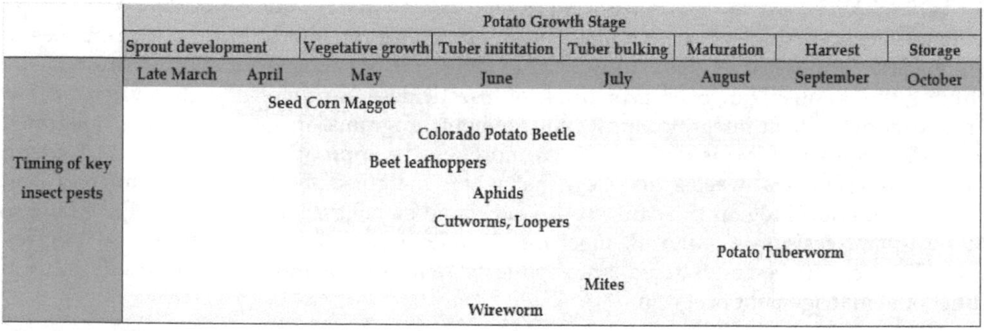

	Potato Growth Stage						
	Sprout development	Vegetative growth	Tuber inititation	Tuber bulking	Maturation	Harvest	Storage
	Late March April	May	June	July	August	September	October
	Seed Corn Maggot						
		Colorado Potato Beetle					
Timing of key insect pests		Beet leafhoppers					
			Aphids				
		Cutworms, Loopers					
						Potato Tuberworm	
				Mites			
			Wireworm				

Fig. 2. Occurrence of potato insect pests in the Pacific Northwest of the United States.

2. Profile of the potato crop

The potato is one of the world's most important food crops, domesticated approximately 8,000-10,000 years ago by native Peruvians in South America (Hawkes, 1990; Ministry of Agriculture, 2008). The potato arrived in Europe in the late 1500s (Salaman, 1985) and since then potatoes have been widely cultivated in Europe, Africa, Asia/Oceania and North America. The top ten potato producers worldwide are China, Russia, India, United States, Ukraine, Poland, Germany, Belarus, Netherlands, and France (http://www.potato2008.org/en/world/index.html; source http://faostat.fao.org/default .aspx).

In the U.S., all 50 states cultivate potatoes; however, including Idaho, the Pacific Northwest produces nearly two thirds of the potatoes grown in North America. Idaho is consistently ranked first in potato production and acreage (28% of total U.S. production), followed by Washington State (23% of total U.S. production). Other top U.S. potato producing areas are Oregon, California, Montana, Colorado, North Dakota, Michigan, Wisconsin, and Maine (Fig. 3) (http://www.agcensus.usda.gov/index.asp). Florida and Texas can be included as leading farm cash receipts (http://usda.mannlib.cornell.edu/MannUsda/view DocumentInfo.do?documentID=1235).

Fig. 3. Top potato producing regions in the U.S.

In the Pacific Northwest, potatoes are typically planted in the spring and harvested in late summer and fall. The growing season in most areas ranges from 120 to 140 days although high elevation areas have a 90 to 100 day season. In the Columbia Basin of Washington and Oregon, one of the most prolific regions in North America, potato production ranges between 160 to 180 days (University of California, 1986). Potatoes are produced as annual plants that grow from specialized underground stems called stolons later known as tubers (Gopal & Khurana, 2006). This vegetative propagation through the planting of tubers or part of tubers (seed pieces) is a common practice in the region (University of California, 1986; Wale et al., 2008). "Russet Burbank" is still a common variety in the region, followed by other Russet varieties such as Ranger Russet and Russet Norkotah (Table 1). Other varieties are available to producers in the region; many originated from the Tri-State potato breeding program (http://potatoes.wsu.edu/varieties/key.html). The Tri-State Program was created in the mid 1980's to intertwine researchers, scientists and stakeholders in the Pacific Northwest to promote and develop new potato varieties.

Thirty-seven Tri State releases has been released since the mid 1980's including the popular "Premier Russet", Western Russet" (both dual purposes fry and fresh packs), "Blazer Russet", "Umatilla Russet", and "Ranger Russet" (processing for fries), and "Alturas" (dehydration/processing) (Potato Variety Management Institute http://www.pvmi.org/varieties/varieties.htm).

Varieties	Year										
	1997	1998	1999	2000	2001	2002	2003	2004	2005	2006	2007
Russet Burbank	80	78	74	75	71	71	69	63	63	66	62
Russet Norkotah	5	5	8	8	8	8	10	14	12	10	10
Ranger Russet	7	6	4	4	4	3	1	2	1	-	1
Shepody	4	7	9	8	11	12	13	13	15	13	14
Umatilla	-	-	-	1	-	-	-	-	-	-	2
Alturas	-	-	-	-	-	-	1	3	3	2	2
Others	4	5	4	4	6	6	5	5	6	9	9

Table 1. Popular potato varieties (%) planted in Idaho, Washington and Oregon (Washington Potato Commission, 2007; National Agricultural Statistics Service, http://www.agcensus.usda.gov/index.asp).

After producers determine the nutritional needs of the crop (nitrogen, phosphorous, potassium), seed potatoes are cut uniformly, treated with fungicides and/or insecticides, and planted. Potatoes emerge 3-5 weeks after planting, depending on weather conditions, location and time of the year. After plant emergence, producers monitor crop needs regarding fertilization, pest control, and general management. Most potatoes in the area are grown under irrigation. Another common practice is desiccation or vine-killing (Hutchinson & Stall, 2007). Timely vine killing is essential for good tuber skin set, and efficient harvest. It is not clear if desiccation (naturally or artificially) of early season potatoes influences pest pressure in other nearby fields awaiting harvest. Johnson (2008) suggests a holistic plan for potato management combining management practices in the years prior to growing the potato crop, during the growing season, at harvest and during tuber storage. Miller and Hopkins (2008) specified that among many things, establishing appropriate long-term crop rotation, applying and alternating pesticides with different mode of action to avoid pesticide resistance, and improving soil microbial, physical and chemical health are necessary to achieve a healthy crop.

The potato plant develops through four clearly defined growth stages: (1) vegetative growth, (2) tuber initiation, (3) tuber growth and (4) plant maturation (Johnson, 2008). Each stage is affected by different groups of insect pests. The degree of the damage will depend on the timing of events, cultivar characteristics, and the intrinsic characteristics of each pest.

3. Managing insect pests in potatoes

The first step for managing insect pests is the timely detection of pest infestation. The goal is to manage the crop as a whole system, keeping pests at acceptable levels by utilizing several harmonizing strategies (Johnson, 2008). Earlier in this chapter, IPM was defined as a comprehensive pest control. Heitefuss (1989) defines IPM as a system in which all economically, ecologically and toxicologically suitable procedures are utilized in maximum harmony to maintain pests below the economic threshold. In the late 1990's, Luckmann and Metcalf (1994) defined IPM as an intelligent selection and use of pest control actions (or tactics) that will ensure favorable economic, ecological, and sociological consequences. IPM techniques include monitoring of pest populations, the judicious use of pesticides, and the effective communication regarding the necessity of implementing a control tactic or not. In a modern definition, the goal is to prevent and suppress pests with minimum impact on human health, the environment and non-target organisms (Dreistadt, 2004). According to Pedigo (1986), in order to use an IPM approach, an understanding of the following factors is required: (1) the biology of the pest and its natural enemies, (2) the response of crops to management practices, (3) the effect of pesticide application on pests and non-target organisms, and (4) the action threshold or level of damage tolerable by the plant.

3.1 Monitoring

Timing detection and monitoring is an essential long-term requirement against pest infestation in potatoes. Plants should be regularly checked for signs and symptoms of pest damage. During the height of the growing season, plants should be checked once or twice per week. The inspection of the undersides of leaves and the inner plant canopy is recommended since many pests prefer sheltered sites. A close look of any plant that has missing, absent or damaged leaves or flowers, or plants whose color, texture, or size looks different than healthy ones, can be a signs of a problem.

Some insect pests can be dislodged by laying a sheet, sturdy cardboard or paper below the infested plants and "beating" the canopy. Insects can be monitored with a hand lens or traps. Both techniques not only allow detection and monitoring of pest problems but also provide estimates of pest population density. The optimal timing of sampling depends upon the life history and behavior patterns of the pest and/or beneficial insects, as well as crop stage and state, and also environmental conditions. Some areas, such as field edges or fields next to main roads are more prone to pest problems (personal observations).

Different sampling procedures can be used depending on the crop, size of field, etc. In the Pacific Northwest, the majority of potatoes are planted in circles, under center-pivot irrigation, thus fields can be divided in quarters (e.g, north, east, west, south). Upon entering the field, quick visual examination of the field is recommended. Uncharacteristic areas with poor stands or patchy growth should be scouted thoroughly. In the Pacific North-western region traps are widely used. Some traps can be baited with a pheromone which is a chemical that usually attracts a single species (e.g., potato tuber moths). Traps can also be coated with adhesive material to "stick" the pest to the trap. Either natural (e.g., virgin females) or more often the synthetic pheromones are used to attract males (Roelofs et al., 1975; Persoons et al,. 1976; Voerman & Rothschild, 1978; Rothschild, 1986; Raman, 1988). Pheromones specifically disrupt the reproductive cycle of harmful insects. Pheromone traps are used extensively in commercial agriculture in the region helping farmers detect the presence of pest species (Merrill et al., 2011). Traps in general will work only for adult insects, as adults have developed wings and are more mobile. Yellow sticky traps attract fruit flies, winged aphids, thrips, psyllids, fungus gnats, wasps, numerous flies; many species collected in sticky cards belong to the same family or group (e.g., several moths in the Gelechiidae family can be found in traps meant to be for tuber moths, a member of the Gelechiidae family), thus correct identification of target pest(s) is needed.

3.2 Management

Before planting, selecting field location can minimize pest damage and improve predictability of pest problems (Hoy et al., 2008). Also circumventing planting dates when insects will emerge and inflict most damage is recommended. Certified seed should be used at all times. If available, the selection of cultivars resistant or tolerant to important pests is also desirable.

3.2.1 Seed quality and certification

High quality seed is essential for the production of a profitable crop (Love et al., 2003). Several diseases, some transmitted by insects and especially aphids, can be transmitted to infected seed such as viruses, bacterial ring rot, blackleg, late blight, scab, and wilt diseases (University of California, 1986). Disease-free seed tubers are available from certified nurseries where seed potatoes are grown for several generations before being commercially offered to producers. This process is regulated by each state's seed certification programs. The use of appropriate production practices is only half of the process of growing high quality potatoes (Love t al. 2003). Samples from seed growing areas are grown in field trial to establish disease problems. There is a range of tolerance from zero to 6% for certain viruses (University of California, 1986). For instance, Idaho inspection tolerance for post-harvest winter tests of seed destined for recertification range from 0.8% for leafroll, 2.0% mosaic, and 5% chemical injury (Love et al., 2003). Thus, (1) clean all equipment thoroughly before entering fields, (2) establish aphid control programs to prevent the development of aphid populations in potato seed fields, (3) control aphids in potato fields to prevent the movement of aphids to seed fields.

3.2.2 Resistant cultivars

Cultivars tolerant or resistant to pests can provide long-term protection for the potato crop. For a long time it has been recognized that the evaluation of potato germplasm for resistance is a valuable tool to developing IPM programs (Horgan et al., 2007). Although no commercial potato cultivar is completely resistant to insect damage, there are variations in the susceptibility of cultivars to the pest (Hoy et al., 2008). For example, tubers of the transgenic clone Spunta G2 are resistant to the potato tuberworm (Douches et al., 2002). Rondon et al., (2009) tested several cultivars including Spunta G2 with excellent results. However the public perception regarding genetically-modified organisms is still "blocking" the widely use of this type of resources.

3.2.3 Cultural practices

Proper management of field preparation, planting, harvesting and storage are essential for maximum yield and tuber quality (University of California, 1986). Pest infestation can be minimized by avoiding contaminated seed tubers, soil, and water, sanitation of machinery brought from infested areas, removing cull piles, and following rational fertilizer and irrigation programs. Crop rotation, which reduces certain pests by breaking their life cycle, should be an integral part of a holistic management program.

3.2.4 Biological control

In a sustainable ecosystem, insect pest populations may be kept in check by natural enemies such as other insects. Parasitoids, predators, pathogens, antagonists, or competitor populations that suppress pest populations are desirable in fields (Driesche & Bellows, 1996). Under current pest management programs in potatoes, especially with an intensive agricultural production system centered on frequent calendar sprays of broad-spectrum insecticides, the impact of natural enemies is relatively unknown (Koss, 2003; Rondon, 2010). In contrast, a lot of information regarding the biology and the potential of natural enemies (a.k.a., biological control agents) can be found in the literature (Rondon, 2010). The advantage of using biological control agents is that they have no pre-harvest intervals, and are safer for application personnel, consumers and non-target organisms.

In the Pacific Northwest region, a number of arthropod species attack insect pests at the egg, immature, and/or adult stage. Species of Heteroptera (former Hemiptera, Family Pentatomidae) such as *Podisus maculiventris* (Say) (spined soldier bug) and the *Perillus bioculatus* (Fabricius) (two-spotted stink bug), *Opolomus dichrosus* L. (no common name) can be found in potato fields feeding on *Leptinotarsa decemlineata* (Say) (Colorado potato beetle) (Ferro, 1994; Lacey et al., 2001). Tamaki et al., (1983) and Lopez et al., (1995) indicated that *Myiopharus doryphorae* L. (Tachinidae fly) and *Edovum puttleri* L. (Eulophidae wasp) are moderately efficient parasitoids of Colorado potato beetles. Nabidae (damsel bugs), Neuroptera (lace wings), Coccinellidae (lady bugs), Carabidae (ground beetles) and spiders are also common inhabitants in potato fields behaving as generalists feeding on aphids, thrips, small larvae and potato beetles.

3.2.5 Chemical methods
Traditional IPM text books recommend considering chemical controls only if other techniques do not result in adequate pest control. However, chemical controls can be effectively used with other techniques (e.g., cultural, physical, biological, etc.). When choosing a chemical, be sure to read the label and choose the right product; chemical selection can be the breaking point for success or failure in controlling target pests. Worldwide, the potato requires more pesticides than any other major food crop (CIP, 1994). Intense pesticide dependency has led to the development of pesticide resistance in cases such as the Colorado potato beetles in almost all potato-growing areas in the U.S. (Weisz et al.,1995). To date, the only exception of documented Colorado potato resistance to chemicals is in the Pacific Northwest (Schreiber et al., 2010). Management tactics that preserve susceptible genotypes and allow them to interbreed with resistant individuals have the greatest potential for resistance management according to Tabashnik (1989) and Weisz et al., (1996).

4. Potato pests in the Pacific Northwest

4.1 Wireworms (Order Coleoptera: Family Elateridae)
Wireworms are one of the most destructive insect pests in the Pacific Northwest. Nearly 40 species from 12 genera attack potato, but only a few are economically important (Hoy et al., 2008). In irrigated production, the most common wireworm species are the Pacific Coast wireworm *(Limonius canus* LeConte), the sugar beet wireworm *(L. californicus* (Mannerheim)), the western field wireworm (*L. infuscatus* Motschulsky*)*, and the Columbia Basin wireworm (*L. subauratus* LeConte). Areas with annual rainfall less than 15 inches may be infested with the Great Basin wireworm (*Ctenicera pruinina* (Horn)). West of the Cascade mountains *Agriotes* spp. is the most common pest. A complex of species may occur (Figure 4) (Andrew et al., 2008).

4.1.1 Pest description
Wireworms are the larval stage of click beetles. Adult click beetles are slender hard-shelled insects. They range in color from chocolate to dark brown and from about 0.8 - 1.9 cm long, depending on species. Click beetles get their name from their ability to snap a spine on their thorax that produces a "clicking" sound and allows them to jump in the air when distressed or disturbed. All beetles in the Elateridae family have this ability. This technique is used to avoid predation or to get back on their feet after falling on their backs. Depending on the

Fig. 4. Wireworms in the Pacific Northwest, from left: *Limonius canus*, *L. californicus*, *L. infuscatus*, *L. subauratus*, *Ctenicera pruinina*, and *Agriotes lineatus* (Photo by C.J. Marshall, Oregon State Arthropod Collection http://osac.science.oregonstate.edu).

species, each female after mating lays an average of 80 eggs, singly or in small clusters in the soil. Immature stages have a hardened and shiny shell and very few hairs (Fig. 5). They have three body regions with a distinct head, a thorax with 3 pairs of legs and a segmented abdomen with a tail-end (Jensen et al., 2011). Characteristics of the tail-end serve for identification purposes. Depending on species and age, wireworm larvae range from about 2 mm after hatching to 4 cm long or more at maturity. Wireworm pupae are first white, but later change to reddish-brown; they pupae in the soil.

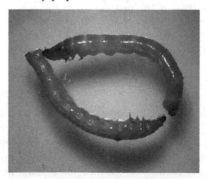

Fig. 5. Wireworm larva (Photo by A. Jensen, Washington Potato Commission).

4.1.2 Damage

Wireworms can cause damage to potatoes by feeding upon potato seed pieces and sprouts in the spring, facilitating infection by pathogens or other insect pests. The latter damage can result in reduction in yield and/or rejection of the entire crop. In the U.S. there is zero tolerance for live larvae in tubers. Wireworms tend to be most damaging in potatoes that follow corn or small grains (wheat, barley) and on ground just entering cultivation. Wireworms damage potatoes both near planting time (damage to seed pieces) and during the growing season (damage to developing tubers). They can also be a problem at harvest and before entering storage.

4.1.3 Hosts

Potatoes, corn, wheat and grass are hosts for several species of wireworms in the Pacific Northwest. Also, beans, carrots, peas, and other annual crops may be infested; while melons, beet roots, and strawberry fruits are affected less frequently.

4.1.4 Biology

Adult click beetles emerge from pupae in the soil from late spring through late summer. In the Pacific Northwest, wireworms overwinter as larvae or adults. They can have up to a 7-year life cycle from egg to adult. Adults can fly, but usually remain in the areas where they developed as larvae. Eggs are laid in grassland or in cereal crops where larvae feed on grass roots and overwinter in the soil. Females tend to lay eggs in grassy areas. Larvae can live from 2-5 years in the soil, depending on the species. They require several years to mature and can overwinter at a depth of 30-100 cm or more in the soil, only to return near the surface in spring to resume feeding when soil temperatures exceed 50°F (10°C). Later in the season when temperatures reach 80°F (26.6°C) and above, the larvae tend to move deeper than 15 cm into the soil to escape the "heat" (Hoy et al., 2008; Schreiber et al., 2010). Wireworm larva can be confused with larvae of the family Tenebrionidae (false wireworms) or Tipulidae (crane fly). The Tenebrionidae group tends to be saprophagous while Tipulidae are associated with grassy crops.

4.1.5 Monitoring

Trapping should start early, especially in areas with history of wireworm problems. In the Pacific Northwest trapping starts mid to late March until April to May. Horton (2006) modeled the relationship between bait trap counts and crop damage by *L. canus* in Wapato, WA. Horton's model predicts tuber damage based on number of wireworms collected. Wireworm presence or absence in a field should be determined before using control measures. Unfortunately, current monitoring methods are time consuming, laborious and often do not accurately reflect field populations of this pest. Historically, wireworms have been monitored by extracting and sifting through soil cores to locate larvae. Since the distribution of wireworms in a field tends to be patchy and unpredictable, large numbers of samples are required to accurately estimate population size. Baits have largely replaced random soil sampling, since they are less labor intensive and may detect low wireworm populations. Baited traps can be constructed by placing 3-4 tablespoons of a mixed of wheat and corn seeds or rolled oats inside a fine mesh bag or nylon. The seed mixture should be soaked in water 24 hours prior to placement in the hole to facilitate germination. Dig a hole about 20-25 cm deep and 3.5- 4 cm wide at the soil surface (Horton, 2006). Bury the mixture at the bottom of the hole. Fill the hole and mound a "soil dome" over the covered bait to serve as a solar collector and to prevent standing water. Cover each mound with a sheet of black plastic and cover the edges with soil to hold the plastic sheet down. The plastic collects solar radiation and speeds germination of the mixture. The germinating seeds attract wireworms. A few days later, remove the plastic and soil covering the bait and count the number of wireworm larvae found at each station. There are not specific recommendations as to how many traps per field should be placed, however, placement of the bait stations should represent different areas of a field (Campbell & Stone, 1939; Simmons et al., 1998)

4.1.6 Control
There are no effective natural enemies for wireworm. If one suspects wireworms are present in a field based on trapping, chemical control is the best management option (http://potatoes.com/Research-IPM.cfm). Fumigants are effective on wireworms that are present at the time of fumigation and within the zone of fumigation (Schreiber et al., 2010). Fumigants are sensitive to soil temperatures. In furrow applications are also effective; however, some rotational restrictions may apply (Schreiber et al., 2010). Use of contemporary chemicals in other crops suggests that stand protection and wireworm reduction are not covered with current chemicals available (Vernon et al., 2009).

4.2 Colorado Potato Beetle (Order Coleoptera: Family Chrysomelidae)
The Colorado potato beetle, *Leptinotarsa decemlineata* (Say), first described in 1824 by Thomas Say, is associated with potato plants and its solanaceous relatives such as nightshade. It is the most important defoliating insect pest of potato. Its remarkable ability to develop insecticide resistance, incredible reproductive potential and sustained feeding by larvae and adults, makes the management of this pest challenging (Hoy et al., 2008).

4.2.1 Pest description
The Colorado Potato Beetle (CPB) is a yellow and black striped beetle, about 1.3 cm long and 0.6 cm wide. They can be found in almost all U.S. potato regions. Larvae are reddish orange, with two rows of black spots on each side. Orange egg clusters are found mainly on the undersides of leaves, mostly in the top third of the plant. Eggs resemble ladybug eggs (Fig. 6).

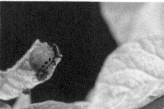

Fig. 6. Colorado potato beetle egg mass (left), larva (center), and adult (right). Photos by R. Marchosky (egg mass), L. Ketchum (larva), and S.I. Rondon (adult), OSU.

4.2.2 Damage
This beetle can cause complete defoliation and nearly complete crop loss if allowed to reproduce unchecked. Both larvae and adults feed on potato foliage throughout the season.

4.2.3 Hosts
Potatoes and other solanaceous plants such as eggplant, nightshade, horsenettle and buffalobur are preferred hosts of this pest.

4.2.4 Biology
Pupation and overwintering occur in the soil. Adults emerge from the soil to lay eggs in the spring. Depending on the region, this insect may have three generations in a season. Adult beetles spend the winter buried 10-25 cm in the soil and emerge in the spring just as the first volunteer potatoes appear. Recently emerged beetles either mate close to the overwintering sites or fly to new potato fields to find a mate. Usually first infestations occur around field margins. Eggs are deposited on potato foliage in masses. CPB eggs resemble lady beetle eggs. Larvae pass through four life stages and then burrow into the soil to pupate.

4.2.5 Monitoring
Start monitoring fields at crop emergence. There are no established treatment thresholds for CPB. Large CPB populations are harder to manage than small ones, thus the goal is to control this pest early in the season.

4.2.6 Control
Crop rotation may help in delaying or reducing CPB pressure. Colonizing beetles need to feed before laying eggs, so controlling volunteer potatoes and solanaceous weeds is important as are rotating crops and planting new potato fields far from the last year's potato fields (Schreiber et al., 2010). These practices will reduce the number of overwintering beetles migrating into the new field. This may not be a practical solution in the Pacific Northwest region since potatoes are use in rotation with other local crops such as wheat or corn. The use of "at planting" and systemic insecticides in early potatoes will contribute to the control of early-season CPB populations. The use of pyrethroid insecticides is not recommended since it has a direct effect on natural enemies. Targeting chemical applications to control eggs and young larvae when possible is recommended.

4.3 Green peach aphid and potato aphid (Order Heteroptera: Family Aphididae)
The aphid population in western North America, north of Mexico, is comprised of 1,020 species in 178 genera in 15 subfamilies (Pike et al., 2003). Several aphid species are known to be pests of potatoes, but the green peach aphid, *Myzus persicae* (Sulzer), and potato aphid, *Macrosiphum euphorbiae* (Thomas), are two of the most important vectors of diseases in the Pacific Northwest. Aphids are important due to their ability to transmit viruses. According to Hoy et al., (2008) there are six commonly found potato viruses transmitted by aphids: *Potato leafroll virus* (PLRV), multiple strains of *Potato virus Y* (PVY), *Potato virus A* (PVA), *Potato virus S* (PVS), *Potato virus M* (PVM), and *alfalfa mosaic virus* (AMV). PLRV and PVY are transmitted by several species of aphids but primarily by green peach aphid. The potato aphid transmits PVY and PVA.

4.3.1 Pest description

Green peach aphids are small, usually less than 0.3 cm long. The body varies in color from pink to green with three darker stripes down the back. The head has long antennae which have an inward pointing projection or tubercle at its base (Fig. 7). Potato aphids are larger than green peach aphids with a body somewhat elongated and wedge-shaped (Fig. 8). The adults of both species may be winged (alatae) or wingless (apterous). Winged forms are usually triggered by environmental changes (e.g., decreasing photoperiod or temperature, deterioration of the host plant or overcrowding) (Branson et al., 1966). On the back of the fifth abdominal segment, a pair of tube-like structures called "siphunculi", "cornicles", or "pipes" are present on most aphid species. The green peach aphid present a "swollen" cornicles with a dark tip, while the cornicles on the potato aphid are 1/3 of the length of the body and are usually curved slightly outward (Alvarez et al., 2003).

Fig. 7. Green peach aphid wingless adult (left) and alatae (right). Photos by A. Jensen, Washington Potato Commission.

Fig. 8. Potato aphid wingless adult and nymphs (left) and alatae (right). Photos by A. Jensen, Washington Potato Commission.

4.3.2 Damage

In general, aphids injure plants directly by removing sap juices from phloem tissues. They also reduce the aesthetic quality of infested plants by secreting a sugary liquid called "honeydew" on which a black-colored fungus called "sooty mold" grows. The "sooty mold" reduces the photosynthetic potential of the plant. Most importantly, aphids transmit plant diseases, particularly viruses. Aphids on potato are serious pests because of their ability to transmit several plant diseases such as PLRV (transmitted mainly by green peach aphid) and PVY (transmitted by several species of aphids). PLRV causes necrosis while strains of PVY can cause internal brown lesions in the tubers. Srinivasan & Alvarez (2007) reported that mixed viral infections of heterologous viruses occur regularly in potatoes.

4.3.3 Hosts

The green peach aphid, also known as tobacco or spinach aphid, survives the winter in the egg stage on peach trees. They can also overwinter on various perennial, biennial, and winter annual weeds, such as tumble mustard, flixweed, shepherd's-purse, chickweed, mallow, horseweed, pennycress and redstem filaree. Besides potatoes and peaches, other hosts include lettuce, spinach, tomatoes, other vegetables and ornamentals (Dickson & Laird, 1967; Wallis, 1967; Tamaki et al., 1980; Barry et al., 1982).

4.3.4 Biology

Green peach aphid migrates to potatoes in the spring from weeds and various crops where it has overwintered as nymphs and adults, or from peach and related trees where it overwinters as eggs. Most aphids reproduce sexually and develop through gradual metamorphosis (overwintering diapause egg, nymphs and winged or wingless adults) but also through a process called 'parthenogenesis' in which the production of offspring occurs without mating (Jensen et al., 2011). Potato aphids also overwinter as active nymphs, adults or eggs; eggs are laid on roses and sometimes other plants. Throughout the growing season aphids produce live young, all of which are female and can be either winged or wingless. In some instances, aphids undergo sexual, oviparous reproduction as a response of a change in photoperiod and temperature, or perhaps a lower food quantity or quality, where females produce sexual females and males. In the fall, winged males are produced which fly to overwintering hosts and mate with the egg-laying females produced on that host. Aphids found in the region undergo multiple overlapping generations per year (Jensen et al., 2011, Schreiber et al., 2010).

4.3.5 Monitoring

Fields should be checked for aphids at least once a week starting after emergence. The most effective scouting method is beating sheets, trays, buckets or white paper. There are no well-established treatment thresholds for aphids in potatoes in the Pacific Northwest but since aphids transmit viruses, producers are encouraged to control aphids early in the season, especially in seed potato producing areas. Schreiber et al., (2010) recommend a minimum sample size of ten locations per 100 acre field. For potatoes that are not to be stored, application of foliar aphidicide should begin when 5 aphids per 100 leaves or 5 aphids/plant are detected. Hoy et al., (2008) suggests some sampling methods and action thresholds for colonizing aphids on processing potatoes, table stock, and seed potato in different productions thresholds.

4.3.6 Control
Weed control and elimination of secondary hosts are critical. Early aphid infestations commonly occur on a number of weeds including species of mustards and nightshade; therefore, those weeds should be kept under control. Research in Idaho indicates that hairy nightshade is an excellent aphid and virus host (Srinivasan & Alvarez, 2007), thus, control of this weed is highly recommended. In some instances, the number of insects available to infest crops in the spring depends upon winter survival (DeBano et al., 2010). Thus, the elimination of overwintering sites is recommended if possible. Peach trees are the most common winter hosts, although apricots and several species of Prunus are sometimes infested (Schreiber et al., 2010). A large numbers of generalist predators feed on aphids including the minute pirate bugs, big-eyed bugs, damsel bugs, lady beetles and their larvae, lacewings, flower fly larvae, and aphid-specific parasitoid wasps. If aphids are present, use of insecticides in commercial fields should occur as soon as non-winged aphids are detected. In seed producing areas, preventive methods are recommended. Application of foliar aphidicide should begin just prior to the decline in performance of seed-treatment insecticides applied at planting (60 days after planting, Rondon unpublished). Schreiber et al., 2010 indicated that complete insect control from planting until aphid flights have ceased is the only means to manage diseases in full season potatoes.

4.4 Beet leafhopper (Order Heteroptera: Family Cicadellidae)
The beet leafhopper, *Circulifer tenellus* Baker, is the carrier of the beet leafhopper-transmitted virescence agent (BLTVA) phytoplasma (a.k.a., Columbia Basin potato purple top phytoplasma) that causes significant yield losses and a reduction in potato tuber quality.

4.4.1 Pest description
The beet leafhopper (BLH) is a wedge-shaped and pale green to gray or brown in color. It has several nymphal instars (Fig. 9). Adults may have dark markings on the upper surface of the body early and late in the season ("darker form") or clear during the season ("clear form") (Fig. 10).

Fig. 9. Beet leafhopper nymphs. Photos by A. Murphy, OSU.

4.4.2 Damage
Beet leafhoppers must feed in the phloem of the plant. Direct feeding can cause relatively minor damage ("hopperburn"); however, BLTVA is a very destructive and detrimental disease affecting potatoes. BLTVA can cause a wide range of symptoms in potatoes, including leaf curling and purpling, aerial tubers, chlorosis, and early senescence. Most BLTVA infection occurs early in the season, during May and June (Munyaneza, 2003; Munyaneza & Crosslin,

2006). Potato is not a preferred host for BLH and will not spend much time on the crop (however it does spend enough time to transmit BLTVA) (Schreiber et al., 2010).

4.4.3 Hosts
Among the favorite hosts are Kochia, Russian thistle, and various weedy mustard species such as tumble mustard. Beet leafhoppers are especially abundant on young, marginal, semi-dry and small weeds plants. They also thrive on radishes, sugar beet (Meyerdirk & Hessein, 1985), and carrots (Munyaneza, 2003).

4.4.4 Biology
The beet leafhopper overwinters on rangeland weeds and migrates to potatoes as early as May. They overwinter as adult females in weedy and native vegetation throughout most of the dry production areas. The beet leafhopper has three life stages: egg, nymph and adult. The adult can have a "darker form", early or late in the season; and a "clear form (during the season) (Fig. 10). Beet leafhoppers can transmit BLTVA as adults and nymphs. Eggs are laid in stems of host plants, and a new spring generation begins developing in March and April. Beet leafhopper begins to move from weeds to potatoes and potentially affect potatoes during the first spring generation, which matures in late May to early June (Jensen et al., 2011). Potatoes are most seriously affected by BLTVA infections that occur early in the growing season (Rondon unpublished). Beet leafhopper remains common through the summer, during which it goes through 2 to 3 overlapping generations. The final generation for the year matures during late October-early November. Total number of beet leafhoppers varies from year to year (Crosslin et al., 2011).

Fig. 10. "Clear form"(left) and "dark form" (right) of the beet leafhopper. Size of adults 2.5-3 mm. Photos by A. Jensen, Washington Potato Commission.

4.4.5 Monitoring
Because potatoes are not a preferred host of the BLH, in-field sampling is problematic. Most recommendations suggest the use of yellow sticky cards around field margins. It is important to keep traps close to the ground where hoppers mostly move. Check and replace

traps at least once a week. Rondon (unpublished data) suggests the use of DVAC (modified leaf blowers) to collect leafhoppers.

4.4.6 Control
Weed control in areas surrounding the potato field can help reduce initial sources of BLTVA inoculum. Due to the nature of the pest, few biological control efforts have been taking place in the Pacific Northwest. However, a species of *Anagrus* (Hymenoptera Mymaridae), has been reported as a common egg parasitoid in California (Meyerdirk & Moratorio, 1987).
Foliar insecticides can reduce BLH populations and ergo, the incidence of the disease. Based on extensive research conducted in the Pacific Northwest, there are several foliar applied insecticides that are effective against BLH. Some evidence suggests that the use of some neonicotinoid insecticides at planting may provide control of BLTVA (Schreiber et al., 2010).

4.5 Potato Tuberworm (Order Lepidoptera: Family Gelechiidae)
The potato tuberworm, *Phthorimaea operculella* Zeller, is one of the most economically significant insect pests of cultivated potatoes worldwide. The first significant economic damage to potato crops in the Columbia Basin region occurred in 2002, when a field in Oregon showed high levels of tuber damage associated with potato tuberworm. By 2003, the pest was a major concern to all producers in the region after potatoes from several fields were rejected by processors because of tuber damage. Since then, potato tuberworm has cost growers in the Columbia Basin millions of dollars through increased pesticide application and unmarketable potatoes (Rondon, 2010).

4.5.1 Pest description
The potato tuberworm has four life stages: adult, egg, larva and pupa. Adults are small moths (approximately 0.94 cm long) with a wingspan of 1.27 cm. Forewings have dark spots (2-3 dots on males; "X" on females). Both pairs of wings have fringed edges (Rondon & Xue, 2010) (Fig. 11). Eggs are ≤ 0.1 cm spherical, translucent, and range in color from white or yellowish to light brown. Eggs are laid on foliage, soil and plant debris, or exposed tubers (Rondon et al., 2007); however, foliage is the preferred oviposition substrate (Varela, 1988). Adult female moths lays 150-200 eggs on the underside of leaves, on stems, and in tubers (Hoy et al., 2008). Larvae are usually light brown with a characteristic brown head. Mature larvae (approximately 0.94 cm long) may have a pink or greenish color (Fig. 12). Larvae close to pupation drop from infested foliage to the ground and may burrow into the tuber. Ultimately, larvae will spin silk cocoons and pupate on the soil surface or in debris under the plant.

Fig. 11. Forewings of potato tuberworm adult females present an "x" pattern (left); while male (right) present 2-3 dark spots. Photos by OSU (Rondon 2010).

Fig. 12. Potato tuberworm larva entering tuber. Photo by L. Ketchum, OSU.

4.5.2 Damage

Tuberworm larvae behave as leaf miners. They can also live inside stems or within groups of leaves tied together with silk. The most important damage is to tubers, also a food source for the larvae, especially exposed tubers, or those within centimeters of the soil surface. Larvae can infest tubers when foliage is vine killed or desiccated right before harvest (Clough et al, 2010). Tunnels left by tuber worms in tubers can be full of droppings or excrement that can be a potential source for secondary infections.

4.5.3 Hosts

Although the potato tuberworm host range includes a wide array of Solanaceous crops such as tomatoes, peppers, eggplants, tobacco, and weeds such as nightshade, the pest has been found only on potatoes in the Pacific Northwest region (Rondon, 2010).

4.5.4 Biology

Potato tuberworm adults emerge as early as April in the Pacific Northwest, and continue to threaten the crop through November. Populations build sharply later in the growing season (September and October). The potato tuberworm has been detected in all potato growing regions of Oregon and throughout the Columbia Basin of Washington. A limited number of adults have been trapped in western Idaho. No tuber damage has been reported in Idaho (Rondon, 2010). A recent study suggests that locations with higher spring, summer, or fall temperatures are associated with increased trapping rates in most seasons (DeBano et al., 2010). Occasionally potato tuberworm pupae can be found on the surface of tubers, most commonly associated with indentations around the tuber eyes, but usually are not found inside tubers (Rondon et al., 2007). Considering the duration period of each instar and its relationship to abiotic factors such as temperature, the potato tuberworm can undergo several generations per year in the Pacific Northwest region.

4.5.5 Monitoring

Pheromone-baited traps to catch adult male moths have been widely used in the region (Rondon et al., 2007). Unfortunately there are no established treatment thresholds. Another way is to check leaf mining. Most mines are found in the upper third of the plant canopy, suggesting that efficient scouting for foliar damage should focus on the top third of the plant (DeBano et al., 2010). The number of mines gives a good indication of the history of potato tuberworm infestation in a plant, but it does not necessarily indicate the severity of larval infestation at a point in time. The study also found that reasonably precise estimates of foliar damage for areas of 23 ft x 30 ft can be made by sampling 9 plants (DeBano et al., 2010).

4.5.6 Control

Control efforts should be directed toward tuberworm populations right before or at harvest. Females prefer to lay eggs on potato foliage, but when potato foliage starts to degrade and change color, or when it is vine-killed, the risk of tuber infestation increases greatly. The greatest risk for tuber infestation occurs between desiccation and harvest (Clough et al., 2010; Rondon, 2010). If tuberworm populations appear to be building prior to late season, additional control measures may be necessary. Other means of control include the elimination of cull piles and the elimination of volunteer potatoes. Daily irrigation that keeps the soil surface moist can also aid in the control of tuberworm populations. Most chemical products aim to reduce larva population in foliage but that technique does not provide 100% protection for the tubers.

4.6 Occasional pests
4.6.1 Mites

The two-spotted spider mite, *Tetranichus urticae* Koch, is the most abundant mite species found in potatoes in the Pacific Northwest. They can occasionally be considered pests of potatoes when crops such as beans, corn, alfalfa or clover seed are planted nearby (Hoy et al., 2008). Mites in general prefer hot and dry conditions; they also prefer stressed plants where irrigation is poorly managed. They damage plants by puncturing the leaf tissue to extract plant juices. Plants respond by changing color from green to brown. Spider mites overwinter in the area as adults in debris around field edges (Jensen et al., 2011). Females are very prolific; after emerging from overwinter, they mate and lay eggs on the underside of leaves. If temperatures are warm (75-80°F or 23.8-26.6C), eggs can hatch in 3-5 days; nymphs to adults can take place in 7-9 days at those temperatures. When leaves get overcrowded, mites climb to the top of the plant and secrete silk that can be used as a "transport" device during light to moderate winds conditions (Fig. 13).

Sampling for mites requires a close visual inspection of leaves from different levels of the plants. Shaking potentially infested leaves above a piece of white paper helps to determine the presence of mites. Applications of miticides should be made upon early detection of mites. All potatoes should be surveyed for the presence of mites and mite eggs starting mid-season (Schreiber et al., 2010). Thorough coverage is essential for good control and it is suggested that foliage should be dry at the time of application. While a single application of a miticide will suffice, if a second application of a miticide is required, the use of a miticide with different chemistry should be considered as a resistance prevention strategy (Jensen et al., 2011).

4.6.2 Cutworm, armyworm and loopers

These are several species of moth larvae that affect potato crops. Cutworms, armyworms and loopers are the immature stages of lepidopteran moths. Moths' typically have four defined life stages: egg, larva, pupa and adult. The most common species in the Pacific

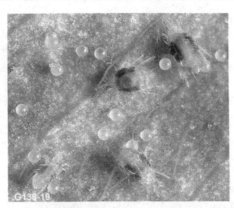

Fig. 13 Two spotted spider mite adults range in size from 0.25 mm to 0.5 mm long; eggs are around 0.1 mm. Adults and nymphs are pale yellow or light green with two dark spots on the abdomen (Photo by R.E. Berry, OSU).

Northwest regions are listed below (Table 2). Cutworms feed on potato seeds, cut stems, and foliage; armyworms and loopers feed on foliage throughout the season. Cutworms and armyworms have three pairs of true legs and five pairs of prolegs behind; loopers have only three pair of true legs and three pair of prolegs behind. At planting insecticides protect potato seed from cutworms; however, after the residual effect is gone, the crop is unprotected; in some years, a foliar chemical application may be needed. Potatoes can tolerate some worm defoliation without loss in marketable yield. The period of full bloom is the most sensitive plant growth stage, but even then defoliation on the order of 10% appears to cause little if any yield loss. Applications should be targeted to control small larvae (1st and 2nd instars), rather than larger larvae (Schreiber et al., 2010, Jensen et al., 2011).

Group	Common name	Scientific name
Cutworms	Spotted cutworms	*Xestia c-nigrum*
Western yellow striped armyworm	Bertha armyworm	*Mamestra configurata* Walker
Looper	Alfalfa looper	*Autographa californica* (Speyer)
	Cabbage looper	*Trichoplusia ni* (Hübner)

Table 2. Most common cutworm, armyworm, looper species in the Pacific Northwest (Zack. et al., 2010).

4.7 Resistance to insecticides

Insecticides are the most powerful tool available for use in pest management (Metcalf, 1994). However the misuse, overuse and historically unnecessary use of insecticides have been some of the most important factors in the increasing interest in integrated pest management (Von Rumker & Horay, 1972; Metcalf, 1994). In the last decades, the Insecticide Resistance Action Committee (IRAC), a group of technical experts that coordinates responses to prevent or delay the development of resistance in insect and mite pests, defined resistance to insecticides as a "heritable change in the sensitivity of a pest population that is reflected in the repeated failure of a product to achieve the expected level of control when used according to the label recommendations for that pest species" (http://www.irac-online.org/). In other words, it is the inherited ability of a pest population to survive a pesticide which is a result of a process of selection (Hamm et al., 2008). Some potato pests have developed resistance to certain groups of pesticides; however, significant insecticide resistance is not yet known to occur in the Pacific Northwest (University of California, 1986). For instance, while spider mite infesting potatoes has demonstrated the ability to readily develop resistance to miticides there appears to be no evidence of this problem developing in the U.S. Pacific region (Schreiber et al., 2010).

Pesticides such as pyrethroids that disturb natural enemies can cause a resurgence of primary or secondary pests, especially when applied mid to late season. In the past few years, package mixes of insecticide, some including pyrethroids have been available for use on potatoes. More research is needed to evaluate the real impact of this pesticide in the Pacific Northwest potato region. Seed and soil treatments with systemic insecticides have become a standard approach to control early "invaders" (Hoy et al., 2008). This approach may be less disruptive to predator and non-target insects than traditional foliar or ground chemical applications.

There are several key components to developing a resistance management program for insect pests: first, producers must employ non-chemical control tactics for control of pest problems, including irrigation, cultivation and proper fertilization management; second, producers must rotate insecticidal modes of action. This integrated pest management approach will lead producers to a sustainable production system with long term economic benefits. Alvarez et al., (2003) suggest keeping good records of chemical applications, rotating insecticide use changing not only the product but also the class of compound, applying insecticides at labeled rates, using newer insecticides with chemistries that are safer for applicators and non-target organisms, and reducing insecticide applications by scouting and making applications only as needed.

5. Conclusions

Potato is one of the most important food crops widely grown over many latitudes and elevations over the world. Increasing potato production in a sustainable manner requires an integrated approach covering a range of strategies. Combating pests is a continuous challenge that producers have to face as they intensify their production techniques to satisfy the increasing demands of the global market.

6. Acknowledgements

The author would like to thank A. Smith, A. Murphy, and R. Marchosky, the author's staff at Oregon State University, for their help providing tables, figures and pictures. Special thanks to A. Goyer, A. Murphy, M. Corp, and G. Clough also from Oregon State University, for peer proofing the manuscript.

7. References

Alvarez, J.M., R.L. Stotlz, C.R. Baird, and L.E. Sandoval. (2003). Insect pest and their management. In *Potato Production Systems* (ed) J.C. Stark and S.L. Love. University of Idaho Extension. Pp 205-239.

Andrews N., M. Ambrosino, G. Fisher, and S.I. Rondon. (2008). Wireworm biology and non-chemical management in potatoes in the Pacific Northwest. *Oregon State University Extension Service Publication*. Dec. PNW 607. Available at http://extension.oregonstate.edu/catalog/pdf/pnw/pnw607.pdf.

Barry, A., R.E. Berry, and G. Tamaki. (1982). Host preference of the green peach aphid, *Myzus persicae* (Hemiptera: Aphididae). *Environmental Entomology*. 11(4): 824-827.

Branson, T., F. Terry, and G. Robert. (1966). Effects of a nitrogen deficient host and crowding on the corn leaf aphid. *Journal of Economic Entomology*. 59(2): 290-294.

Cambell, R.E. & M.W Stone. (1939). Trapping elaterid beetles as a control measure against wireowmrs. *Journal of Economic Entomology*. 32: 47-53.

CIP (Centro Internacional de la Papa or International Potato Center). (1994). *International Potato Center Circular* 20(1). CIP, Lima, Perú.

Clough, G.H., S.I. Rondon, S.J. DeBano, N. David, and P.B. Hamm. (2010). Reducing tuber damage by the potato tuberworm (Lepidoptera: Gelechiidae) with cultural practices and insecticides. *Journal of Economic Entomology* 103(4): 1306-1311.

Crosslin, J.M., S.I. Rondon and P.B. Hamm. (2011). Population dynamics of the beet leafhopper in northern Oregon and incidence of the Columbia Basin potato purple top phytoplasma. American Journal of Potato Research (In press).

DeBano, S.J., P.B. Hamm, A. Jensen, S.I. Rondon, and P.J. Landolt. (2010). Spatial and temporal dynamics of potato tuberworm (Lepidoptera: Gelechiidae) in the Columbia Basin of the Pacific Northwest. *Journal of Economic Entomology* 39(1): 1-14.

Dickson, R., & E.F. Laird. (1967). Fall dispersal of green peach aphids to desert valleys. *Annals of the Entomological Society of America*. 60(5): 1088-1091.

Douches, D.S, W. Li, K. Zarka, J. Coombs, W. Pett, E. Grafius, T. El-Nasr. (2002). Development of Bt-cry5 insect-resistant potato lines 'Spunta-G2' and 'Spunta-G3'. HortScience 37: 1103-1107.

Dreistadt, S.H., J.K. Clark and M.L. Flint. (2004). *Pests of Landscape Trees and Shrubs: An Integrated Pest Management Guide*. 2nd edition, Publication 3359. University of California Division of Agriculture and Natural Resources, Davis. Pp 501.

Driesche, R.G., & T.S. Bellows. (1996). *Biological control*. Chapman and Hall. International Thomson Publishing Company. New York. Pp 539.

Ferro, D. N. (1994). Biological control of the Colorado potato beetle. *In*: G. W. Zehnder, M. L. Powelson, R. K. Jansson and K. V. Raman (eds), *Advances in Potato Pest Biology and Management*. APS Press, St. Paul. Pp. 357-375.

Gopal, J. & S.M.P. Khurana. (2006). Potato production, improvement and postharvest management. The Haworth press. Pp 587.

Hamm, P.B.., C,W. Hy, P.J. Hutchinson, W.R. Stevenson, R.A. Boydston, J.M. Alvarez, A. Alyokin, G. Dively, N.C. Gudmestad, and W.W. Kirk. (2008). Managing pesticide resistance. *In Potato Health Management. Plant Health Management Series* (Ed. D.A. Johnson). American Phytopathological Society, St Paul, Minnesota, USA. Second Edition. Pp 123-131.

Hawkes, J.G. (1990). *The potato, evolution, biodiversity, and genetic resources*. Belhaven Press, London. Pp 259.

Heitefuss, R. (1989). *Crop and plant protection: the practical foundations*. Ellis Horwood Ltd, Chichester. Pp 261.

Hoy, C.W., G. Boiteau, A. Alyokhin, G. Dively, and J.M. Alvarez. (2008). Managing insect and mite pests. In (Ed D. Johnson) *Potato Health Management. Plant Health*

Management Series. American Phytopathological Society, St Paul, Minnesota, USA. Second Edition. Pp 133-147.

Horgan, F.G., D.T. Quiring, A. Lagnaoui, A. Salas, and Y. Pelletier. (2007). Mechanism of resistance to tuber-feeding *Phthorimaea operculella* (Zeller) in two wild potato species. *Entomological Experimental Applicata* 125: 249-258.

Horton, D. (2006). Quantitative relationship between potato tuber damage and counts of Pacific coast wireworm (Coleoptera: Elateridae) in baits: seasonal effects. *Journal of the Entomological Society of British Columbia* 103: 37-48.

Hoy, C.W., G. Boiteau, A. Alyokhin, G. Dively, J.M. Alvarez. (2008). Managing insects and mites. *In Potato Health Management. Plant Health Management Series* (Ed. D.A. Johnson). American Phytopathological Society, St Paul, Minnesota, USA. Second Edition. Pp 133-147.

Hutchinson, C.M., & W.M. Stall. (2007). *Potato vine killing or desiccation.* University of Florida IFAS Extension. HS925. Available at http://edis.ifas.ufl.edu/pdffiles/HS/HS18100.pdf.

Jensen, A. , A. Schreiber and N. Bell. (2011). Irish potato pests. In Pacific Northwest Insect Management Handbook. Ed. C. Holligsworth. U. Massachusetts Extension. Available at http://uspest.org/pnw/insects?00INTR05.dat.

Johnson, D.A. (2008). *Potato Health Management.* Plant Health Management Series. American Phytopathological Society, St Paul, Minnesota, USA. Second Edition. Pp 259.

Koss, A. (2003). *Integrating chemical and biological control in Washington State potato fields.* M.S. Thesis, Washington State University, Pullman, WA, USA.

Lacey, L.A., D.R. Horton, T.R. Unruh, K. Pike, and M. Marquez. (2001). Biological control of insect pests of potato in north america. *In Washington State Potato Conference and Trade Show proceedings.* Spanish seminar. Tri Cities, WA. Pp 123

Lopez, R., & D.N. Ferro. (1995). Larviposition of Myiopharus doryphorae (Diptera: Tachinidae) to Colorado potato beetle (Coleoptera: Chrysomelidae) larvae treated with lethal and sublethal doses of *Bacillus thuringensis* Berliner subsp. tenebrionis. *Journal of Economic Entomology* 88(4); 870-874.

Love, S.L., P. Nolte, D.L. Corsini, J.C. Whitmore, L.L. Ewing, and J.L. Witworth. 2003. Seed production and certification. In *Potato production systems.* (ed) J.C. Stark and S.L. Love. University of Idaho Extension. Pp 49

Luckmann, W.H., & R.L. Metcalf. &. (1994). Pest management concept. *In Introduction to Insect Pest Management.* 3rd edition. A Wiley-Interscience publication. John Wiley & Son. Pp 1-34.

Metcalf, R.L., & W.H. Luckmann. (1994). *Introduction to Insect Pest Management.* 3rd edition. A Wiley-Interscience publication. John Wiley & Son. Pp 650.

Metcalf, R.L. (1994). Insecticides in Pest Management. In *Introduction to Insect Pest Management* (ed R.L. Metcalf and W.H. Luckmann). 3rd edition. A Wiley-Inter science publication. John Wiley & Son. Pp 245-314.

Merrill, S.C., S.M. Walter, F.B. Peairs, and J.A. Hoeting. (2011). Spatial variability of western bean cutworm (Lepidoptera: Noctuidae) pheromone trap captures in sprinkler irrigated corn in eastern Colorado. *Environmental Entomology* 40(3): 654-660.

Meyerdirk, D.E. & N.A. Hessein. (1985). Population dynamics of the beet leafhopper, *Circulifer tenellus* (Baker), and associated *Empoasca* spp. (Homoptera: Cicadellidae) and their egg parasitoids on sugar beets in southern California. *Journal of Economic Entomology.* 78: 346-353.

Meyerdirk, D.E. & M.S. Moratorio. (1987). Biology of *Anagrus giraulti* (Hymenoptera: Mymaridae), and egg parasitod of the beet leafhopper, *Circulifer tenellus* (Homoptera: Cicadellidae). *Annals of the Entomological Society of America*. 80: 272-277.

Miller, J.S., & B.G. Hopkins. (2008). Checklist for a holistic potato health management plan. In *Potato Health Management. Plant Health Management Series* (Ed. D.A. Johnson). American Phytopathological Society, St Paul, Minnesota, USA. Second Edition. Pp 7-10.

Ministry of Agriculture. (2008). *Native potatoes of Perú*. J & J Proyects Editoriales S.A.C. Pp 115.

Munyaneza, J.E. (2003). Leafhopper identification and biology. In *Pacific Northwest Vegetable Association Proceedings*. Pp 89-91.

Munyaneza, J.E., and J.M. Crosslin. (2006). The beet leafhopper (Hemiptera: Cicadellidae) transmits the Columbia basin potato purple top phytoplasma to potatoes, beets, and weeds. *Journal of Economic Entomology*. 99(2): 268-272.

Pedigo, L.P. 1996. *Entomology and Pest Management*. Second Edition. 1996. Prentice-Hall Pub., Englewood Cliffs, NJ. Pp 679.

Persoons, C.J., S. Voerman, P.E.J. Verwiel, F.J. Ritter, W.J. Nooyen, and A.K. Minks. (1976). Sex-pheromone of potato tuberworm moth, *Phthorimaea operculella:* isolation, identification and field evaluation. *Entomological Experimental Applicata*. 20: 289-300.

Pike, K.S., L.L. Boysdston, and D.W. Allison. (2003). *Aphids of western North America north of Mexico*. Washington State University Extension. MISC0523. Pp 282.

Raman, K.V. (1988). Control of potato tuber moth *Phthorimaea operculella* with sex pheromones in Perú. *Agriculture, Ecosystems and Environment* 21: 85-99.

Roelofs, W.L., J.P. Kochansky, R.T. Carde, G.G. Kennedy, C.A. Henrick, J.N. Labovitz, and V.L. Corbin. (1975). Sex-pheromone of potato tuberworm moth, *Phthorimaea operculella*. *Life Sciences* 17: 699-706.

Rondon, S.I., S.J. DeBano, G.H. Clough, P.B. Hamm, A. Jensen, A. Schreiber, J.M. Alvarez, M. Thornton, J. Barbour, and M. Dögramaci. (2007). Biology and management of the potato tuberworm in the Pacific Northwest. *Oregon State University Extension Service Publication*. Apr. PNW 594.
http://extension.oregonstate.edu/catalog/pdf/pnw/pnw594.pdf.

Rondon, S.I., D. Hane, C.R. Brown, M.I. Vales, and M. Dögramaci. (2009). Resistance of potato germplasm to the potato tuberworm (Lepidoptera: Gelechiidae). *Journal of Economic Entomology*. 102(4): 1649-1653.

Rondon, S. I., S. J. DeBano, G. H. Clough, P. H. Hamm, and A. Jensen. (2008). Ocurrence of the potato tuber moth in the Columbia Basin of Oregon and Washington. J. Kroschel and L. Lacey (eds) *Integrated pest management for the potato tuber moth, Phthorimaea operculella Zeller: a potato pest of global importance*. Tropical Agriculture 20, Advances in Crop Research 10 Margraf Publishers, Weikersheim, Germany: 9-13.

Rondon, S.I. (2010). The potato tuberworm: a literature review of its biology, ecology, and control. *American Journal of Potato Research*. 87:149–166.

Rondon, S.I. & L. Xue. (2010). Practical techniques and accuracy for sexing the potato tuberworm, *Phthorimaea operculella* (Lepidoptera: Gelechiidae). *Florida Entomologist*. 93(1): 113-115.

Rothschild, G.H.L. (1986). The potato moth: an adaptable pest of short term cropping systems. In: Kitching R. L. (Ed.), *The ecology of exotic plants and animals*. J. Wiley, Brisbane. Pp 144-162.

Salaman, R. (1985). *The history and social influence of the potato.* Cambridge University Press. Pp 685.

Schreiber, A., A. Jensen, K. Pike, J. Alvarez, and S.I. Rondon. (2010). *Integrated Pest Management guidelines for insects and mites in Idaho, Oregon, and Washington Potatoes.* http://oregonstate.edu/potatoes/ipm/publications.htm.

Simmons, C., L. Pedigo, and M.E. Rice. (1998). Evaluation of seven sampling techniques for wireworms. *Environmental Entomology* 27(5): 1062-1068.

Srinivasan, R. & J.M. Alvarez. (2007). Effect of mixed viral infections (potato virus Y-potato leafroll virus) on biology and preference of vectors *Myzus persicae* and *Macrosiphum euphorbiae* (Hemiptera: Aphididae). *Journal of Economic Entomology.* 100(3): 646-655.

Tabashnick, B.E. (1989). Managing resistance with multiple pesticides tactics: theory, evidence, and recommendations. *Journal of Economic Entomology.* 92: 1263-1269.

Tamaki, G., L. Fox, and R.L. Chauvin. (1980). Green peach aphid: orchard weeds are host to fundatrix. *Environmental Entomology.* 9(1): 62-65.

Tamaki, G., R.L. Chauvin and A.K. Burditt, Jr. (1983). Field evaluation of *Doryphorophaga doryphorae* (Diptera: Tachinidae), a parasite, and its host the Colorado potato beetle (Coleoptera: Chrysomelidae). *Environmental Entomology.* 12: 386-389.

University of California. (1986). *Integrated Pest Management for potatoes in the western United States.* University of California, Division of Agriculture and natural Resources. Publication 3316. Pp 146.

Varela, L.G., and E.A. Bernays. (1988). Behavior of newly hatched potato tuber moth larvae, *Phthorimaea operculella* Zell. (Lepidoptera: Gelechiidae), in relation to their host plants. *Journal of Insect Behavior.* 1: 261-275.

Vernon, R.S., W.G. van Herk, M. Clodius, and C. Harding. (2009). Wireworm management I: Stand protection versus wireworm mortality with wheat seed treatments. *Journal of Economic Entomology* 102(6): 2126-2136.

Voerman, S., and G.H.L. Rothschild. (1978). Synthesis of 2 components of sex-pheromone system of potato tuberworm moth, *Phthorimaea operculella* (Zeller) (Lepidoptera: Gelechiidae) and field experience with Them. *Journal of Chemical Ecology.* 4: 531-542.

Von Rumker, R. & F. Horay. (1972). Pesticide manual. *U.S. Agency for International Development.* Pp. 126.

Wale, S., H.W. Platt, and N. Cattlin. (2008). *Potatoes a color handbook.* Manson publishing, Ltd. London. Pp

Wallis, R.L. (1967). Some host plants of the green peach aphid and beet western yellow virus in the Pacific Northwest. *Journal of Economic Entomology.* 60(4): 904-907.

Washington Potato Commission. (2007). *Potato Varieties in the Northwest.* (Ed.) A. Jensen. Vol VII, No15. Available at www.potatoes.com

Weisz, P.R., Z. Smilowitz, M.C. Saunders, and B. Christ. (1995). *Integrated pest management for potatoes.* Pennsylvania State Unviersity Cooperative Extension, Unviersity Park, PA. Pp 63.

Weinsz, P.R., S. Fleischer, and Z. Smilowitz. (1996). Site-specific integrated pest management for high value crops: impact on potato pests management. *Journal of Economic Entomology.* 59(2) 501-509.

Zack, R.S., P.J. Landolt, A. Jensen, and A.Schreiber. (2010). Lepidopterous "worms" on, but not in potatoes. In *Annual progress report-Washington Potato Commission.* Pp 1-5. Available at www.potatoes.com.

Management Strategies for Western Flower Thrips and the Role of Insecticides[1]

Stuart R. Reitz[1] and Joe Funderburk[2]

[1]*United States Department of Agriculture, Agricultural Research Service, Center for Medical, Agricultural and Veterinary Entomology, Tallahassee, FL,*
[2]*North Florida Research and Education Center, University of Florida, Quincy, FL,*
USA

1. Introduction

Today, the western flower thrips, *Frankliniella occidentalis* (Pergande) (Thysanoptera: Thripidae) is one of the most significant agricultural pests globally because of the damage it is able to inflict on a wide range of crops. Adults and larvae feed by piercing plant tissues with their needle-shaped mandible and draining the contents of punctured cells (reviewed in Kirk, 1997b). Feeding by adults and larvae produces scarring on foliage, flowers and fruits, which results in aesthetic crop damage and disrupts plant growth and physiology. Also, oviposition can produce a wound response in fruiting structures, which reduces the marketability of certain horticultural produce (Childers, 1997). Most importantly, western flower thrips is able to transmit several species of destructive plant viruses in the genus *Tospovirus* (Bunyaviridae). It is the most important vector of *Tomato spotted wilt virus* and *Impatiens necrotic spot virus* worldwide, and it is also known to vector Chrysanthemum stem necrosis virus, *Groundnut ringspot virus* and *Tomato chlorotic spot virus* (Pappu et al., 2009; Webster et al., 2011).

The actual amounts of economic losses attributable to any pest are difficult to determine, but Goldbach and Peters (1994) estimated that *Tomato spotted wilt virus* alone caused over US$1 billion in losses annually on a global basis. This estimate did not include the direct damage caused by western flower thrips, and it still would further underestimate present day losses, as the western flower thrips has continued to spread throughout the world (Kirk & Terry, 2003; Reitz et al., 2011). The state of Georgia, USA compiles estimates of the economic costs of pests to its crops. These estimates include both crop losses and the costs of control measures. From 2001 – 2006, costs from thrips and TSWV for tomatoes (*Solanum lycopersicum*) and peppers (*Capsicum annuum*) have averaged over 12% of the harvested value of those crops per year (Sparks, 2003, 2004, 2005, 2006; Sparks & Riley, 2001, 2002). Over 60% of the total for economic losses from pests and control costs in tomato and pepper are from thrips and tomato spotted wilt virus (Sparks, 2003). In addition, losses in Georgia

---

[1] The use of trade, firm, or corporation names in this publication is for the information and convenience of the reader. Such use does not constitute an official endorsement or approval by the United States Department of Agriculture or the Agricultural Research Service of any product or service to the exclusion of others that may be suitable.

in 2006 to ornamentals caused primarily by western flower thrips damage exceeded US$ 15 million. Clearly, these economic assessments show that the western flower thrips is one of the most destructive agricultural pests globally, but its emergence as a major pest has only occurred relatively recently.

The species "*Frankliniella occidentalis*" was first described as a member of the genus *Euthrips* by Theodore Pergande in 1895 from specimens collected in California, where its widespread distribution and abundance across many flowering plants was noted (Pergande, 1895). Subsequently, the species was described under several other names, which have since been synonymized (Mound, 2011). During the early to mid 20th century, the western flower thrips was regularly mentioned as a member of complexes of pestiferous thrips in western North America, but it was generally regarded as a less significant problem than species such as *Taeniothrips inconsequens* (Uzel), *Thrips tabaci* Lindeman, *Scirtothrips citri* (Moulton) and *Heliothrips haemorrhoidalis* (Bouché) (Moulton, 1931). For most of the early 20th century, western flower thrips remained a localized problem in California and other areas of the western USA and Canada, with sporadic problems reported from southern Texas to British Columbia.

One of the first cases of damage attributed to western flower thrips occurred on potatoes grown in the San Gabriel Valley of southern California. Feeding damage from large populations of the western flower thrips was implicated as the cause of curling of new foliage on potato *(Solanum tuberosum)*. The condition, termed potato curly leaf, would lead to significant reductions in tuber size and yield (Crawford, 1915). The recommended control for these thrips infestations was early season applications of Bordeaux mixture combined with extracts of tobacco, which was able to reduce losses when timed appropriately.

Also early in the 20th century, the western flower thrips was recorded as a pest of other vegetable, fruit and nut crops. Riherd (1942) observed oviposition damage on peas grown in the Rio Grande Valley of Texas, which reduced the aesthetic quality of fresh market peas *(Pisum sativum)*. Similar spotting patterns, termed "pansy spots", on apples *(Malus domestica)* grown in the northwestern USA were attributed to western flower thrips oviposition (Childs, 1927; Moulton, 1931), as was damage to table grapes *(Vitis vinifera)* grown in California (Jensen, 1973; Yokoyama, 1977). Another significant type of damage to fruit crops resulted from feeding on floral tissues that would cause flower loss before fruit set. Occasionally, larval feeding was known to produce scarring of citrus fruit (Essig, 1926; Woglum & Lewis, 1935). During this period, western flower thrips also was implicated as one of thrips causing damage to alfalfa flowers *(Medicago sativa)*, which reduced set seed (Borden, 1915; Seamaxs, 1923).

Later, during the 1950's the western flower thrips began to receive more attention as a pest of seedling cotton *(Gossypium hirsutum)* in the southwestern USA. Extensive use was made of all major classes of insecticides, including organochlorines, organophosphates, and carbamates, during the 1950's – 1960's, to combat this early season threat. Recommendations were made to growers to make multiple applications of insecticides to prevent loss of seedlings from thrips feeding damage. However, control failures with organochlorine insecticides were observed as early as 1960 (Race, 1961). The control failures with organochlorines then led growers to make preventative applications of organophosphates to protect seedlings. These were often made as systemic treatments at planting. However, Race (1965) recognized a significant disadvantage with this approach of intense preventative insecticide use, namely that growers could be making unnecessary applications when the thrips populations did not warrant treatment. Further complicating management issues were the findings by Shorey et al. (1962) that carbamates actually led to rapid resurgence of western flower thrips populations after applications in cotton, which were likely related to adverse effects of the insecticides on beneficial insects.

Harding (1961a) observed that western flower thrips was the predominant thrips species infesting onions (*Allium cepa*) grown in southern Texas. It was also observed to be the predominant thrips in California onions (Hale & Shorey, 1965). Multiple applications of insecticides, predominately organophosphates, were found to reduce thrips abundance but not to improve yields significantly (Harding, 1961b).

Western flower thrips were noted as a minor pest of other vegetable crops in California. It was confirmed as a vector of *Tomato spotted wilt virus* in the 1930's (see Sakimura, 1962), but viral epidemics caused by western flower thrips apparently were not common at this time, and most damage was a result direct feeding rather than virus transmission (Sakimura, 1961). Similar to the situation with seedling cotton, western flower thrips feeding on young tomato plants was known to reduce photosynthesis and lead to defoliation and plant death (Shorey & Hall, 1963). Thrips feeding on lettuce (Latuca sativa) produces scarred, corky tissue on leaves. Shorey found that organophosphates would reduce populations for a short time after applications, but that populations would rebound soon after applications (Shorey & Hall, 1962). One of the causes identified for the lack of long term control of western flower thrips by insecticides in open field crops was that crops were subject to repeated ongoing dispersal of adults from outside crop fields (Shorey & Hall, 1963). Although tomato spotted wilt epidemics were uncommon in California at this time, Shorey and Hall (1963) also noted the inability of conventional insecticides to reduce Tomato spotted wilt virus transmission within fields because of the repeated dispersal of viruliferous thrips into fields from external sources.

Notably in major review articles concerning floriculture crops, western flower thrips was seldom discussed as a significant pest before the 1970's (Bryan & Smith, 1956; Naegele & Jefferson, 1964; Price et al., 1961), although the damage it caused to greenhouse-grown cut flowers was well recognized in California by the 1930's (Bohart, 1943). At this time, greenhouse growers relied on methods such as intensive sprays of nicotine sulfate or tartar emetic, and fumigation with nicotine, napthalene or calcium cyanide. Fumigation was found to kill adults but was far less effective against immature stages. Therefore, populations could rebound quickly after a fumigation treatment, thus exacerbating overall pest management concerns for growers.

2. Western flower thrips as an invasive species

It is likely that Naegele and Jefferson (1964) did not discuss western flower thrips because of its limited distribution at the time of their review article on floriculture pests. However, the pest status of western flower thrips began to expand rapidly in the late 1970's when growers in California began to experience more extensive damage to cut flower crops from thrips feeding and virus transmission (Robb, 1989). Because of the exceedingly low damage thresholds for these crops, growers responded to the threat with intensive insecticide treatments, leading to the rapid development of resistance to all major classes of insecticides available at the time, including pyrethroids, carbamates, organophosphates and abamectin (Immaraju et al., 1992).

This insecticide-resistant strain(s) of western flower thrips that originated in California is thought to then have spread around the world in association with the globalization of the cut flower and horticulture industries (Bonarriva, 2003; Huang, 2004). The western flower thrips was established throughout agroecosystems of eastern North America by the mid 1990's, and its spread through Europe was even more rapid (Kirk & Terry, 2003). It was first recorded in the Netherlands in 1983 (Mantel & Van de Vrie, 1988), and large outbreaks were observed in almost all European countries by 1990 (Kirk & Terry, 2003). In northern areas of Europe, western flower thrips is largely restricted to glasshouses[2] because climatic conditions prevent persistent populations from establishing (McDonald et al., 1997). In more southerly areas around the Mediterranean basin, populations have become established in open field crops, which facilitates the repeated colonization of protected crops (Brødsgaard, 1993; Kontsedalov et al., 1998). Since the distribution review by Kirk and Terry (2003), western flower thrips has continued to spread to new regions. Significantly, it is now a widely established pest in China (Reitz et al., 2011). As a reflection of its current cosmopolitan distribution and pest status, western flower thrips, a native of North America, is one of the most frequently intercepted insect species at USA ports of entry (Nickle, 2004).

Recently, the taxonomic status of *Frankliniella occidentalis* has been called into question. Molecular evidence indicates that *"Frankliniella occidentalis"* is a complex of two cryptic, sympatric species (Rugman-Jones et al., 2010). Rugman-Jones et al. (2010) have designated the two species as the lupin (L) and glasshouse (G) species. Neither of these genetic species corresponds to the historic morphological descriptions of the species of *F. moultoni* (Hood) or *F. occidentalis* (Pergande). The "lupin" and "glasshouse" designations are based on the similarity of certain individuals with a "western flower thrips" population that has been associated with *Lupinus arboreus* in New Zealand since the 1930's (Martin & Workman, 1994) and other individuals corresponding to pest strains found in European glasshouses (Brødsgaard, 1994). This recognition of species diversity within *"F. occidentalis"* adds uncertainty as to which species is under consideration in earlier literature. However, circumstantial evidence suggests that recent pest problems are largely the result of the highly invasive "glasshouse" type (Martin & Workman, 1994; Rugman-Jones et al., 2010).

Ironically perhaps, responses to these new, expanding problems from invasive western flower thrips populations were similar to historic control efforts attempted in the western USA; namely, there was a heavy reliance on insecticide use for control. Despite pervasive attempts to control western flower thrips with the widespread, intense use of insecticides, there are many factors that limit the efficacy of insecticides. These include ecological, behavioral and physiological factors, and these need to be appreciated and understood to place insecticide use in a proper context. Ecologically, western flower thrips is highly polyphagous and capable of reproducing on numerous host plants (Northfield et al., 2008; Paini et al., 2007). As large populations can develop on non-crop hosts, mass dispersal into crops occurs, whether open field crops (Pearsall & Myers, 2001; Puche et al., 1995; Ramachandran et al., 2001) or crops in protected environments (Antignus et al., 1996). The potential for continual recolonization of crops limits the observed field efficacy of insecticides (Eger et al., 1998; Reitz et al., 2003). This ongoing dispersal means that even repeated insecticide applications have little utility in reducing pest damage, especially in high value crops with low damage thresholds (Bauske, 1998; Kontsedalov et al., 1998). Once

---

[2] The terms "greenhouse" and "glasshouse" are used interchangeably in this article for convenience, although they are not necessarily structurally equivalent.

having landed on plants, western flower thrips preferentially reside within flowers or other concealed, protected places on plants (Hansen et al., 2003; Kirk, 1997a). This thigmotactic behavior of thrips limits their exposure to many foliar applied insecticides. Also, the anthophilous nature of western flower thrips limits their exposure to systemic insecticides, which are not readily transported into floral tissues (Cloyd & Sadof, 1998; Daughtrey et al., 1997). Therefore, some of the most effective materials are those with translaminar properties, which increase the probability of thrips concealed in flowers actually ingesting toxins (Kay & Herron, 2010).

3. Insecticide use and insecticide resistance

While delivering toxins to western flower thrips can be problematic, and the species behavior and ecology can minimize exposure to insecticides, the species is well suited to evolve resistance to multiple classes of insecticides. Since the first reported case of control failures with insecticides (toxaphene, an organochlorine, Race, 1961), there have been numerous incidences of resistance reported to all major classes of insecticides from all regions of the world (Bielza et al., 2007b; Brødsgaard, 1994; Dağli & Tunç, 2007; Immaraju et al., 1992; Jensen, 2000a; Kay & Herron, 2010; Morishita, 2001; Robb et al., 1995; Weiss et al., 2009; Zhao et al., 1995). Resistance has not only developed against insecticides targeting western flower thrips, but also insecticides used to treat other pest species. Correspondingly, a number of different resistance mechanisms have been characterized to date, including metabolic detoxification, reduced penetration, altered target site resistance, and knockdown resistance (Bielza, 2008).

One class of insecticides where resistance problems have been particularly acute is the pyrethroids. Robb (1989) recorded control failures with pyrethroids in ornamental greenhouses in California. Management became so difficult, with a lack of alternative insecticides that growers returned to using legacy non-synthetic insecticides, such as nicotine sulfate. Despite these early reports regarding difficulties with pyrethroids, they have continued to be used against western flower thrips extensively, with the same outcome of resistance development (Broadbent & Pree, 1997; Espinosa et al., 2002b; Frantz & Mellinger, 2009; Immaraju et al., 1992; Seaton et al., 1997; Thalavaisundaram et al., 2008; Zhao et al., 1995). In these cases, the development of resistance to pyrethroids has tended to occur rapidly. Resistance to pyrethroids led Australia to abandon their use for western flower thrips management less than eight years after the pest was first detected (Herron & Gullick, 2001). Likewise, pyrethroids are no longer recommended for use in Turkish greenhouses because of the rapid development of resistance and cross resistance to other chemicals (Dağlı & Tunç, 2008).

Resistance to pyrethroids is primarily derived from metabolic detoxification pathways (Broadbent & Pree, 1997; Espinosa et al., 2005; Maymó et al., 2006; Zhao et al., 1995). A broad range of enzymatic detoxification pathways to detoxify pyrethroids have been identified, including cytochrome P450 monooxygenases, glutathione S-transferases and esterases. The predisposition of western flower thrips to evolve resistance based on metabolic detoxification is likely a product of its polyphagous nature. Because individuals move from host to host, they are likely to encounter a variety of plant defensive chemicals. Therefore, it is adaptive to have multiple means to contend with the unpredictable suite of host plant defenses that they may encounter (Rosenheim et al., 1996).

These inherent metabolic detoxification pathways predispose the western flower thrips to overcome pyrethroids and other classes of insecticides. Adding to the factors that make western flower thrips amenable to developing insecticide resistance are its rapid development rate so that populations may pass through several generations within a single cropping system, its high fecundity, so that resistant females can produce many offspring, and the haplodiploid sex determination characteristic of thrips (Reitz, 2009). In this type of sex determination, females are diploid, but males are haploid. Because males are haploid, their alleles are exposed directly to selection, which enable alleles for resistance to become fixed rapidly in a population (Denholm et al., 1998).

Insecticide resistance in western flower thrips is a complex phenomenon. It is important to note that resistance to a particular insecticide may not derive from a single trait and can involve multiple pathways (Jensen, 1998). For example, resistance to diazinon, an organophosphate, and methiocarb, a carbamate, has been linked to both metabolic detoxification and altered target site sensitivity (Jensen, 2000b; Zhao et al., 1994). In addition, distinct populations can evolve resistance to a particular insecticide through different mechanisms (Jensen, 2000b; Thalavaisundaram et al., 2008). However, a single metabolic mechanism may confer cross resistance between different classes of insecticides. Espinosa et al. (2002a) found that single metabolic pathway appears to confer resistance to pyrethroids and carbamates. In turn, Bielza et al. (2007a) proposed that such a single detoxification mechanism for the pyrethroid acrinathrin and for carbamates could be exploited, by using carbamates as synergists to increase the activity of acrinathrin through competitive substrate inhibition. These variable pathways complicate the prediction of resistance development and the potential for cross resistance to multiple classes of insecticides in any population.

Individuals carrying resistance alleles are often considered to be at a fitness disadvantage in a population when the particular insecticide to which they are resistant is not used (i.e., in the absence of particular selection pressures) (Georghiou & Taylor, 1986). When such a fitness disadvantage is present in resistant individuals, reversion to a susceptible population should occur soon after removal of the specific insecticide (selective pressure). However, this outcome may not always be the case for western flower thrips. There are cases of long term maintenance of resistance in the absence of insecticide exposure. Kontsedalov et al. (1998) reported that resistance to the pyrethroid cypermethrin did not decline after more than seven years of non-exposure for a laboratory colony. Likewise, resistance to organophosphates can be maintained in field and laboratory populations for several years without exposure (Brødsgaard, 1994; Robb, 1989). Bielza et al. (2008) compared fecundity, fertility, longevity, and egg-to-adult developmental time in populations of western flower thrips from Spain that were resistant and susceptible to acrinathrin (pyrethroid) and spinosad (a spinosyn formulation, see below), and found that resistance to either material did not carry significant fitness costs, at least in terms of the parameters they measured. Therefore, resistance may be expected to develop rapidly and be maintained for long periods of time in the absence of the use of an insecticide, and these scenarios must be taken into account in developing insecticide resistance management programs, which are integral to integrated pest management (IPM) programs.

Given the ongoing issues with management failures and resistance development with synthetic insecticides, and the cancellation of registrations for many other synthetic insecticides as a result of the Food Quality and Protection Act of 1996 (US EPA, 1996), there is a limited pool of efficacious insecticides for use against western flower thrips. Consequently, there is intense interest in developing new, alternative insecticides. Some of the most efficacious insecticides against western flower thrips in recent years have been spinosyns, which are metabolites derived from fermentation of the actinomycete bacterium, *Saccharopolyspora spinosa* (Sparks et al., 1999). Spinosyns are in a group of insecticides with a novel mode of action, the nicotinic acetylcholine receptor (nAChR) allosteric activators (Group 5 – Insecticide Resistance Action Committee) (IRAC International MoA Working Group, 2011). The unique mode of action of spinosyn-based insecticides and their translaminar properties have made them highly effective against western flower thrips. The first products with spinosad as the active ingredient (a combination of spinosyns A and D; Dow AgroSciences, Indianapolis, IN) were registered for use in the late 1990's (Thompson et al., 2000). This was followed later by the release of spinetoram in 2008. Given the effectiveness of spinosyns and the lack of effective alternatives, growers tended to place an overreliance on spinosad, making them a victim of their own success. The first evidence of resistance to spinosyns was detected in western flower thrips populations in Australia by 2002 (Herron & James, 2005), Spain by 2003 (Bielza et al., 2007b) and the US by 2006 (Weiss et al., 2009). Spinosyn resistance in western flower thrips appears to be based on altered target site resistance, with spinosad resistance in a Spanish population based a single locus, autosomal recessive trait (Bielza et al., 2007c). The evidence that spinosad resistance is a recessive trait means that this resistance may not be stable, which would facilitate reversion to susceptibility in populations (Weiss et al., 2009). However, as with other resistance cases, mechanisms and genetics of spinosyn resistance likely vary across populations. Zhang et al. (2008) reported that spinosad resistance in a population from Japan is likely polygenic. Perhaps, more troubling is the potential cross resistance to other insecticides that Zhang et al (2008) found in their spinosad resistant strain. This potential will necessitate further caution in managing the use of spinosyn products.

In efforts to broaden the range of insecticides available for western flower thrips management, there has been interest in adapting other existing chemistries not previously labeled for use against western flower thrips. Neonicotinoids, which have been used since the 1990's to manage various types of other sucking insect pests, have recently received attention for a potential role in western flower thrips management. In experimental trials in Australia, Broughton and Herron (2009) found that two neonicotinoids, acetamiprid and thiamethoxam, were as effective as spinosad against larvae and adults of western flower thrips in pepper and lettuce. However, none of the tested insecticides, including spinosad, were effective in reducing the abundance of western flower thrips adults in tomato. Coutts and Jones (2005) found that drenching lettuce seedlings with neonicotinoids, in particular imidacloprid, just before transplanting reduced the incidence of tomato spotted wilt significantly in field trials. However, during their field trial, the predominant vector species were *Frankliniella schultzei* (Trybom) and *Thrips tabaci*, with very few western flower thrips found. Interestingly, their results showed that there were minimal effects of the neonicotinoids on the abundance of any of the thrips species. Likewise, western flower thrips populations showed little effect from applications of another neonicotinoid, dinotefuran, in pepper and strawberry (Dripps et al., 2010). It is possible that in these cases, mortality induced by the insecticide was counteracted by the loss of natural enemies of thrips from the insecticide applications. Alternatively, disease reductions observed by Coutts and Jones (2005) may have resulted from sublethal effects of the insecticides on thrips. Imidacloprid was found to actually enhance feeding of western flower thrips on tomato foliage while it reduced feeding by another TSWV vector, *Frankliniella fusca* (Hinds) (Joost & Riley, 2005). Likewise, the anthranilic diamide, cyantraniliprole, which disrupts insect feeding activity, reduced TSWV transmission by *F. fusca*, but had no effect on transmission by western flower thrips (Jacobson & Kennedy, 2011). Therefore, more thorough evaluations should be made of the use of neonicotinoids specifically for western flower thrips management in different cropping systems. Given these results, it is important also not to extrapolate from the effects of insecticides on one thrips species, even congeneric ones, to *F. occidentalis*.

Despite indications of resistance to abamectin developing (Immaraju et al., 1992), it and related materials in the avermectin class have continued to receive interest as a means to manage western flower thrips in greenhouse and open field crops. Spiers et al. (2006) recently found that abamectin was as effective as spinosad in reducing western flower thrips feeding damage to gerbera daisies (*Gerbera jamesonii*). Overall flower quality was high not only because of the low amount of damage but also because of the lack of phytotoxic effects. The abamectin derivative, emamectin benzoate, has also been found to have efficacy against western flower thrips (Ishaaya et al., 2002). This material was found to be more potent than the parent material abamectin, and it was more effective against larvae than against adults. An advantageous property of abamectin and emamectin benzoate is that they are translaminar, increasing the likelihood that toxins will reach concealed thrips adults and larvae. However, other results have shown that these avermectins are not effective against western flower thrips in field trials (Kay & Herron, 2010). They found that their test population of western flower thrips was highly susceptible to abamectin and emamectin benzoate in direct exposure assays so that the field failures could result from difficulties in applying adequate doses of avermectins to plants in the field (Broughton & Herron, 2007).

Other newly developed insecticides with novel modes of action continue to be evaluated for their potential role in western flower thrips management. Pyridalyl, which has not been classified as yet for its mode of action, is more toxic to larvae than to adults of the western flower thrips, but it is compatible with biological control agents such the predatory bug *Orius strigicollis* (Poppius) (Isayama et al., 2005). This feature makes it an attractive insecticide for rotational use in an overall integrated pest management program. Pyridalyl is registered for greenhouse ornamentals in the USA and certain other crops elsewhere in the world. Likewise, the pyrrrol chlorfenapyr is registered for greenhouse ornamental and vegetables in the USA, but is approved for use in other situations globally. It has shown efficacy comparable to spinosad in greenhouse trials conducted in Australia (Broughton & Herron, 2009). Both pyridalyl and chlorfenapyr have translaminar properties. The systemic insecticide fipronil, a phenylpyrazole, is effective against western flower thrips larvae, and to a lesser extent, adults (Kay & Herron, 2010). In an interesting approach to synergizing insecticides, Cook et al. (2002) found that the addition of dodecyl acetate, a component of the western flower thrips alarm pheromone, increased the efficacy of fipronil in field trials against western flower thrips in strawberry (*Fragaria* × *ananassa*).

There has also been recent interest in the use botanically derived insecticides for use against the western flower thrips. Certain essential oils can help reduce the incidence of tomato spotted wilt in tomato (Reitz et al., 2008), and other *Chenopodium* based materials have shown efficacy under greenhouse conditions (Chiasson et al., 2004). *Chenopodium* based products, while not highly toxic under open field conditions, provide sufficient suppression of western flower thrips larvae to warrant inclusion in insecticide rotation schemes (Funderburk, 2009). These essential oil products tend to have little negative impact on natural enemies, so they may be compatible in overall IPM programs (Bostanian et al., 2005). One drawback in the use of plant essential oils has been that concentrations of oils needed to have lethal insecticidal properties to pests can be phytotoxic (Cloyd et al., 2009, S. R. Reitz, unpublished).

There has been considerable interest in the use of microbial insecticides against western flower thrips (Butt & Brownbridge, 1997). Although experimental work has demonstrated the effectiveness of these materials and natural epizootics have been recorded (Vacante et al., 1994), there has been limited commercial success with them. Several products are available for commercial use (Shah & Goettel, 1999), and new formulations are still being developed (e.g., Zhang et al., 2009).

There has been a long history of insecticide use against the western flower thrips. Yet, there has only been a limited number of efficacious available at any given time, and none have been able to serve as a stand-alone management tactic. Although several new insecticides have been reported to have efficacy against western flower thrips, there is still a limited suite of insecticides that are effective. This limitation is likely to be an ongoing constraint because of the cost of developing and registering new insecticides. Therefore, to maintain the utility of efficacious insecticides as a part of IPM programs for western flower thrips, it is critical to take conservative and judicious approaches to the use of these insecticides.

4. Insecticide resistance management

Given the history of insecticide use against western flower thrips, resistance development is more than likely to occur to any insecticide, regardless of mode of action. Therefore, it is critical to develop strategies to employ them effectively. A key element in this regard is proper insecticide resistance management programs to maintain efficacy for as long as possible. In general, insecticide resistance management programs for western flower thrips do not differ conceptually from those designed for other pests. The basic concept is to rotate among insecticides with different modes of action at appropriate intervals to delay or inhibit the evolution of resistance within a pest population.

Most current insecticide resistance management plans recommend rotation of chemical classes after every generation of thrips (Broadbent & Pree, 1997; Herron & Cook, 2002; Robb et al., 1995). In Australia, the initial recommendations for insecticide resistance management were for growers to alternate among insecticides from different chemical classes with each application, a practice recommended in other regions (Funderburk, 2009). However, Herron and Cook (2002) proposed that this simple strategy would not be effective because of the long term persistence of resistance in populations to cypermethrin, a pyrethroid. They argued that reversion to a susceptible population would not occur before an insecticide was used again, rendering that material ineffective. It is also possible that alternating chemical classes too frequently (within a generation) could more readily select for individuals with resistance to multiple insecticides. In a subsequent study, Broughton and Herron (2007) advocated a three-consecutive spray program of a particular insecticide before rotating to a different chemical class. A key component to this strategy was that the three-consecutive applications needed to be made within a single thrips generation to gain the maximum effectiveness for that treatment.

The actual implementation of such an approach is limited by the continuous, overlapping generations present within a crop (Reitz, 2009), and may best be interpreted as rotating chemistries on an appropriate time interval (approximately 3 weeks). Bielza (2008) further cautioned that it is not simply enough to rotate among different chemistries. Rather rotation schedules should be based on known resistance mechanisms to avoid problems with cross resistance. An example would be to rotate from chemistries in which metabolic resistance is likely to develop to chemistries in which target site resistance is likely to develop. The more types of resistance modes that can be built into a rotation plan, the more effective each material would be expected to be. Unfortunately in some cropping systems, growers may be faced with having only one or two efficacious classes of insecticides, which increases the risk of resistance development (Broughton & Herron, 2007). Many growers will make applications that are mixtures of more than one insecticides (Cloyd, 2009a). This is done either to combat more than one pest at a particular time, or in the belief that better control of a particular pest can be achieved with mixtures. However, Bielza (2008) also cautioned against using mixtures of insecticides because the structure of western flower thrips populations and resistance mechanisms may actually increase rates of resistance development when mixtures are used.

Bielza (2008) outlined a general resistance management protocol that also serves as a foundation for a sound IPM program. The four recommendations are to: 1) apply insecticides only when required; 2) make accurate and precise insecticide applications; 3) diversify the types of management methods that are used in a crop; and 4) conserve natural enemies. In addition, resistance monitoring needs to be conducted on an ongoing basis so that insecticides can be quickly removed from use before complete failures occur, and so that susceptibility to those materials can be restored. The proper stewardship of insecticide use will help to forestall the development of resistance in western flower thrips populations, as well as populations of other pest species inhabiting particular crops. Even with sound insecticide resistance management programs in effect, it is clear that insecticides cannot function as a stand-alone control method for western flower thrips, and most authors have advocated that insecticides not be used as a stand-alone management tactic. In fact, there cannot be a reliance on any single tactic, and truly integrated management approaches need to be employed.

5. Western flower thrips IPM in open field vegetables in the Southeastern USA

IPM programs developed in Florida for open field vegetable crops are an example of the evolution management programs for the western flower thrips and tomato spotted wilt virus. The development of these IPM programs has relied on a thorough understanding of western flower thrips biology and ecology. Perhaps, the most important aspect for successful management is the recognition that complete control of western flower thrips and elimination of damage is not attainable. Rather, the goal should be to manage thrips within acceptable limits that do not result in economically significant damage, and this goal has become the focus of current western flower thrips management programs (Funderburk, 2009).

Northern Florida, and the rest of the southeastern USA, is a major producer of fresh market tomatoes and peppers, although farms in this region tend to be relatively small (10 – 100 ha) and dispersed throughout the landscape (Bauske, 1998). Vegetable crops in the region are grown on beds covered with plastic mulches (Castro et al., 1993). Crops are started from transplants, with the typical crop growing in the field for 12 – 14 weeks. Tomato and pepper crops in the region did not experience pest problems from thrips until the western flower thrips invaded in the 1980's (Beshear, 1983; Olson & Funderburk, 1986). Initially, damage from western flower thrips was observed from oviposition and direct feeding of adults and larvae on developing tomato fruits, which reduce their aesthetic quality and marketability (Ghidiu et al., 2006; Salguero-Navas et al., 1991). Similar scarring damage from feeding can occur on pepper fruit (Funderburk et al., 2009). However, soon after the invasion of the western flower thrips, epidemics of tomato spotted wilt began to occur throughout the southeast (Csinos et al., 2009) and crops remain at risk if proper management is not employed (Reitz et al., 2008). It is important to note that the most prevalent *Frankliniella* species in Florida and the southeastern USA are *F. tritici* (Fitch) and *F. bispinosa* (Morgan) (in southern Florida, Hansen et al., 2003), but these species do not cause the damage that western flower thrips do in vegetable crops.

The extensive crop losses caused by western flower thrips and *Tomato spotted wilt virus* have spurred considerable research to develop effective management programs. Understandably though, when western flower thrips and tomato spotted wilt first emerged as problems, tomato and pepper growers in the region responded with intensive insecticide treatments in attempts to prevent disease spread. By the 1990's growers in northern Florida were making an average of 16 separate applications of insecticides, with each application often being a mixture of multiple insecticides (Bauske, 1998). Despite such intense insecticide treatments, attempts at vector control through insecticides did not substantially reduce the problems. This lack of success results from the fact that most virus transmission in these crops is a result of primary spread of the pathogen – that is, infection comes from viruliferous individuals that disperse into the crop from external sources (Gitaitis et al., 1998; Puche et al., 1995). Furthermore, *Tomato spotted wilt virus* transmission occurs in as little as 5 minutes of an adult thrips feeding on a plant (Wijkamp et al., 1996). The following is a description of key pest management tactics that have been developed to successfully manage thrips and *Tomato spotted wilt virus* in open field fruiting vegetables.

5.1 Scouting

Because thrips species vary in their pest status, and because insecticides can differentially impact populations of native flower thrips and the invasive western flower thrips, it is necessary to accurately identify the species in order to make and evaluate management decisions. Thrips can be easily sampled by collecting flower samples into containers with alcohol. These samples can then be examined under a microscope with at least 40X magnification to determine the species. Various identification keys are available to assist with species identifications at this level (e.g., Frantz & Fasulo, n.d.). Periodic sampling can be used to assess shifts in the relative abundance of species of thrips throughout the growing season. When done systematically, this sampling is invaluable for determining the need for any insecticide application and the effects of such applications.

Sampling for crop management scouting purposes can be accomplished by counting the thrips from samples of ten flowers collected from each of several locations throughout a field. The number of samples needed to collect depends, in part, on field size. Because of the potential for scarring damage to fruit, it is also important to examine small, medium, and large fruits for thrips, with care taken to look under the calyx because of the thigmotactic behavior of thrips. Small fruits especially need to be inspected frequently as the eggs generally are laid during the flower stage, and larvae on the small fruit are the first indication of a developing problem. Again, it is important to sample fruit from several locations in each field.

5.2 Economic thresholds

Economic thresholds have been developed for thrips management in fruiting vegetable crops, including tomato, pepper and eggplant (Funderburk, 2009). These thresholds primarily apply to oviposition damage by female western flower thrips and to the feeding damage caused by adult western flower thrips and larvae of this and other thrips. Consequently, species identification in scouting is critical for the use of thresholds. Thresholds can guide growers for making therapeutic insecticide applications to mitigate these types of damage. However, growers must be aware that therapeutic treatments can do little to mitigate virus transmission. Therefore, preventative tactics must be employed to manage the primary spread of disease in crop fields. Secondary spread may be managed by scouting for larvae and treating, as appropriate.

Tomato: Although adults of all *Frankliniella* species that occur in Florida feed on pollen, petals and other floral structures in tomato, this feeding injury does not result in economic damage. However, once feeding by adults of western flower thrips and larvae of all species commences on immature fruits, it can produce "flecking" damage, which becomes apparent as the tomatoes ripen (Ghidiu et al., 2006). Oviposition in developing tomato fruit from western flower thrips also causes aesthetic damage (Salguero-Navas et al., 1991). Whereas 25 adults of native thrips (*F. tritici* and/or *F. bispinosa*) per bloom do not cause damage, one western flower thrips adult per flower is the threshold at which growers need to take action. An average of up to two larvae per small, medium or large fruit can be tolerated, but growers should take action at these thresholds (Funderburk, 2009; Funderburk et al., 2011).

Pepper and eggplant: As with tomato, adults of *F. tritici* and *F. bispinosa* cause little, if any, damage to pepper and eggplant, even with densities of 25 per bloom, and they beneficially outcompete western flower thrips and melon thrips, *Thrips palmi* Karny (Funderburk, 2009). Direct feeding damage from adults of western flower thrips and melon thrips is less severe than in tomato, and oviposition in immature pepper and eggplant fruit by western flower thrips does not cause the damage that is typical in tomato and some other crops. Therefore, higher thresholds can be tolerated in pepper and eggplant than in tomato. Up to six (6) western flower thrips and/or melon thrips adults per flower can be tolerated without damage. Once fruits begin to develop, growers need to be aware of scarring damage that adults of the western flower thrips and the melon thrips and larvae of all species may cause. Up to two larvae per small, medium, or large fruit on average in a field are tolerable. Growers should be prepared to take action if larval populations exceed two per fruit (Funderburk 2009). Because of the critical role that *Orius* spp. play in suppressing thrips populations and secondary virus spread, their populations should be monitored in scouting programs, and considered when assessing the need for insecticide treatments.

5.3 Biological control

Despite certain similarities between crops of tomato and pepper, there are fundamental differences in interactions between thrips and these two plant species. These differences mean that management programs must be designed for each crop. Adult western flower thrips readily colonize pepper and tomato (Baez et al., 2011). Tomato, though, is not a significant reproductive host for western flower thrips (Funderburk, 2009), but pepper can be a good reproductive host for these thrips (van den Meiracker & Ramakers, 1991). Consequently, there is the potential for secondary virus spread from within the crop (Gitaitis et al., 1998). However, Funderburk et al. (2000) demonstrated that the predator

Orius insidiosus (Say) colonizes peppers and can effectively suppress thrips populations in the crop (Funderburk et al., 2000; Ramachandran et al., 2001). In particular, *O. insidiosus* preferentially preys on western flower thrips over the native species *F. tritici* and *F. bispinosa* (Baez et al., 2004; Reitz et al., 2006). Therefore, conservation of these valuable naturally occurring biological control agents can significantly reduce pest problems in pepper and related crops (e.g., eggplant, *Solanum melongena*) and has become a cornerstone of IPM for pepper production (Funderburk et al., 2009). One of the keys to conservation of *Orius* species is to use insecticides that are minimally toxic to *Orius* spp., whether for thrips management or for management of other pests (Reitz et al., 2003). In contrast to pepper, *Orius* species do not have an affinity for tomato (Baez et al., 2011; Pfannenstiel & Yeargan, 1998), and so this naturally occurring biological control is not available for tomato.

5.4 Interspecific competition

Biotic limitations on western flower thrips populations can come from other species of thrips as well as predators, such as *O. insidiosus*. Recent studies have shown that interspecific competition from native thrips limits the larval survivorship of western flower thrips (Paini et al., 2008). In a survey conducted in northern Florida, Northfield et al. (2008) observed that over 75% of thrips collected from a range of uncultivated hosts were the native species *F. tritici*, with only 1% being western flower thrips. Similar results have been observed in crop fields where two-thirds or more of thrips in untreated or spinosad-treated pepper are the native species. Yet, the demographics differ in pyrethroid treated plots, where western flower thrips predominant (Hansen et al., 2003; Reitz et al., 2003). Because the native species *F. tritici* and *F. bispinosa* do not cause the economic damage that western flower thrips do, and because they outcompete western flower thrips, their conservation contributes to overall pest management. This difference in pest status among the species is why species identifications are an essential component of scouting in IPM programs.

5.5 Host plant location and ultraviolet reflective mulches

Thrips locate host plants primarily through a combination of visual cues, with anthophilous thrips tending to be attracted to colors of flowers. Western flower thrips are attuned to spectral radiation in the ultraviolet range (~365 nm) and in the yellow-green range (~540 nm) (Matteson et al., 1992). The yellow-green sensitivity is thought to play a role in long distance orientation to plants, and the ultraviolet sensitivity is part of the visual system to distinguish flowers. Anthophilous thrips, such as western flower thrips, are attracted to colors of flowers, especially white, blue and yellow flowers with low ultraviolet reflectance (Antignus, 2000). Therefore, increasing the reflectivity in ultraviolet range of the spectrum can repel thrips.

The ultraviolet reflective mulches available for the raised-bed plastic mulch production system of Florida are effective in repelling migrating adults of the western flower thrips, and this repellency reduces the primary and secondary spread of tomato spotted wilt. The use of ultraviolet reflective mulch also reduces the influx of the native thrips, *F. tritici* and *F. bispinosa*, but not disproportionately to reductions in western flower thrips (Momol et al., 2004; Reitz et al., 2003). Ultraviolet reflective mulches are most effective early in the crop season before the plant canopy begins to cover the mulch and reduce the surface area available for reflectance. Application of certain bactericides/fungicides and other pesticides also reduces the ultraviolet reflectance and hence the efficacy of the mulch. A single application of copper and mancozeb for bacterial or fungal control can reduce the reflectance by nearly 50%. Repeated applications can consequently lead to higher incidences of tomato spotted wilt (S. R. Reitz, unpublished data). Therefore, using alternatives to copper and mancozeb early in the season for foliar pathogen management is advisable.

Ultraviolet reflective mulches also deter other pests, especially whiteflies and aphids, which can vector other plant viruses (Fanigliulo et al., 2009; Stapleton & Summers, 2002; Summers et al., 2010). Consequently, these materials are a good overall IPM tactic to employ where insect vectors are of concern. However, growers need to balance these benefits with the potential delay in plant growth in the spring because these mulches do not warm the soil as readily as standard black plastic mulches (Harpaz, 1982; Maynard & Olson, 2000). Newly developed mulches have helped to mitigate this effect by excluding the reflective metalized layer where transplants are placed.

5.6 Host plant fertilization

Soils in the southeastern USA tend to be nutrient deficient, so that growers need to add up to 200 kg of nitrogen per hectare (the recommended rate for Florida, Olson & Simmonne, 2009). However, growers have often overfertilized crops by up to 70% (Castro et al., 1993). This extra nitrogen fertilization can actually increase densities of western flower thrips. Female western flower thrips, in particular, preferentially settle on plants with higher nitrogen content (Baez et al., 2011; Brodbeck et al., 2001). This association seems to be most closely related to the phenylalanine content of tomato. From a pest management perspective, as vector populations increase with increasing fertilization there is an increase in the incidence of tomato spotted wilt. In north Florida tomatoes, the incidence of tomato spotted wilt was 50% lower for plants grown at recommended nitrogen levels compared with plants grown with supraoptimal nitrogen (Stavisky et al., 2002). Interestingly, F. tritici and F. bispinosa do not respond in the same manner to nitrogen fertilization as western flower thrips (Baez et al., 2011; Stavisky et al., 2002). Besides the increasing pest problems, excess nitrogen fertilization does not increase per plant yield. Therefore, growers can improve overall crop production of tomatoes and pepper by maintaining optimal fertilization levels.

5.7 Systemic acquired resistance

Many plants possess traits for systemic acquired resistance, which are induced defensive mechanisms against pathogens (Sticher et al., 1997). Certain chemicals have been found to stimulate these natural plant defenses against pathogens when applied to plants before infection occurs. Acibenzolar-S-methyl is a systemic acquired resistance inducer that stimulates the salicylic acid pathway for disease resistance in tomato and other crops. Commercial formulations of acibenzolar-S-methyl have been shown to reduce the incidence of tomato spotted wilt (Momol et al., 2004). Its use has minimal impacts on populations of the flower thrips. When tomatoes are grown on ultraviolet reflective mulches, Momol et al. (2004) concluded that acibenzolar-S-methyl provided little additional disease protection because of the large effect of the mulch. Nevertheless, acibenzolar-S-methyl is highly effective against bacterial pathogens that afflict tomatoes (Obradovic et al., 2005; Pradhanang et al., 2005), making it an excellent replacement for copper and mancozeb sprays on ultraviolet reflective mulches as well as standard black plastic mulch.

5.8 Host plant resistance

The single best defense against insect-vectored pathogens is host plant resistance. Numerous cultivars of tomato and pepper are resistant or tolerant to *Tomato spotted wilt virus* are currently commercially available (for a partial listing, see Funderburk et al., 2011). These cultivars have resistance to the virus, but not to thrips feeding or oviposition. All of the commercially available cultivars of tomato share a single source of resistance from the *Sw-5* gene. In pepper, all resistance is conferred by the *Tsw* gene. Both of the *Sw-5* and *Tsw* genes appear to be single dominant genes (Boiteux & de Avila, 1994; Stevens et al., 1992), and thus susceptible to being compromised by resistance breaking strains of the virus. In fact, such resistance breaking strains have commonly developed around the world (Roselló et al., 1996; Sharman & Persley, 2006). Presently in Florida, tomato spotted wilt resistant cultivars can maintain tomato spotted wilt incidences at economically acceptable levels. However, the threat of epidemics from resistance breaking strains is real. Coupled with the potential damage from western flower thrips feeding and oviposition, growers must maintain an integrated approach to thrips and tomato spotted wilt management.

5.9 Insecticides

Insecticides continue to have an important role to play in western flower thrips management. However, the use of insecticides must be done judiciously. Decisions regarding which insecticides to use and when need to be made in the context of both short-term and long-term management goals. Minimizing resistance development and avoiding the flaring of western flower thrips populations by their release from natural enemies need to be critical factors in insecticide use decisions. Populations of the invasive western flower thrips likely arrived in Florida with resistance to most classes of broad-spectrum insecticides (Immaraju et al., 1992). Further, flaring of the populations of the western flower thrips and other pests is possible when any broad-spectrum synthetic insecticide is used (Funderburk et al., 2000; Reitz et al., 2003). For this reason, growers are encouraged to move to newer, safer, and more selective insecticides in different chemical classes that are becoming available. Although growers are encouraged to use more selective materials when needed, the use of certain organophosphate and carbamate insecticides against western flower thrips may be warranted in certain circumstances. These should only be used in particular instances when nontarget effects would be minimal, for instance near the end of the production season to prevent scarring damage to fruit.

As discussed above, the most efficacious insecticides for western flower thrips, at present, are in the spinosyn class. No other insecticide class provides a similar level of effectiveness against western flower thrips. However, as resistance to spinosyns has been documented in Florida and elsewhere, limits are being placed on the number of applications that can be made in each crop to forestall further resistance development. A number of other insecticides are registered or in the process of being registered by the United States Environmental Protection Agency (EPA) that are able to suppress western flower thrips adults and larvae. Lists of currently available insecticides for western flower thrips management and their role in overall IPM programs for fruiting vegetable crops are available (Funderburk, 2009; Funderburk et al., 2011; Funderburk et al., 2009).

The fact that these materials are not as efficacious as spinosyns should not deter their inclusion in IPM programs. The focus of management should not be placed on killing the maximum number of thrips. Rather, the focus of management should be in minimizing damage below economically injurious levels. Because economic damage from oviposition and scarring from feeding only occur at high levels (see above), even limited suppression of western flower thrips adults and larvae can maintain these types of damage well within tolerable limits. Secondary spread of tomato spotted wilt in tomato can also be limited by suppressing populations rather than attempting complete control (Momol et al., 2004). In pepper, conservation of *O. insidiosus* significantly reduces both primary and secondary spread of tomato spotted wilt (Funderburk et al., 2000; Reitz et al., 2003). We have found repeatedly that avoiding treatments that induce outbreaks of western flower thrips populations by killing natural enemies and competing species of native thrips within crop fields is an effective approach to minimizing losses to western flower thrips. Most broad-spectrum synthetic insecticides, including pyrethroids, organophosphates, and carbamates kill the native species of thrips that outcompete western flower thrips (Hansen et al., 2003; Reitz et al., 2003; Srivistava et al., 2008), leading to dramatic large scale shifts in thrips demographics (Frantz & Mellinger, 2009). These synthetic broad-spectrum insecticides not only can disrupt western flower thrips management, they also can disrupt management of other pests including spider mites, whiteflies, and leafminers, by eliminating natural enemies of those pests.

5.10 Vertical integration of the management program
One of the most important keys to successful crop production is not to consider problems in isolation. It is critical to understand how one management tactic may affect other production aspects. For example, in northern Florida, western flower thrips and *Tomato spotted wilt virus* are clearly the most important pest-complex facing tomato and pepper production. The use of ultraviolet reflective mulches has been effective in reducing populations of western flower thrips and the incidence of tomato spotted wilt (e.g., Momol et al., 2004; Reitz et al., 2003). Still, sweetpotato whitefly (*Bemisia tabaci* [Gennadius]) and whitefly-vectored viruses are occasionally important pests in northern Florida tomatoes (Momol et al., 1999). Ultraviolet reflective mulches used to manage thrips and tomato spotted wilt are also efficacious in reducing whitefly-caused damage (Antignus, 2000; Csizinszky et al., 1999).

In contrast, in southern Florida, western flower thrips and tospoviruses have only recently emerged as damaging problems requiring management consideration. Whiteflies and whitefly-vectored viruses have historically been the key insect pest and disease problems in tomato there. Growers use an wide range of insecticides to manage whitefly vectored viruses, particularly *Tomato yellow leaf curl virus*, which can devastate entire crops (Moriones & Navas-Castillo, 2000). Most tomatoes in southern Florida are treated with neonicotinoid insecticides at planting for management of immature whiteflies, with imidacloprid, thiamethoxam, or dinotefuran being most commonly used (Schuster et al., 2010). As the season progresses, growers may rotate "soft" insecticides, such as azadirachtin-based products, microbial insecticides such as *Beauveria bassiana* and insect growth regulators, into management programs against whitefly nymphs. While such materials are compatible with thrips management, they have little effect in suppressing primary virus transmission by whitefly adults that disperse into fields. As a result, growers still place a heavy reliance on broad spectrum insecticides, such as organophosphates and pyrethroids, for management of primary virus spread by whiteflies. The unintended consequence of this approach to whitefly management has been to release populations of western flower thrips from their natural controls, which greatly complicates overall crop management (Frantz & Mellinger, 2009; Weiss et al., 2009).

To facilitate overall crop management, growers are advised to anticipate key pests such as whiteflies or western flower thrips and to employ preventive tactics to minimize their impact. For tomato, one such preventative tactic would ultraviolet reflective mulches. It is also important from an areawide management perspective to maintain crop free periods and remove crops immediately after harvest so that crop residues do not serve as reservoirs for later infestations. In pepper grown in southern Florida, pepper weevil, *Anthonomus eugenii* Cano, is another significant pest. Its management also can be facilitated by crop free periods, and the destruction of crops immediately after harvest. Further sanitation, including the control of solanaceous weeds that serve as alternative hosts helps to reduce future populations. Using such preventative measures would minimize the need for insecticide applications for this pest. Consequently, there would be an overall benefit crop management because many of the available insecticides for pepper weevil management are pyrethroids or other disruptive broad spectrum synthetic insecticides. These are just some of the many pest problems that growers must contend with. Consequently, there is a clear need to integrate management programs for the diverse pests attacking crops. It is also critical to provide growers with realistic economic thresholds for different pests, and proper scouting techniques to assess pest abundances and the need, if any, to apply pesticides.

6. Conclusions and future directions

Only recently has the reliance on insecticides for western flower thrips management been challenged. Yet, failures to control western flower thrips with insecticides have become so severe that Cloyd (2009b) suggested we have reached an impasse in the use of insecticides against western flower thrips in ornamental greenhouse production and diverse management tactics must be employed. The successful management programs for western flower thrips and thrips-vectored viruses developed for solanaceous crops in Florida have been based on an understanding of thrips ecology and how different species interact with different crops (Funderburk, 2009). These strategies involve: an emphasis on scouting and the identification of thrips species present in a crop; the optimal nitrogen fertilization inputs to reduce the attractiveness of crops to western flower thrips without adversely affecting yields; the use of ultraviolet reflective mulches to deter thrips entrance into crop fields; the use of acibenzolar-S-methyl to suppress development of tomato spotted wilt symptoms in fruit of susceptible varieties; and the use of economic thresholds to determine the need and timing of insecticide applications; and the use of select insecticides to help suppress reproduction of thrips in the field, and thus manage secondary spread of tomato spotted wilt from within fields. The understanding of the importance of biotic resistance against western flower thrips provided by competing species of native thrips and natural enemies, such as *O. insidiosus*, has been of fundamental importance to improving western flower thrips management. This low input approach helps to avoid other pest problems as well. For example, the conservation biological control program for western flower thrips and *Tomato spotted wilt virus* in peppers has been used by growers in north Florida since the late 1990's, and growers have experienced far fewer problems from whiteflies, aphids, and other pests than when they were following a prophylactic, calendar-based spray program that included frequent use of broad-spectrum insecticides to control pests.

If western flower thrips were the only pest concern for growers, they would, perhaps, feel relieved. However, there are many other insect and pathogen threats to crops. We have now begun to appreciate how management of one pest can interact with and affect management of other pests. Therefore, the need for sustainable, truly integrated pest management programs that do not consider individual pests in isolation is clear. There must be a focus on pest complexes rather than considering individual pest species. For example, insecticide resistance can be exacerbated when western flower thrips are exposed to insecticide treatments aimed at other pests (Immaraju et al., 1992).

The development of more selective insecticides that do not adversely affect natural enemies will improve overall crop management. These materials would reduce the risk of releasing non-target pests from control when treatments are necessary for another pest. Additional management tools that would reduce the need for insecticide applications will continue to be beneficial. Improved host plant resistance is one of those areas. The recent development of cultivars of tomatoes and peppers that show some degree of resistance to *Tomato spotted wilt virus* has eased concerns among growers. However, because the resistance conferred by the *Sw-5* gene in tomato and the *Tsw* gene in pepper is not durable, resistance-breaking strains can readily develop (Aramburu & Marti, 2003; Ciuffo et al., 2005; Latham & Jones, 1998; Margaria et al., 2004; Roggero et al., 2002; Thomas-Carroll & Jones, 2003; Thompson & van Zijl, 1995). In addition, virus resistance does not protect against physical damage. Therefore, it is still important for growers to maintain a multifaceted integrated management program.

Currently there are major research efforts underway to identify new germplasm sources of resistance to *Tomato spotted wilt virus*. One promising source of resistance appears to be the *Sw-7* gene from *Solanum chilense* (Price et al., 2007). Although this resistance is another single gene dominant trait, it is not linked to the *Sw-5* gene. Another exciting prospect for the future of host plant resistance is the identification of germplasm that is resistant to western flower thrips. Several accessions of different *Capsicum* species have been identified that have significant resistance against western flower thrips feeding (Maharijaya et al., 2011). If these traits can be incorporated into commercial cultivars, they may offer some degree of protection against virus transmission and aesthetic damage.

The management of western flower thrips will continue to be an ongoing challenge. Management programs cannot be static as they will need to be continually refined and updated with the advent of new invasive pests or as other conditions change. For example, *Scirtothrips dorsalis* Hood invaded Florida within the past five years and poses a threat to crops including pepper and eggplant. Recently, a new form of *Tospovirus*, a genetic reassortant of *Groundnut ringspot virus* and *Tomato chlorotic spot virus* has been found in Florida (Webster et al., 2010). Although western flower thrips can transmit this virus (Webster et al., 2011), it is not yet known if it or another thrips species is the most important vector. Despite these challenges, we believe that western flower thrips can be successfully managed, given a thorough understanding of its ecology and pest status.

7. Acknowledgments

We appreciate the collaborative efforts of Steve Olson, Pete Andersen, Brent Brodbeck, Julianne Stavisky, Mrittunjai Srivistava, M. Timur Momol, and Norm Leppla (University of Florida), Anthony Weiss and Jim Dripps (Dow AgroSciences), Charles Mellinger and Galen Frantz (Glades Crop Care), and Scott Adkins (USDA-ARS) for their insights and contributions to these research projects and to the management programs that have developed from this research. We appreciate the comments by Katherine Luhring, which have improved this manuscript. We are especially indebted to Laurence Mound (CSIRO) for many invaluable discussions regarding thrips taxonomy, biology and ecology. We recognize the contributions of Larry Pedigo (Professor Emeritus, Iowa State University) to the theory and practice of integrated pest management generally and to stressing the importance of understanding pest status in particular.

8. References

Antignus, Y. (2000) Manipulation of wavelength-dependent behaviour of insects: An IPM tool to impede insects and restrict epidemics of insect-borne viruses. *Virus Research* 71: 213-220.

Antignus, Y., Mor, N., Joseph, R.B., Lapidot, M., & Cohen, S. (1996) Ultraviolet-absorbing plastic sheets protect crops from insect pests and from virus diseases vectored by insects. *Environmental Entomology* 25: 919-924.

Aramburu, J., & Marti, M. (2003) The occurrence in north-east Spain of a variant of Tomato spotted wilt virus (TSWV) that breaks resistance in tomato *(Lycopersicon esculentum)* containing the Sw-5 gene. *Plant Pathology* 52: 407.

Baez, I., Reitz, S.R., & Funderburk, J.E. (2004) Predation by *Orius insidiosus* (Heteroptera: Anthocoridae) on species and life stages of *Frankliniella* flower thrips (Thysanoptera: Thripidae) in pepper flowers. *Environmental Entomology* 33: 662-670.

Baez, I., Reitz, S.R., Funderburk, J.E., & Olson, S.M. (2011) Variation within and between *Frankliniella* thrips species in host plant utilization. *Journal of Insect Science* 11: 41.

Bauske, E.M. (1998) Southeastern tomato growers adopt integrated pest management. *HortTechnology* 8: 40-44.

Beshear, R.J. (1983) New records of thrips in Georgia (Thysanoptera, Terebrantia, Tubulifera). *Journal of Georgia Entomological Society* 18: 342-344.

Bielza, P. (2008) Insecticide resistance management strategies against the western flower thrips, *Frankliniella occidentalis*. *Pest Management Science* 64: 1131-1138.

Bielza, P., Espinosa, P.J., Quinto, V., Abellán, J., & Contreras, J. (2007a) Synergism studies with binary mixtures of pyrethroid, carbamate and organophosphate insecticides on *Frankliniella occidentalis* (Pergande). *Pest Management Science* 63: 84-89.

Bielza, P., Quinto, V., Contreras, J., Torne, M., Martin, A., & Espinosa, P.J. (2007b) Resistance to spinosad in the western flower thrips, *Frankliniella occidentalis* (Pergande), in greenhouses of south-eastern Spain. *Pest Management Science* 63: 682-687.

Bielza, P., Quinto, V., Fernandez, E., Gravalos, C., & Contreras, J. (2007c) Genetics of spinosad resistance in *Frankliniella occidentalis* (Thysanoptera: Thripidae). *Journal of Economic Entomology* 100: 916-920.

Bielza, P., Quinto, V., Grávalos, C., Abellán, J., & Fernández, E. (2008) Lack of fitness costs of insecticide resistance in the western flower thrips (Thysanoptera: Thripidae). *Journal of Economic Entomology* 101: 499-503.

Bohart, R.M. (1943) Calcium cyanide fumigation for the western thrips. *Journal of Economic Entomology* 36: 442-444.

Boiteux, L.S., & de Avila, A.C. (1994) Inheritance of a resistance specific to tomato spotted wilt Tospovirus in *Capsicum chinense* 'PI 159236'. *Euphytica* 75: 139-142.

Bonarriva, J. (2003) *Cut Flowers: Industry and Trade Summary*, United States International Trade Commission, Available from: http://www.usitc.gov/publications/332/pub3580.pdf.

Borden, A.D. (1915) The mouthparts of the Thysanoptera and the relation of thrips to the non-setting of certain fruits and seeds. *Journal of Economic Entomology* 8: 354-360.

Bostanian, N.J., Akalach, M., & Chiasson, H. (2005) Effects of a *Chenopodium*-based botanical insecticide/acaricide on *Orius insidiosus* (Hemiptera: Anthocoridae) and *Aphidius colemani* (Hymenoptera: Braconidae). *Pest Management Science* 61: 979-984.

Broadbent, A.B., & Pree, D.J. (1997) Resistance to insecticides in populations of *Frankliniella occidentalis* (Pergande) (Thysanoptera: Thripidae) from greenhouses in the Niagara region of Ontario. *Canadian Entomologist* 129: 907-913.

Brodbeck, B.V., Stavisky, J., Funderburk, J.E., Andersen, P.C., & Olson, S.M. (2001) Flower nitrogen status and populations of *Frankliniella occidentalis* feeding on *Lycopersicon esculentum*. *Entomologia Experimentalis et Applicata* 99: 165-172.

Brødsgaard, H.F. (1993) Cold hardiness and tolerance to submergence in water in *Frankliniella occidentalis* (Thysanoptera: Thripidae). *Environmental Entomology* 22: 647-653.

Brødsgaard, H.F. (1994) Insecticide resistance in European and African strains of western flower thrips (Thysanoptera: Thripidae) tested in a new residue-on-glass test. *Journal of Economic Entomology* 87: 1141-1146.

Broughton, S., & Herron, G.A. (2007) *Frankliniella occidentalis* (Pergande) (Thysanoptera: Thripidae) chemical control: insecticide efficacy associated with the three consecutive spray strategy. *Australian Journal of Entomology* 46: 140-145.

Broughton, S., & Herron, G.A. (2009) Potential new insecticides for the control of western flower thrips (Thysanoptera: Thripidae) on sweet pepper, tomato, and lettuce. *Journal of Economic Entomology* 102: 646-651.

Bryan, D.E., & Smith, R.F. (1956) The *Frankliniella occidentalis* (Pergande) complex in California (Thysanoptera: Thripidae). *University of California Publications in Entomology* 10: 359-410.

Butt, T.M., & Brownbridge, M. (1997) Fungal pathogens of thrips, In: T. Lewis (Ed.), *Thrips as crop pests*, CAB International, New York. pp. 399-433.

Castro, B.F., Durden, B.R., Olson, S.M., & Rhoads, F.M. (1993) Telogia Creek irrigation energy conservation demonstration on mulched staked tomatoes. *Proceedings of the Florida State Horticultural Society* 106: 219-222.

Chiasson, H., Vincent, C., & Bostanian, N.J. (2004) Insecticidal properties of a *Chenopodium*-based botanical. *Journal of Economic Entomology* 97: 1378-1383.

Childers, C.C. (1997) Feeding and oviposition injuries to plants, In: T. Lewis (Ed.), *Thrips as crop pests*, CAB International, New York. pp. 505-537.

Childs, L. (1927) Two Species of Thrips injurious to Apples in the Pacific Northwest. *Journal of Economic Entomology* 20: 805-808 pp.

Ciuffo, M., Finetti-Sialer, M.M., Gallitelli, D., & Turina, M. (2005) First report in Italy of a resistance-breaking strain of *Tomato spotted wilt virus* infecting tomato cultivars carrying the *Sw5* resistance gene. *Plant Pathology* 54: 564.

Cloyd, R.A. (2009a) Getting mixed-up: Are greenhouse producers adopting appropriate pesticide mixtures to manage arthropod pests? *HortTechnology* 19: 638-646.

Cloyd, R.A. (2009b) Western flower thrips *(Frankliniella occidentalis)* management on ornamental crops grown in greenhouses: Have we reached an impasse? *Pest Technology* 3: 1-9.

Cloyd, R.A., Galle, C.L., Keith, S.R., Kalscheur And, N.A., & Kemp, K.E. (2009) Effect of commercially available plant-derived essential oil products on arthropod pests. *Journal of Economic Entomology* 102: 1567-1579.

Cloyd, R.A., & Sadof, C.S. (1998) Flower quality, flower number, and western flower thrips density on transvaal daisy treated with granular insecticides. *HortTechnology* 8: 567-570.

Cook, D.F., Dadour, I.R., & Bailey, W.J. (2002) Addition of alarm pheromone to insecticides and the possible improvement of the control of the western flower thrips, *Frankliniella occidentalis* (Pergande) (Thysanoptera: Thripidae). *International Journal of Pest Management* 48: 287-290.

Coutts, B.A., & Jones, R.A.C. (2005) Suppressing spread of Tomato spotted wilt virus by drenching infected source or healthy recipient plants with neonicotinoid insecticides to control thrips vectors. *Annals of Applied Biology* 146: 95-103.

Crawford, D.L. (1915) Potato curly leaf (caused by *Euthrips occidentalis*). *California State Commission of Horticulture, Monthly Bulletin* 4: 389-391, Number 8, August.

Csinos, A., Martinez-Ochoa, N., & Bertrand, P. (2009) History, In: S.W. Mullis (Ed.), *Tospoviruses in Solanaceae and Other Crops in the Coastal Plain of Georgia*, University of Georgia, Athens, GA. pp. 7-8.

Csizinszky, A.A., Schuster, D.J., & Polston, J.E. (1999) Effect of ultraviolet-reflective mulches on tomato yields and on the silverleaf whitefly. *Hortscience* 34: 911-914.

Dağli, F., & Tunç, I. (2007) Insecticide resistance in *Frankliniella occidentalis* (Pergande) (Thysanoptera: Thripidae) collected from horticulture and cotton in Turkey. *Australian Journal of Entomology* 46: 320-324.

Dağlı, F., & Tunç, I. (2008) Insecticide resistance in *Frankliniella occidentalis*: Corroboration of laboratory assays with field data and cross-resistance in a cypermethrin-resistant strain. *Phytoparasitica* 36: 352-359.

Daughtrey, M.L., Jones, R.K., Moyer, J.W., Daub, M.E., & Baker, J.R. (1997) Tospoviruses strike the greenhouse industry: INSV has become a major pathogen on flower crops. *Plant Disease* 81: 1220-1230.

Denholm, I., Cahill, M., Dennehy, T.J., & Horowitz, A.R. (1998) Challenges with managing insecticide resistance in agricultural pests, exemplified by the whitefly *Bemisia tabaci*. *Philosophical Transactions of the Royal Society of London Series B Biological Sciences* 353: 1757-1767.

Dripps, J.E., Gomez, L.E., Weiss, A.W., Funderburk, J., Castro, B.A., & Paroonagian, D.L. (2010) Insecticide rotation as a component of thrips resistance management programs. *Resistant Pest Management Newsletter* 19: 32-35.

Eger, J.E., Stavisky, J., & Funderburk, J.E. (1998) Comparative toxicity of spinosad to *Frankliniella* spp. (Thysanoptera: Thripidae), with notes on a bioassay technique. *Florida Entomologist* 81: 547-551.

Espinosa, P.J., Bielza, P., Contreras, J., & Lacasa, A. (2002a) Field and laboratory selection of *Frankliniella occidentalis* (Pergande) for resistance to insecticides. *Pest Management Science* 58: 920-927.

Espinosa, P.J., Bielza, P., Contreras, J., & Lacasa, A. (2002b) Insecticide resistance in field populations of *Frankliniella occidentalis* (Pergande) in Murcia (south-east Spain). *Pest Management Science* 58: 967-971.

Espinosa, P.J., Contreras, J., Quinto, V., Grávalos, C., Fernández, E., & Bielza, P. (2005) Metabolic mechanisms of insecticide resistance in the western flower thrips, *Frankliniella occidentalis* (Pergande). *Pest Management Science* 61: 1009-1015.

Essig, E.O. (1926) *Insects of Western North America* Macmillan, New York.

Fanigliulo, A., Comes, S., Pacella, R., Crescenzi, A., Momol, M.T., Olson, S.M., & Reitz, S. (2009) Integrated management of viral diseases in field-grown tomatoes in southern Italy. *Acta Horticulturae* 808: 387-392.

Frantz, G., & Fasulo, T.R. (n.d.) Thrips KnowledgeBase, Accessed: 20 June 2011, Available from: http://www.gladescropcare.com/pg1.html.

Frantz, G., & Mellinger, H.C. (2009) Shifts in western flower thrips, *Frankliniella occidentalis* (Thysanoptera: Thripidae), population abundance and crop damage. *Florida Entomologist* 92: 29-34.

Funderburk, J. (2009) Management of the western flower thrips (Thysanoptera: Thripidae) in fruiting vegetables. *Florida Entomologist* 92: 1-6.

Funderburk, J., Reitz, S., Olson, S., Stansly, P., Smith, H., McAvoy, G., Demirozer, O., Snodgrass, C., Paret, M., & Leppla, N. (2011) Managing thrips and tospoviruses in tomato, University of Florida, Gainesville, FL.

Funderburk, J., Reitz, S., Stansly, P., Schuster, D., Nuessly, G., & Leppla, N. (2009) Managing thrips in pepper and eggplant, ENY-658, Accessed: 26 July 2011, Available from: http://edis.ifas.ufl.edu.

Funderburk, J., Stavisky, J., & Olson, S. (2000) Predation of *Frankliniella occidentalis* (Thysanoptera: Thripidae) in field peppers by *Orius insidiosus* (Hemiptera: Anthocoridae). *Environmental Entomology* 29: 376-382.

Georghiou, G.P., & Taylor, C.E. (1986) Factors influencing the evolution of resistance, In: Committee on Strategies for the Management of Pesticide Resistant Pest Populations: National Research Council (Ed.), *Pesticide resistance: strategies and tactics for management*, National Academies Press, Washington, DC. pp. 157-169.

Ghidiu, G.M., Hitchner, E.M., & Funderburk, J.E. (2006) Goldfleck damage to tomato fruit caused by feeding of *Frankliniella occidentalis* (Thysanoptera: Thripidae). *Florida Entomologist* 89: 279-281.

Gitaitis, R.D., Dowler, C.C., & Chalfant, R.B. (1998) Epidemiology of tomato spotted wilt in pepper and tomato in southern Georgia. *Plant Disease* 82: 752-756.

Goldbach, R., & Peters, D. (1994) Possible causes of the emergence of tospovirus diseases. *Seminars in Virology* 5: 113-120.

Hale, R.L., & Shorey, H.H. (1965) Systemic insecticides for the control of western flower thrips on bulb onions. *Journal of Economic Entomology* 58: 793-794.

Hansen, E.A., Funderburk, J.E., Reitz, S.R., Ramachandran, S., Eger, J.E., & McAuslane, H. (2003) Within-plant distribution of *Frankliniella* species (Thysanoptera: Thripidae) and *Orius insidiosus* (Heteroptera: Anthocoridae) in field pepper. *Environmental Entomology* 32: 1035-1044.

Harding, J.A. (1961a) Effect of migration, temperature, and precipitation on thrips infestations in south Texas. *Journal of Economic Entomology* 54: 77-79.

Harding, J.A. (1961b) Studies on the control of thrips attacking onions. *Journal of Economic Entomology* 54: 1254-1255.

Harpaz, I. (1982) Nonpesticidal control of vector-borne viruses, In: K.F. Harris, & K. Maramorosch (Eds.), *Pathogens, Vectors and Plant Diseases: Approaches to Control*, Academic Press, New York. pp. 1-21.

Herron, G.A., & Cook, D.F. (2002) Initial verification of the resistance management strategy for *Frankliniella occidentalis* (Pergande) (Thysanoptera: Thripidae) in Australia. *Australian Journal of Entomology* 41: 187-191.

Herron, G.A., & Gullick, G.C. (2001) Insecticide resistance in Australian populations of *Frankliniella occidentalis* (Pergande) (Thysanoptera: Thripidae) causes the abandonment of pyrethroid chemicals for its control. *General and Applied Entomology* 30: 21-26.

Herron, G.A., & James, T.M. (2005) Monitoring insecticide resistance in Australian *Frankliniella occidentalis* (Pergande) (Thysanoptera: Thripidae) detects fipronil and spinosad resistance. *Australian Journal of Entomology* 44: 299-303.

Huang, S.W. (2004) *Global Trade Patterns in Fruits and Vegetables*, USDA Economic Research Service, Available from: http://www.ers.usda.gov/publications/WRS0406/WRS0406.pdf.

Immaraju, J.A., Paine, T.D., Bethke, J.A., Robb, K.L., & Newman, J.P. (1992) Western flower thrips (Thysanoptera: Thripidae) resistance to insecticides in coastal California greenhouses. *Journal of Economic Entomology* 85: 9-14.

IRAC International MoA Working Group. (2011) IRAC MoA (Insecticide Resistance Action Committee Mode of Action) Classification Scheme, version 7.1, June 2011, Accessed: 3 July 2011, Available from: http://www.irac-online.org/wp-content/uploads/2009/09/MoA_Classification.pdf.

Isayama, S., Saito, S., Kuroda, K., Umeda, K., & Kasamatsu, K. (2005) Pyridalyl, a novel insecticide: Potency and insecticidal selectivity. *Archives of Insect Biochemistry and Physiology* 58: 226-233.

Ishaaya, I., Kontsedalov, S., & Horowitz, A.R. (2002) Emamectin, a novel insecticide for controlling field crop pests. *Pest Management Science* 58: 1091-1095.

Jacobson, A.L., & Kennedy, G.G. (2011) The effect of three rates of cyantraniliprole on the transmission of tomato spotted wilt virus by *Frankliniella occidentalis* and *Frankliniella fusca* (Thysanoptera: Thripidae) to *Capsicum annuum*. *Crop Protection* 30: 512-515.

Jensen, F. (1973) Timing of halo spotting by flower thrips on table grapes. *California Agriculture* 31: 6-8.

Jensen, S.E. (1998) Acetylcholinesterase activity associated with methiocarb resistance in a strain of western flower thrips, *Frankliniella occidentalis* (Pergande). *Pesticide Biochemistry and Physiology* 61: 191-200.

Jensen, S.E. (2000a) Insecticide resistance in the western flower thrips, *Frankliniella occidentalis*. *Integrated Pest Management Reviews* 5: 131-146.

Jensen, S.E. (2000b) Mechanisms associated with methiocarb resistance in *Frankliniella occidentalis* (Thysanoptera: Thripidae). *Journal of Economic Entomology* 93: 464-471.

Joost, P.H., & Riley, D.G. (2005) Imidacloprid effects on probing and settling behavior of *Frankliniella fusca* and *Frankliniella occidentalis* (Thysanoptera: Thripidae) in tomato. *Journal of Economic Entomology* 98: 1622-1629.

Kay, I.R., & Herron, G.A. (2010) Evaluation of existing and new insecticides including spirotetramat and pyridalyl to control *Frankliniella occidentalis* (Pergande) (Thysanoptera: Thripidae) on peppers in Queensland. *Australian Journal of Entomology* 49: 175-181.

Kirk, W.D.J. (1997a) Distribution, abundance and population dynamics, In: T. Lewis (Ed.), *Thrips as crop pests*, CAB International, New York. pp. 217-257.

Kirk, W.D.J. (1997b) Feeding, In: T. Lewis (Ed.), *Thrips as crop pests*, CAB International, New York. pp. 119-174.

Kirk, W.D.J., & Terry, L.I. (2003) The spread of the western flower thrips *Frankliniella occidentalis* (Pergande). *Agricultural and Forest Entomology* 5: 301-310.

Kontsedalov, S., Weintraub, P.G., Horowitz, A.R., & Ishaaya, I. (1998) Effects of insecticides on immature and adult western flower thrips (Thysanoptera: Thripidae) in Israel. *Journal of Economic Entomology* 91: 1067-1071.

Latham, L.J., & Jones, R.A.C. (1998) Selection of resistance breaking strains of tomato spotted wilt tospovirus. *Annals of Applied Biology* 133: 385-402.

Maharijaya, A., Vosman, B., Steenhuis-Broers, G., Harpenas, A., Purwito, A., Visser, R.G.F., & Voorrips, R.E. (2011) Screening of pepper accessions for resistance against two thrips species (*Frankliniella occidentalis* and *Thrips parvispinus*). *Euphytica*: 1-10.

Mantel, W.P., & Van de Vrie, M. (1988) The western flower thrips, *Frankliniella occidentalis*, a new thrips species causing damage in protected cultures in The Netherlands. *Entomologische Berichten (Amsterdam)* 48: 140-142.

Margaria, P., Ciuffo, M., & Turina, M. (2004) Resistance breaking strain of *Tomato spotted wilt virus* (*Tospovirus; Bunyaviridae*) on resistant pepper cultivars in Almería, Spain. *Plant Pathology* 53: 795-795.

Martin, N.A., & Workman, P.J. (1994) Confirmation of a pesticide-resistant strain of Western flower thrips in New Zealand. *Proceedings of the NZ Plant Protection Conference* 47: 144-148.

Matteson, N., Terry, I., Ascoli-Christensen, A., & Gilbert, C. (1992) Spectral efficiency of the western flower thrips, *Frankliniella occidentalis*. *Journal of Insect Physiology* 38: 453-459.

Maymó, A.C., Cervera, A., Garcerá, M.D., Bielza, P., & Martínez-Pardo, R. (2006) Relationship between esterase activity and acrinathrin and methiocarb resistance in field populations of western flower thrips, *Frankliniella occidentalis*. *Pest Management Science* 62: 1129-1137.

Maynard, D.N., & Olson, S.M. (2000) *Vegetable production guide for Florida* University of Florida, IFAS, Florida Cooperative Extension Service, Gainesville, FL.

McDonald, J.R., Bale, J.S., & Walters, K.F.A. (1997) Low temperature mortality and overwintering of the western flower thrips *Frankliniella occidentalis* (Thysanoptera: Thripidae). *Bulletin of Entomological Research* 87: 497-505.

Momol, M., Simone, G., Dankers, W., Sprenkel, R., Olson, S., Momol, E., Polston, J., & Hiebert, E. (1999) First report of tomato yellow leaf curl virus in tomato in South Georgia. *Plant Disease* 83: 487-487.

Momol, M.T., Olson, S.M., Funderburk, J.E., Stavisky, J., & Marois, J.J. (2004) Integrated management of tomato spotted wilt on field-grown tomatoes. *Plant Disease* 88: 882-890.

Moriones, E., & Navas-Castillo, J. (2000) Tomato yellow leaf curl virus, an emerging virus complex causing epidemics worldwide. *Virus Research* 71: 123-134.

Morishita, M. (2001) Toxicity of some insecticides to larvae of *Frankliniella occidentalis* (Pergande) (Thysanoptera: Thripidae) evaluated by the petri dish-spraying tower method. *Applied Entomology and Zoology* 36: 137-141.

Moulton, D. (1931) Western Thysanoptera of economic importance. *Journal of Economic Entomology* 24: 1031-1036.

Mound, L.A. (2011) Thysanoptera (Thrips) of the World – a checklist, Accessed: 26 July 2011, Available from: http://www.ento.csiro.au/thysanoptera/worldthrips.html.

Naegele, J.A., & Jefferson, R.N. (1964) Floricultural entomology. *Annual Review of Entomology* 9: 319-340.

Nickle, D.A. (2004) Commonly intercepted thrips (Thysanoptera) from Europe, the Mediterranean, and Africa at U.S. ports-of-entry. Part II. *Frankliniella* Karny and *Iridothrips* Priesner (Thripidae). *Proceedings of the Entomological Society of Washington* 106: 438-452.

Northfield, T.D., Paini, D.R., Funderburk, J.E., & Reitz, S.R. (2008) Annual cycles of *Frankliniella* spp. (Thysanoptera: Thripidae) thrips abundance on north Florida uncultivated reproductive hosts: Predicting possible sources of pest outbreaks. *Annals of the Entomological Society of America* 101: 769-778.

Obradovic, A., Jones, J.B., Momol, M.T., Olson, S.M., Jackson, L.E., Balogh, B., Guven, K., & Iriarte, F.B. (2005) Integration of biological control agents and systemic acquired resistance inducers against bacterial spot on tomato. *Plant Disease* 89: 712-716.

Olson, S.M., & Funderburk, J.E. (1986) A new threatening pest in Florida - Western flower thrips, In: W.M. Stall (Ed.), *Proceedings of the Florida Tomato Institute*, University of Florida Extension Report, Gainesville, FL. pp. 43-51.

Olson, S.M., & Simmonne, E. (2009) *Vegetable Production Handbook for Florida*, Univ. of Florida, IFAS, Florida Cooperative Extension Service, Gainesville, FL. pp. 247.

Paini, D.R., Funderburk, J.E., Jackson, C.T., & Reitz, S.R. (2007) Reproduction of four thrips species (Thysanoptera: Thripidae) on uncultivated hosts. *Journal of Entomological Science* 42: 610-615.

Paini, D.R., Funderburk, J.E., & Reitz, S.R. (2008) Competitive exclusion of a worldwide invasive pest by a native. Quantifying competition between two phytophagous insects on two host plant species. *Journal of Animal Ecology* 77: 184-190.

Pappu, H.R., Jones, R.A.C., & Jain, R.K. (2009) Global status of tospovirus epidemics in diverse cropping systems: Successes achieved and challenges ahead. *Virus Research* 141: 219-236.

Pearsall, I.A., & Myers, J.H. (2001) Spatial and temporal patterns of dispersal of western flower thrips (Thysanoptera: Thripidae) in nectarine orchards in British Columbia. *Journal of Economic Entomology* 94: 831-843.

Pergande, T. (1895) Observations on certain Thripidae. *Insect Life* 7: 390 – 395

Pfannenstiel, R.S., & Yeargan, K.V. (1998) Association of predaceous Hemiptera with selected crops. *Environmental Entomology* 27: 232-239.

Pradhanang, P.M., Ji, P., Momol, M.T., Olson, S.M., Mayfield, J.L., & Jones, J.B. (2005) Application of acibenzolar-S-methyl enhances host resistance in tomato against *Ralstonia solanacearum*. *Plant Disease* 89: 989-993.

Price, D.L., Memmott, F.D., Scott, J.W., Olson, S.M., & Steven, M.R. (2007) Identification of molecular markers linked to a new *Tomato spotted wilt virus* resistance source in tomato. *Tomato Genetics Cooperative* 57: 35-36, Available from: http://tgc.ifas.ufl.edu/vol57/vol57.pdf.

Price, R.G., Walton, R.R., & Drew, W.A. (1961) Arthropod pests collected in Oklahoma greenhouses. *Journal of Economic Entomology* 54: 819-821.

Puche, H., Berger, R.D., & Funderburk, J.E. (1995) Population dynamics of *Frankliniella* species (Thysanoptera: Thripidae) thrips and progress of spotted wilt in tomato fields. *Crop Protection* 14: 577-583.

Race, S.R. (1961) Early-season thrips control on cotton in New Mexico. *Journal of Economic Entomology* 54: 974-976.

Race, S.R. (1965) Predicting thrips populations on seedling cotton. *Journal of Economic Entomology* 58: 1013-1014.

Ramachandran, S., Funderburk, J., Stavisky, J., & Olson, S. (2001) Population abundance and movement of *Frankliniella* species and *Orius insidiosus* in field pepper. *Agricultural and Forest Entomology* 3: 1-10.

Reitz, S.R. (2009) Biology and ecology of the western flower thrips (Thysanoptera: Thripidae): The making of a pest. *Florida Entomologist* 92: 7-13.

Reitz, S.R., Funderburk, J.E., & Waring, S.M. (2006) Differential predation by the generalist predator *Orius insidiosus* on congeneric species of thrips that vary in size and behavior. *Entomologia Experimentalis et Applicata* 119: 179-188.

Reitz, S.R., Gao, Y.L., & Lei, Z.R. (2011) Thrips: Pests of concern to China and the United States. *Agricultural Sciences in China* 10: 867-892.

Reitz, S.R., Maiorino, G., Olson, S., Sprenkel, R., Crescenzi, A., & Momol, M.T. (2008) Integrating plant essential oils and kaolin for the sustainable management of thrips and tomato spotted wilt on tomato. *Plant Disease* 92: 878-886.

Reitz, S.R., Yearby, E.L., Funderburk, J.E., Stavisky, J., Momol, M.T., & Olson, S.M. (2003) Integrated management tactics for *Frankliniella* thrips (Thysanoptera: Thripidae) in field-grown pepper. *Journal of Economic Entomology* 96: 1201-1214.

Riherd, P.T. (1942) An unusual injury to peas by thrips. *Journal of Economic Entomology* 35: 453-453.

Robb, K.L. (1989) *Analysis of Frankliniella occidentalis (Pergande) as a pest of floricultural crops in California greenhouses*, University of California, Riverside, Riverside, CA. pp. 135.

Robb, K.L., Newman, J., Virzi, J.K., & Parella, M.P. (1995) Insecticide resistance in western flower thrips, In: B.L. Parker, M. Skinner, & T. Lewis (Eds.), *Thrips biology and management, NATO ASI series. Series A, Life sciences ; v. 276.*, Plenum Press, New York. pp. 341-346.

Roggero, P., Masenga, V., & Tavella, L. (2002) Field isolates of *Tomato spotted wilt virus* overcoming resistance in pepper and their spread to other hosts in Italy. *Plant Disease* 86: 950-954.

Roselló, S., Díez, M.J., & Nuez, F. (1996) Viral diseases causing the greatest economic losses to the tomato crop. I. The tomato spotted wilt virus - A review. *Scientia Horticulturae* 67: 117-150.

Rosenheim, J.A., Johnson, M.W., Mau, R.F.L., Welter, S.C., & Tabashnik, B.E. (1996) Biochemical preadaptations, founder events, and the evolution of resistance in arthropods. *Journal of Economic Entomology* 89: 263-273.

Rugman-Jones, P.F., Hoddle, M.S., & Stouthamer, R. (2010) Nuclear-mitochondrial barcoding exposes the global pest Western flower thrips (Thysanoptera: Thripidae) as two sympatric cryptic species in its native California. *Journal of Economic Entomology* 103: 877-886.

Sakimura, K. (1961) Field observations on the thrips vector species of the tomato spotted wilt virus in the San Pablo area, California. *Plant Disease Reporter* 45: 772-776.

Sakimura, K. (1962) *Frankliniella occidentalis* (Thysanoptera: Thripidae), a vector of the tomato spotted wilt virus, with special reference to the color forms. *Annals of the Entomological Society of America* 55: 387-389.

Salguero-Navas, V.E., Funderburk, J.E., Olson, S.M., & Beshear, R.J. (1991) Damage to tomato fruit by the western flower thrips (Thysanoptera: Thripidae). *Journal of Entomological Science* 26: 436-442.

Schuster, D.J., Mann, R.S., Toapanta, M., Cordero, R., Thompson, S., Cyman, S., Shurtleff, A., & Morris, R.F. (2010) Monitoring neonicotinoid resistance in biotype B of *Bemisia tabaci* in Florida. *Pest Management Science* 66: 186-195.

Seamaxs, H.L. (1923) The alfalfa thrips and its effect on alfalfa seed production. *Canadian Entomologist* 4: 101-105.

Seaton, K.A., Cook, D.F., & Hardie, D.C. (1997) The effectiveness of a range of insecticides against western flower thrips (*Frankliniella occidentalis*) (Thysanoptera: Thripidae) on cut flowers. *Australian Journal of Agricultural Research* 48: 781-787.

Shah, P.A., & Goettel, M.S. (1999) *Directory of microbial control products and services* Society for Invertebrate Pathology, Gainseville, Florida, USA.

Sharman, M., & Persley, D.M. (2006) Field isolates of Tomato spotted wilt virus overcoming resistance in capsicum in Australia. *Australasian Plant Pathology* 35: 123-128.

Shorey, H.H., & Hall, I.M. (1962) Effect of chemical and microbial insecticides on several insect pests of lettuce in southern California. *Journal of Economic Entomology* 55: 169-174.

Shorey, H.H., & Hall, I.M. (1963) Toxicity of chemical and microbial insecticides to pest and beneficial insects on poled tomatoes. *Journal of Economic Entomology* 56: 813-817.

Shorey, H.H., Reynolds, H.T., & Anderson, L.D. (1962) Effect of Zectran, Sevin, and other new carbamate insecticides upon insect populations found on vegetable and field crops in southern California. *Journal of Economic Entomology* 55: 6-11.

Sparks, A.N., Jr. (2003)Vegetable pests, In: *Summary of Losses from Insect Damage and Cost of Control in Georgia - 2003*, P. Guillebeau, N. Hinkle, & P. Roberts (Eds.), pp. 41-47, Department of Entomology, University of Georgia, Retrieved from http://www.ent.uga.edu/pubs/SurveyLoss03.pdf.

Sparks, A.N., Jr. (2004)Vegetable pests, In: *Summary of Losses from Insect Damage and Cost of Control in Georgia - 2004*, P. Guillebeau, N. Hinkle, & P. Roberts (Eds.), pp. 42-48, Department of Entomology, University of Georgia, Retrieved from http://www.ent.uga.edu/pubs/SurveyLoss04.pdf.

Sparks, A.N., Jr. (2005)Vegetable pests, In: *Summary of Losses from Insect Damage and Cost of Control in Georgia - 2005*, P. Guillebeau, N. Hinkle, & P. Roberts (Eds.), pp. 31-37, Department of Entomology, University of Georgia, Retrieved from http://www.ent.uga.edu/pubs/SurveyLoss05.pdf.

Sparks, A.N., Jr. (2006)Vegetable pests, In: *Summary of Losses from Insect Damage and Cost of Control in Georgia - 2006*, P. Guillebeau, N. Hinkle, & P. Roberts (Eds.), pp. 38-43, Department of Entomology, University of Georgia, Retrieved from http://www.ent.uga.edu/pubs/SurveyLoss06.pdf.

Sparks, A.N., Jr., & Riley, D. (2001)Vegetable pests, In: *Summary of Losses from Insect Damage and Cost of Control in Georgia - 2001*, P. Guillebeau, N. Hinkle, & P. Roberts (Eds.), pp. 42-47, Department of Entomology, University of Georgia, Retrieved from http://www.ent.uga.edu/pubs/SurveyLoss01.pdf.

Sparks, A.N., Jr., & Riley, D. (2002)Vegetable pests, In: *Summary of Losses from Insect Damage and Cost of Control in Georgia - 2002*, P. Guillebeau, N. Hinkle, & P. Roberts (Eds.), pp. 38-43, Department of Entomology, University of Georgia, Retrieved from http://www.ent.uga.edu/pubs/SurveyLoss02.pdf.

Sparks, T.C., Thompson, G.D., Kirst, H.A., Hertlein, M.B., Mynderse, J.S., Turner, J.R., & Worden, T.V. (1999) Fermentation-derived insect control agents, In: F. Hall, & J.J. Menn (Eds.), *Biopesticides: use and delivery*, Humana, Totowa, NJ. pp. 171-188.

Spiers, J.D., Davies Jr, F.T., He, C., Bográn, C.E., Heinz, K.M., Starman, T.W., & Chau, A. (2006) Effects of insecticides on gas exchange, vegetative and floral development, and overall quality of gerbera. *Hortscience* 41: 701-706.

Srivistava, M., Bosco, L., Funderburk, J., Olson, S., & Weiss, A. (2008) Spinetoram is compatible with the key natural enemy of *Frankliniella* species thrips in pepper. *Plant Health Progress* doi:10.1094/PHP-2008-0118-02-RS.

Stapleton, J.J., & Summers, C.G. (2002) Reflective mulches for management of aphids and aphid-borne virus diseases in late-season cantaloupe (*Cucumis melo L. var. cantalupensis*). *Crop Protection* 21: 891-898.

Stavisky, J., Funderburk, J.E., Brodbeck, B.V., Olson, S.M., & Andersen, P.C. (2002) Population dynamics of *Frankliniella* spp. and tomato spotted wilt incidence as influenced by cultural management tactics in tomato. *Journal of Economic Entomology* 95: 1216-1221.

Stevens, M.R., Scott, S.J., & Gergerich, R.C. (1992) Inheritance of a gene for resistance to *Tomato spotted wilt virus* Tswv from *Lycopersicon peruvianum* Mill. *Euphytica* 59: 9-17.

Sticher, L., Mauch-Mani, B., & Métraux, J.P. (1997) Systemic acquired resistance. *Annual Review of Phytopathology* 35: 235-270.

Summers, C.G., Newton, A.S., Mitchell, J.P., & Stapleton, J.J. (2010) Population dynamics of arthropods associated with early-season tomato plants as influenced by soil surface microenvironment. *Crop Protection* 29: 249-254.

Thalavaisundaram, S., Herron, G.A., Clift, A.D., & Rose, H. (2008) Pyrethroid resistance in *Frankliniella occidentalis* (Pergande) (Thysanoptera: Thripidae) and implications for its management in Australia. *Australian Journal of Entomology* 47: 64-69.

Thomas-Carroll, M.L., & Jones, R.A.C. (2003) Selection, biological properties and fitness of resistance-breaking strains of Tomato spotted wilt virus in pepper. *Annals of Applied Biology* 142: 235-243.

Thompson, G.D., Dutton, R., & Sparks, T.C. (2000) Spinosad – a case study: an example from a natural products discovery programme. *Pest Management Science* 56: 696-702.

Thompson, G.J., & van Zijl, J.J.B. (1995) Control of Tomato spotted wilt virus in tomatoes in South Africa. *Acta Horticulturae* 431: 379-384.

US EPA. (1996) *Food Quality Protection Act of 1996, Public Law 104–170,* Available from: http://www.epa.gov/pesticides/regulating/laws/fqpa/gpogate.pdf.

Vacante, V., Cacciola, S.O., & Pennisi, A.M. (1994) Epizootiological study of *Neozygites parvispora* (Zygomycota: Entomophthoraceae) in a population of *Frankliniella occidentalis* (Thysanoptera: Thripidae) on pepper in Sicily. *Entomophaga* 39: 123-130.

van den Meiracker, R.A.F., & Ramakers, P.M.J. (1991) Biological control of the western flower thrips *Frankliniella ocidentalis* in sweet pepper with the anthocorid predator *Orius insidiosus*. *Mededelingen van de Faculteit Landbouwwetenschappen Rijksuniversiteit Gent* 56: 241-249.

Webster, C., Perry, K., Lu, X., Horsman, L., Frantz, G., Mellinger, C., & Adkins, S. (2010) First report of *Groundnut ringspot virus* infecting tomato in south Florida. *Plant Health Progress* 10.1094/PHP-2010-0707-01-BR.

Webster, C.G., Reitz, S.R., Perry, K.L., & Adkins, S. (2011) A natural M RNA reassortant arising from two species of plant- and insect-infecting bunyaviruses and comparison of its sequence and biological properties to parental species. *Virology* 413: 216-225.

Weiss, A., Dripps, J.E., & Funderburk, J. (2009) Assessment of implementation and sustainability of integrated pest management programs. *Florida Entomologist* 92: 24-28.

Wijkamp, I., Van De Wetering, F., Goldbach, R., & Peters, D. (1996) Transmission of tomato spotted wilt virus by *Frankliniella occidentalis:* median acquisition and inoculation access period. *Annals of Applied Biology* 129: 303-313.

Woglum, R.S., & Lewis, H.C. (1935) Notes on citrus pests new or seldom injurious in California. *Journal of Economic Entomology* 28: 1018-1021.

Yokoyama, V.Y. (1977) *Frankliniella occidentalis* and scars on table grapes. *Environmental Entomology* 6: 25-30.

Zhang, S. -Y., Kono, S., Murai, T., & Miyata, T. (2008) Mechanisms of resistance to spinosad in the western flower thrip, *Frankliniella occidentalis* (Pergande) (Thysanoptera: Thripidae). *Insect Science* 15: 125-132.

Zhang, S.H., Lei, Z.R., Fan, S.Y., & Wen, J.Z. (2009) Pathogenicity of four *Beauveria bassiana* strains against *Frankliniella occidentalis* at different temperatures. *Plant Protection* 35: 64-67.

Zhao, G., Liu, W., Brown, J.M., & Knowles, C.O. (1995) Insecticide resistance in field and laboratory strains of western flower thrips (Thysanoptera: Thripidae). *Journal of Economic Entomology* 88: 1164-1170.

Zhao, G., Liu, W., & Knowles, C.O. (1994) Mechanisms associated with diazinon resistance in western flower thrips. *Pesticide Biochemistry and Physiology* 49: 13-23.

3

Bioactive Natural Products from Sapindaceae Deterrent and Toxic Metabolites Against Insects

Martina Díaz and Carmen Rossini
Laboratorio de Ecología Química, Facultad de Química, Universidad de la República,
Uruguay

1. Introduction

The Sapindaceae (soapberry family) is a family of flowering plants with about 2000 species occurring from temperate to tropical regions throughout the world. Members of this family have been widely studied for their pharmacological activities; being *Paullinia* and *Dodonaea* good examples of genera containing species with these properties. Besides, the family includes many species with economically valuable tropical fruits, and wood (Rodriguez 1958), as well as many genera with reported anti-insect activity.

Antioxidant, anti-inflammatory and anti-diabetic properties are the pharmacological activities most commonly described for this family (Sofidiya et al. 2008; Simpson et al. 2010; Veeramani et al. 2010; Muthukumran et al. 2011). These activities are in some cases accounted for isolated phenolic compounds such as prenylated flavonoids (Niu et al. 2010), but in many cases, it is still unknown which are the active principles. Indeed, there are many studies of complex mixtures (crude aqueous or ethanolic extracts) from different species in which several other pharmacological activities have been described without characterization of the active compounds [*e.g.* antimigrane (Arulmozhi et al. 2005), anti-ulcerogenic (Dharmani et al. 2005), antimalarial (Waako et al. 2005), anti-microbial (Getie et al. 2003)].

Phytochemical studies on Sapindaceae species are abundant and various kinds of natural products have been isolated and elucidated. Examples of these are flavonoids from *Dodonaea* spp. (Getie et al. 2002; Wollenweber & Roitman 2007) and *Koelreuteria* spp. (Mahmoud et al. 2001), linear triterpenes from *Cupaniopsis* spp. (Bousserouel et al. 2005) and caffeine, xanthenes and cathequines from *Paullinia* spp. (Benlekehal et al. 2001; Sousa et al. 2009). All these compounds are naturally occurring in almost every plant family, however, the Sapindaceae do produce an unusual group of secondary metabolites: the cyanolipids (Avato et al. 2005). Eventhough these compounds exhibit a potential health hazard for humans and animals, for the plants, cyanolipids may have a protective physiological role. However, not many investigations have been developed involving the study of the ecological interactions among the plants producing them and other sympatric organisms. On the other hand, the toxicity of these cyanocompounds might be a potential source of pesticides. Indeed, not only cyanocompounds, but also a wide range of species of Sapindaceae have been tested on their anti-insect activity. Several extracts, fractions or pure compounds of different phenological stages, have been tested against diverse species of lepidopterans, dipterans and coleopterans of major importance in agriculture as well as in veterinary and medical applications. Examples of this include the larvicidal activity of *Magonia pubescens* against *Aedes aegypti*

(Diptera: Culicidae) (Arruda et al. 2003a,b); the toxicity of *Sapindus* spp. against *Sitophilus oryzae* (Coleoptera: Curculionidae) (Zidan et al. 1993; Rahman et al. 2007); and the toxicity of *Dodonaea* spp. against *Spodoptera* spp. (Lepidoptera: Noctuidae) (Abdel et al. 1995; El-Din & El-Gengaihi 2000; Deepa & Remadevi 2007; Malarvannan et al. 2008). The activity against different pest models of extracts from various South American species within the Sapindaceae was recently described. These species included *Allophylus edulis*, *Dodonaea viscosa* and *Serjania meridionalis*, from which isolated metabolites with anti-insect capacities have not yet been described (Castillo et al. 2009). This chapter will examine the available information on extracts and secondary metabolites from Sapindaceae focused on their defensive role for the plant against herbivory; and consequently this appraisal will also present a compilation of potential anti-insect agents from Sapindaceae.

2. Sapindaceae: its anti-insect potential

Anti-insect activity has been described in at least 15 of the 202 genera (Anonymous 2011) belonging to this family. Among these findings, the cases in which the bioactive compounds were isolated represent the least. It has been in general tested the activity of aqueous or ethanolic extracts from different organs and from plants of different phenological stages against a variety of insect targets using different bioassay designs. As a consequence, different modes of action have been described. Extracts have revealed to be potentially deterrent agents, growth inhibitors and even toxic agents against different genera of insects. The following appraisal comprises the main Sapindaceae genera from which extracts or isolated compounds with anti-insect activity have been described.

2.1 *Sapindus*

A vast number of species showing great potential as anti-insect agents belong to this genus. *Sapindus saponaria*, a tree widely distributed in Central and South America, is also frequently used as ornamental (Lorenzi 2004). Brazilian people commonly prepare homemade soap from this tree; and use its seeds to make handcrafts. Its wood is broadly used in construction. Its fruits and roots are popularly used as painkillers, astringents, expectorants and diuretics (Ferreira Barreto et al. 2006). Besides, its medicinal potential as healing and anti thrombotic agents has been studied. Research on that area has revealed that flavonoids in the leaf extracts are responsible for those activities (Meyer Albiero et al. 2002). On the other hand, much research has been devoted to the anti-insect capacity of extracts from different organs of this plant. Boiça Junior et al (2005), on their search for activity against larvae of the cabbage pest *Plutella xylostella* (Lepidoptera: Plutellidae) investigated eighteen plant species from a variety of families, finding that aqueous leaf extract of *S. saponaria* was one of the most active products. The extract produced 100 % of mortality in tests where the larvae were offered cabbage foliage disks coated with the extracts to be evaluated as a sole food (Boiça Junior et al. 2005). In another study, the aqueous fruit extract of this tree showed deterrent properties against another cabbage pest, *Ascia monuste orseis* (Lepidoptera: Pieridae). In this case the activity was comparable to that showed by aqueous extracts of the neem tree, *Azadirachta indica*, the newest botanical pesticide in the market (Isman et al. 1996; Medeiros et al. 2007). The aqueous seeds´ extracts were evaluated against another lepidopteran, *Spodoptera frugiperda* (Lepidoptera: Noctuidae), showing strong effect on larvae development and midgut trypsin activity (dos Santos et al. 2008). Not only against lepidopterans has this tree revealed anti-insect potential, but also against other insect orders.

For instance, a saponin extract from fruits from this species showed toxicity against adults of the greenhouse whitefly *Trialeurodes vaporariorum* (Hemiptera: Aleyrodidae) (Porras & Lopez-Avila 2009); and complete ethanolic extracts from fruits have shown larvicidal and morphological alterations effects on the mosquito *Aedes aegypti* (Diptera: Culicidae) (Ferreira Barreto et al. 2006). Some other saponins presenting other kinds of biological activity, isolated from the fruits of this species, are shown in Figure 1 (Lemos et al. 1992; Ribeiro et al. 1995).

Sapindus emarginatus, another tree from this genus, widely distributed in India, has also demonstrated larvicidal activity of its fruit extract against three important vector mosquitoes: *A. aegypti, Anopheles stephensi* and *Culex quimquefasciatus* (Diptera: Culicidae) (Koodalingam et al. 2009). Later, this group has also investigated the impact of the extracts on the activity of mosquito phosphatases and esterases to gain an insight into the extent of disturbance in metabolic homeostasis inflicted upon exposure to the extract (Koodalingam et al. 2011). Previous reports on this species have shown that the pericarps contain triterpene saponins (Figure 1), which are commonly used as antifertility, antipruritic and anti-inflammatory agents in traditional Indian and Thai medicine (Jain 1976; Kanchanapoom et al. 2001). Perhaps, anti insect activity may be due to saponins in this plant similarly to the case of *Sapindus saponaria* (Porras et al. 2009).

Activity of members of this genus against other insect orders, further than dipterans and lepidopterans, has also been evaluated, including coleoptera and lice. That is the case of the extract from *Sapindus trifoliatus* fruit cortex which showed activity against the red flour beetle *Tribolium castaneum* (Herbst) (Coleoptera: Tenebrionidae) (Mukherjee & Joseph 2000). In this case, weight gain was significantly reduced when larvae were fed on diets including the extract at different doses; and females topically treated -upon emergence- with the extract laid fewer viable eggs. (Mukherjee & Joseph 2001). Another ethanolic fruit extracts, in this case from *Sapindus mukorossi*, also showed anti coleopteran activity against another pest of stored grains, *Sitophilus oryzae* (Coleoptera: Curculionidae) and also against *Pediculus humanus* (Phthiraptera: Pediculidae) (Rahman et al. 2007). Finally, from the methanolic extract of fruits of this species triterpenic saponins (Figure 1) have also been isolated and these natural products demonstrated their potential as growth regulators and antifeedants against *Spodoptera littura* (Lepidoptera: Noctuidae), both as glycosides and as free genines (Saha et al. 2009). In this particular study, it was verified that upon hydrolysis of the saponins, the growth regulatory activity was improved, whereas very little difference was found in regard to the antifeedant activity.

All in all, the genus *Sapindus* contains a variety of species which have been studied on their activity for some insects from different orders. In spite of the fact that not many reports do exist on the action of isolated compounds, the previous ethnobotanical uses of *Sapindus* spp. and the isolation of some active saponins from this genus, may suggest that this group of secondary metabolites might be related to the anti-insect activity. Saponins -glycosides of sapogenins containing a monosaccharide or a polysaccharide unit- reduce the surface tension becoming biological detergents. They are widely distributed secondary plant metabolites, found among almost 100 plant families (Bruneton 1995). Being effective defences for some insects (Plasman et al. 2001; Prieto et al. 2007), saponins have been implied in mechanisms of plant resistance against potential herbivores (Nielsen et al. 2010). The genus *Sapindus*, rich in this kind of compounds, may therefore be promissory raw material to develop plant pest control products. Further information can be found at recent works reviewing saponins from *Sapindus* spp. and their activity (Pelegrini et al. 2008; Sharma et al. 2011).

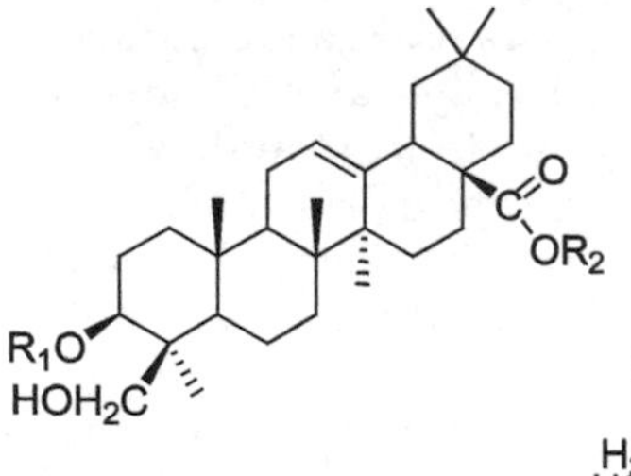

R1: Xyl (OAc)-Ara-Rha
R2: Glc-Glc-Rha
Sapindus mukorossi
(Saha et al., 2010)

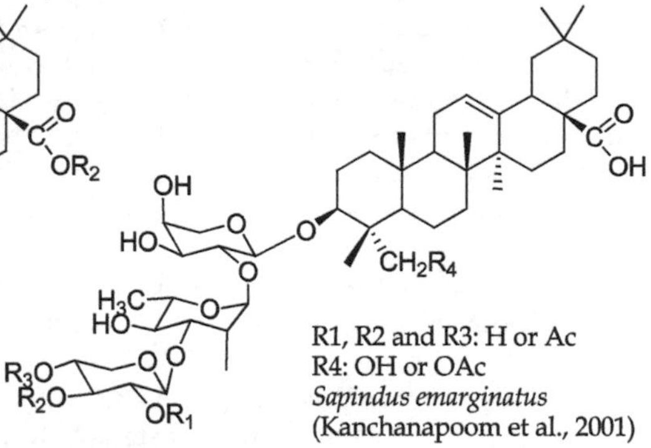

R1, R2 and R3: H or Ac
R4: OH or OAc
Sapindus emarginatus
(Kanchanapoom et al., 2001)

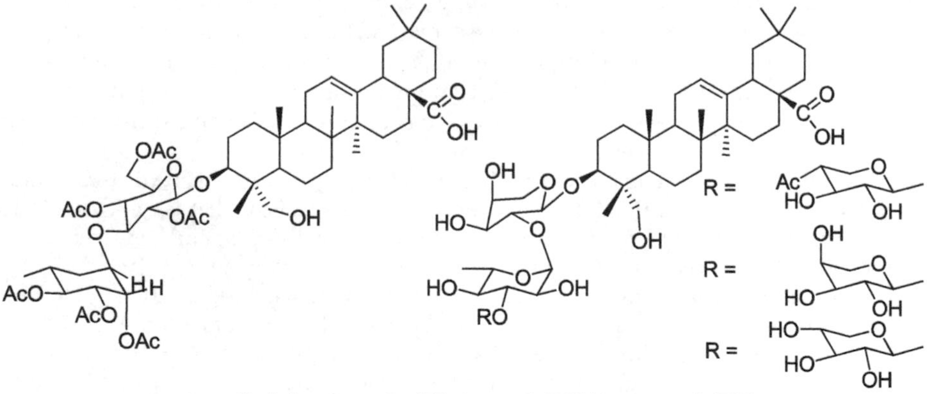

Sapindus saponaria (Ribeiro et al, 1995; Lemos et al, 1992)

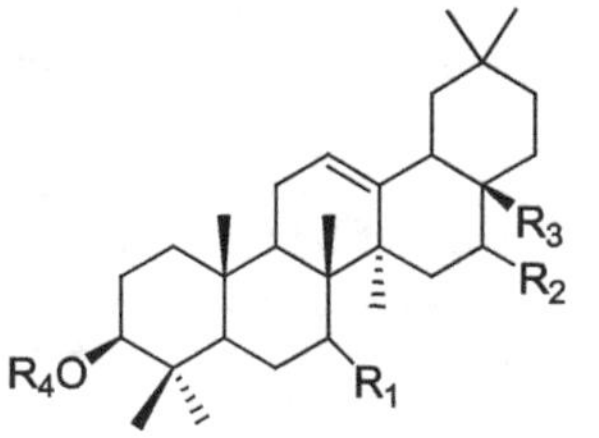

	R1	R2	R3	R4
1	H	H	CH₂OH	-Xyl$\overset{2}{-}$Glc $\overset{3}{\diagdown}$Ara
2	H	H	CH₂OH	-Glc$\overset{2}{-}$Glc $\overset{3}{\diagdown}$Ara
3	H	H	CH₂OH	-Xyl^3-Glc^6-Ac
4	α-OCH₃	H	CH₂OH	-Glc$\overset{2}{-}$Glc $\overset{3}{\diagdown}$Ara
5	H	β-OH	CH₃	-Xyl$\overset{2}{-}$Glc $\overset{3}{\diagdown}$Ara
6	H	β-OH	CH₃	-Glc$\overset{2}{-}$Glc $\overset{3}{\diagdown}$Ara

Nepheliosides **1-6** *Nephelium lappaceum* (Ito et al., 2003)

Fig. 1. Saponins isolated from *Sapindus* and *Nephelium* spp.

2.2 *Dodonaea*

From this genus, there are two species to which almost all research has been devoted: *Dodonaea angustifolia* and *Dodonaea viscosa*. These two species are considered by some taxonomists to be synonymous, while others recognize *D. angustifolia* as a sub-species of *D. viscosa* (cited in Omosa et al. 2010). *D. angustifolia* is widely distributed in Australia, Africa, Asia and South America; and it has been employed until present days in traditional medicine all over the world. It is traditionally used as analgesic, laxative, antipyretic, and to treat rheumatism, eczema, and skin ailments (Malarvannan et al. 2009; Omosa et al. 2010). *Dodonaea viscosa* is a shrub, rarely a small tree, widely distributed in tropical and subtropical areas of both hemispheres. It is used in folk medicine as a febrifuge, a diaphoretic drug, and also for the treatment of rheumatism, gout, inflammations, swelling and pain (Niu et al. 2010).

Anti-insect activity has been described for extracts from both plant species mostly against lepidopterans (Malarvannan & Subashini 2007; Malarvannan et al. 2008; 2009; Sharaby et al. 2009). For instance, extracts of leaves of *D. angustifolia* (obtained with hexane, petroleum ether, chloroform, acetone and water) were tested in field bioassays, showing to be effective biocontrol agents for the larvae of *Earias vitella* (Lepidoptera: Noctuidae) (Malarvannan et al. 2007). Besides, those extracts also showed ovicidal activity against *Helicoverpa armigera* (Lepidoptera: Noctuidae) (Malarvanan 2003). However, while in the case of extracts from different organs (fruits, leaves and twigs) of *D. viscosa* coming from Uruguay, none of these products proved to be active against the polyphagous *Spodoptera littoralis* (Lepidoptera: Noctuidae) (Castillo et al. 2009), insects from other orders were deterred by *D. viscosa* extracts. Interestingly, while extracts from leaves and twigs exhibited good activity against aphids (*Rhopalosiphum padi* and *Myzus persicae*) and a coleopteran (*Epilachna paenulata*), they were innocuous to beneficial insects (*Apis mellifera*) (Castillo et al. 2009). This selectivity makes *D. viscosa* a good candidate from which to develop botanical pesticides. Another independent study also showed a strong contact activity of the seed extracts against the coleopteran, *S. oryzae* (Zhao et al. 2006).

D. angustifolia is known to contain essential oils, flavonoids, terpenoids, phenols, coumarins, sterols and unidentified alcohols (Malarvannan et al. 2008). Meanwhile several flavonoids, diterpenoid acids and saponins have been isolated from *D. viscosa* (Niu et al. 2010). However, the chemical basis for the pesticide and antifeedant activities remains unclear as tests on individual compounds have not been performed. Nevertheless, a series of clerodane diterpenoids (Figure 2) and prenylated flavonoids (Figure 3) were isolated from the aerial parts of *D. viscosa* from China, having them not shown larvicidal activity against two mosquito species tested (Niu et al. 2010), however the authors stated that previous studies showed activity of these clerodanes against two lepidopterans (*Plutella xylostella* and *Pieris rapae*) and against the coleopteran *Sitophilus oryzae*.

At the same time, an investigation on this family of compounds from *D. angustifolia* from Kenya showed that the extracts from the leaf surface of this plant is composed mainly by clerodanes (Figure 2) and also by methylated flavones and flavonols (Figure 3) (Omosa et al.). Clerodanes isolated from *D. viscosa* and *D. angustifolia* belong to the *neo*-clerodane group. As it is well known, these secondary metabolites have a structure based on the carbon skeleton and absolute stereochemistry of clerodin (Klein Gebbinck et al. 2002) isolated first from *Clerodendron infortunatum* (Lamiaceae) (Banerjee 1936). This large group of plant secondary metabolites have been described mainly from Lamiaceae and Asteraceae, and they have exhibited a wide range of anti-insect properties as it has been reviewed

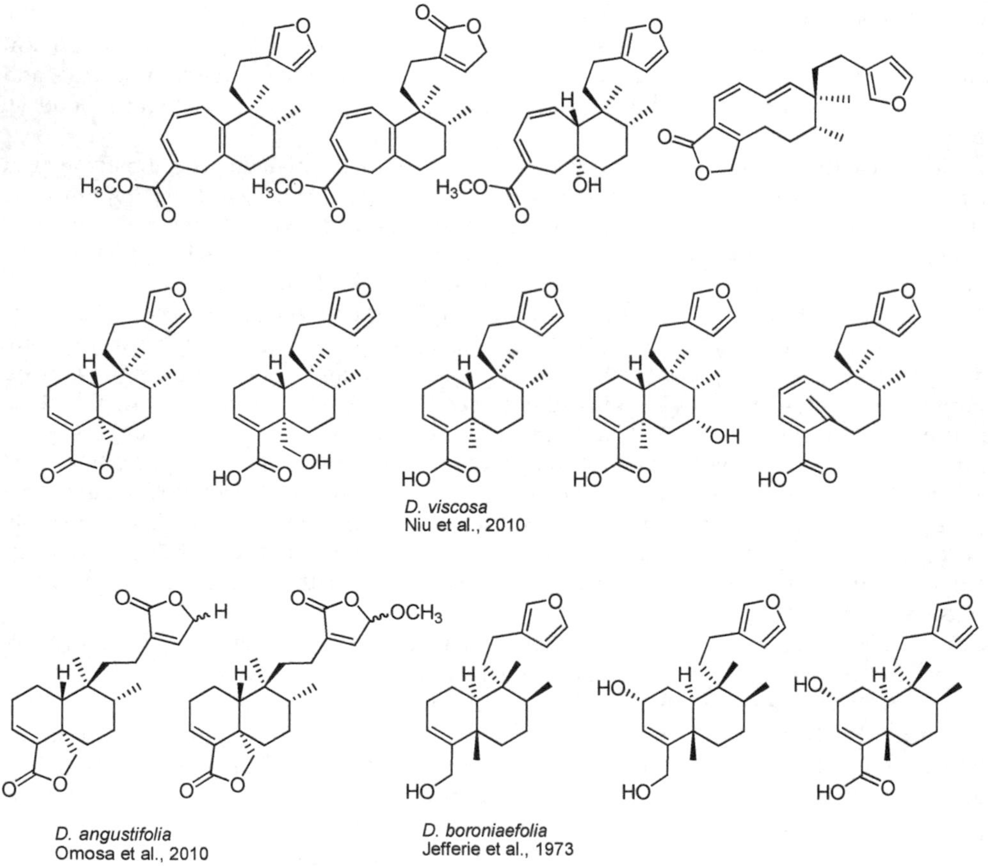

Fig. 2. Clerodanes from *Dodonaea* spp. Structures shown were isolated from *D. viscosa* (Niu et al. 2010) with the exception of the indicated ones.

previously (Klein Gebbinck et al. 2002; Sosa & Tonn 2008). Worth to be noticed, an earlier work by Jefferies et al. (1973) reported the occurrence of various diterpenes in another species, *Dodonaea boroniaefolia*, of the opposite configuration in the main skeleton (Figure 3), that is *ent*-clerodanes.

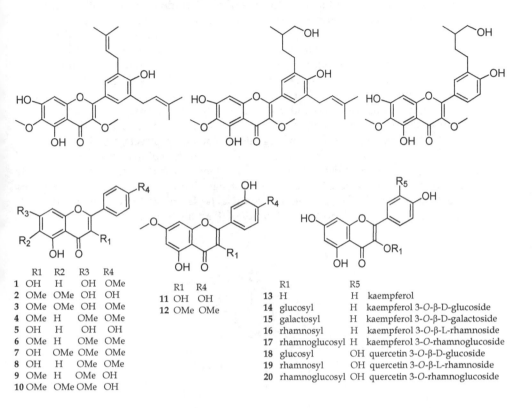

	R1	R2	R3	R4
1	OH	H	OH	OMe
2	OMe	OMe	OH	OH
3	OMe	OMe	OH	OMe
4	OMe	H	OMe	OMe
5	OH	H	OH	OH
6	OMe	H	OMe	OMe
7	OH	OMe	OMe	OMe
8	OH	H	OMe	OMe
9	OMe	H	OMe	OH
10	OMe	OMe	OMe	OH

	R1	R4
11	OH	OH
12	OMe	OMe

	R1	R5	
13	H	H	kaempferol
14	glucosyl	H	kaempferol 3-O-β-D-glucoside
15	galactosyl	H	kaempferol 3-O-β-D-galactoside
16	rhamnosyl	H	kaempferol 3-O-β-L-rhamnoside
17	rhamnoglucosyl	H	kaempferol 3-O-rhamnoglucoside
18	glucosyl	OH	quercetin 3-O-β-D-glucoside
19	rhamnosyl	OH	quercetin 3-O-β-L-rhamnoside
20	rhamnoglucosyl	OH	quercetin 3-O-rhamnoglucoside

Fig. 3. Flavonoids from *Dodonaea* spp. (Niu et al. 2010; Omosa et al. 2010; Teffo et al. 2010).

2.3 *Magonia and Paullinia species*

Paullinia spp. is one of the few purine alkaloid-containing genera used to prepare stimulant drinks worldwide (Weckerle et al. 2003). *Paullinia clavigera* grows in primary forests and shores of the South American Amazonic aquatic ecosystems, and it has been studied mostly in Perú for the control of different insect pests (cited in Pérez et al. 2010). The toxicity of aqueous extracts of lianas from this species against larvae of *Anopheles benarrochi* (Diptera: Culicidae) (Pérez & Iannacone 2004) and afterwards the mortality and repellence of such extracts against larvae of *Rhynchophorus palmarum* (Coleoptera: Curculionidae) (Pérez & Iannacone 2006) were tested. Besides, the activity of aqueous extracts against *Eupalamides cyparissias* (Lepidoptera: Castniidae), and the activity of hydroalcoholic extracts against *Tuthillia cognata* (Hemiptera: Psyllidae) (Pérez et al. 2008) were also reported, showing the potential of this vegetal species as a biocontrol agent against different insect orders. However, it is still unclear which compounds are responsible for the anti-insect effects of *P. clavigera*. Flavonoids, phenols, triterpenes and saponins were detected in a phytochemical study in extracts from the stem cortex of *P. clavigera.* (Pérez et al. 2010); and other species from the same genus (*P. cururu*) contain saponins, tannins and polyphenols (Wilbert & Haiek 1991).

Magonia pubescens, widely distributed in the Brazilian Cerrado, is commonly used in the construction industry. It has been mostly studied for its larvicidal activity against *A. aegypti* (Arruda et al. 2003; da Silva et al. 2003; Rodrigues et al. 2006). In this case, one of the most active fractions from the ethanolic extract of stem barks was shown to be rich in tannins, and specially in a proanthocyanidin (catequic tannin which structure is shown in Figure 4) (Silva et al. 2004). Tannins are largely distributed in nature, usually being the active principles of plants used in traditional medicine. Condensed tannins have a great ability to interact with metallic ions and macromolecules and to form soluble complexes with electron-donor groups such as the ones found in alkaloids and proteins. That may be one of the reasons explaining their toxicity against different organisms, including insects, fungi and bacteria. Morphological alterations caused by this active fraction on the epithelium of the midgut of larvae of *A. aegypti* resembled the ones recorded for tannic acid (Rey et al. 1999). In a side note, it is worth to notice that this vegetal species has also demonstrated potential on its ethanolic extract of stem barks, as acaricide against the larvae of the common cattle tick, *Rhipicephalus (Boophilus) sanguineus* (Acari: Ixodidae) (Fernandes et al. 2008).

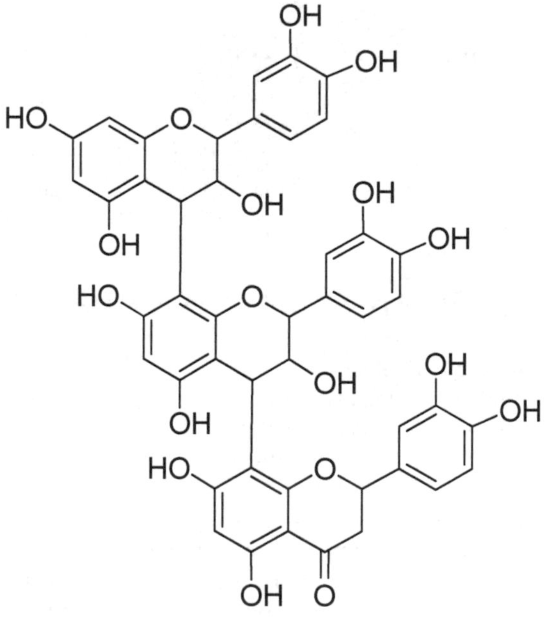

Fig. 4. Catequic tannin with larvicidal activity against *A. aegypti* isolated from *Magonia pubescens* (Silva et al. 2004).

2.4 Miscellaneous
Talisia esculenta, locally known as pitomba occurs in northern and northeastern Brazil. Its fruits are edible to humans and birds which disperse the seeds. However, popular information also mentions that chickens die after ingesting the fruit (cited in Macedo et al.

2002). *Koelreuteria paniculata* is popularly grown as an ornamental tree in temperate regions all across the world (Kamala-Kannan et al. 2009). From these two species of Sapindaceae, lectins have been isolated from their seeds (Macedo et al. 2002; Macedo et al. 2003). Lectins from *T. esculenta* inhibited larval growth of two bruchids *(Callobroschus maculatus* and *Zabrotes subfasciatus)* (Macedo et al. 2002). And in the case of *K. paniculata*, lectins not only showed insecticide activity against *C. maculatus* but also against *Anagasta kuehniella* (Lepidoptera: Pyralidae) (Macedo et al. 2003). Plant lectins are a large group of proteins defined as "plant proteins that possess at least one non-catalytic domain that binds reversibly to a specific mono- or oligosaccharide" (Peumans & Van Damme 1995). Plant lectines have been implied in many ecological roles, being their action as defences against insects one of the latest described (Murdock & Shade 2002; Van Damme et al. 2008). Their mechanisms of action as anti-insect agents are yet poorly understood, with many emerging hypotheses proposed (Van Damme et al. 2008). According to Macedo et al. (2003), the action of *K. paniculata* lectins on *C. maculatus* and *A. kuehniella* larvae may involve *(1) binding to glycoconjugates on the surface of epithelial cells along the digestive tract, (2) binding to glycosylated digestive enzymes, thereby inhibiting their activity, and (3) binding to the chitin component of the peritrophic membrane (or equivalent structures) in the insect midgut.* Finally, regarding *T. esculenta*, its aqueous seeds extracts were studied on its effect on *S. frugiperda* larvae which development was negatively affected, but the activity of the midgut trypsin was not inhibited (dos Santos et al. 2008).

The red maple, *Acer rubrum*, is another Sapindaceae from which bioactive compounds have been isolated. This prominent maple occurs in hardwood forests, being avoided by several potential sympatric consumers [for instance, larvae of *Malacossoma disstria* (Lepidoptera: Lasiocampidae), and the North American beavers, *Castor canadensis*] (Abou-Zaid et al. 2001). The main constituents of an aqueous leaf extract have been phytochemically characterized as ellagic acid, gallate derivatives (structures **1-7** in Figure 5) and glycosides of flavonoids (quercetin and structures **13-20** in Figure 3). When these compounds were tested by themselves against *M. disstria* larvae, it was found that all gallate derivatives exhibited deterrent activity, but not the flavonoids. Among gallate derivatives, compounds **2** and **4-7** in Figure 5 were the five most active compounds. Perhaps in this case, the feeding deterrence effect of the extracts may be traced to the galloyl moiety in its secondary metabolites.

Blighia sapida, commonly known as Ackee, is an evergreen tree, native from West African wild forests. In the late 18th century, the plant was introduced in Jamaica, where nowadays its fruit has been adopted as the national fruit (cited in Gaillard et al. 2011). Its bark is used as fish poison and also in folk medicine in the treatment of malaria, ulcers, back aches and headaches (Kayode 2006). The plant contains triterpenic and steroidal saponins, alkaloids, polyphenols and aminoacidic secondary metabolites (Mazzola et al. 2011). By ingestion, the unripe fruits can cause vomiting and circulatory collapse in humans due to the presence of hypoglycin-A (seeds and flesh) and hypoglycin-B (seeds) (Figure 6) (Hassall et al. 1954; Hassall & Reyle 1955; Gaillard et al. 2011). Acetone and ethanolic extracts of the fruits showed repellent properties against stored-product pests, namely, *C. maculatus, Cryptolestes ferrugineus* (Coleoptera: Cucujidae), and *Sitophilus zeamais* (Coleoptera: Curculionidae) (Khan & Gumbs 2003). Furthermore, ethanol, acetone, hexane, methanol, chloroform, and water extracts from the seeds were evaluated on their repellence against *T. castaneum*, demonstrating the aqueous extract to be the most active (Khan et al. 2002).

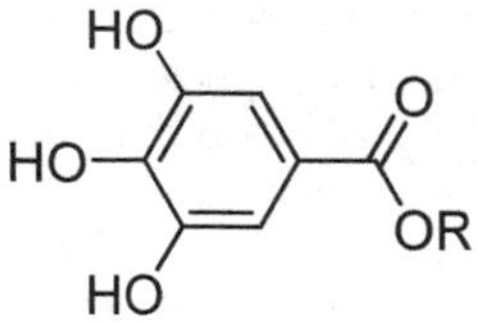

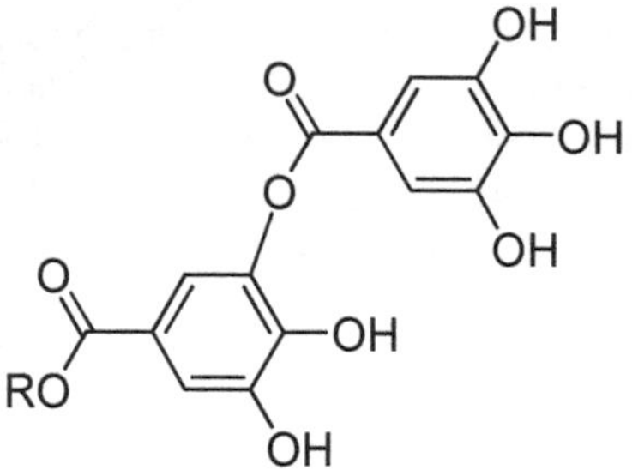

1 R = H   Gallic acid
2 R = Me   Methyl gallate
3 R = Et   Ethyl gallate
4 R = β-D-Glc   1-O-Galloyl-b-D-glucosa
5 R = α-L-Rha   1-O-Galloyl-a-L-rhamnose

6 R = OH   m-Digallate
7 R = OC₂O₅   Ethyl m-digallate

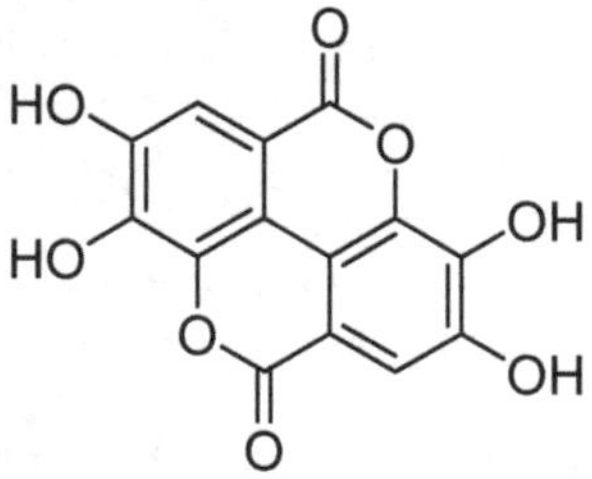

8 ellagic acid

Fig. 5. Ellagic acid and gallate derivatives isolated from *A. rubrum*. (Abou-Zaid et al. 2001).

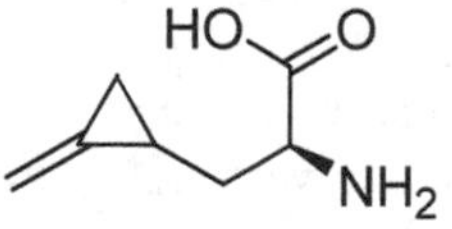

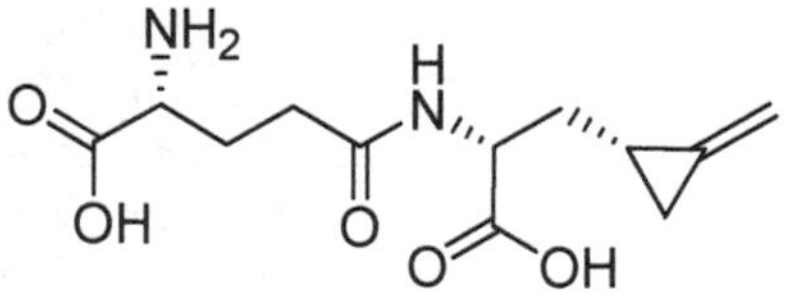

Hypoglycin-A   Hypoglycin-B

Fig. 6. Hypoglycins isolated from the fruits of *B. sapida*.

Serjania lethalis is another species that has been studied in its anti-insect activity. The ethanolic stem bark extract showed larvicidal activity against *A. aegypti* (Rodrigues et al. 2006). In this study the active compounds were not identified. However, the presence in this species of tannins, flavonoids and of saponins (serjanosides) has been reported (Teixeira et al. 1984; de Sousa Araújo et al. 2008). *S. meridionalis,* a species phytochemically not described, exhibited deterrent activity against *E. paenulata* and *M. persicae* (Hemiptera: Aphididae) when its ethanolic leaf extracts were assayed. Disappointingly, this extract was also toxic against beneficial insects (honey bees) (Castillo et al. 2009).

From the genus *Nephelium,* the species *N. lappaceum* is commonly known for its edible fruit "rambutan". It is native from Southeast Asia where the fruits are an important commercial crop (Palanisamy et al. 2008). These fruits have shown potential on its ethanol seed extract, against *S. oryzae,* revealing to reduce esterase and glutathione-S-transferase activities from such insect (Bullangpoti et al. 2004). *N. maingayi,* native from Malasya and Indonesia, has also edible fruits and it has not been studied in its anti-insect activity. However, six saponins, namely, nepheliosides **1-6** (Figure 1) were isolated from a chloroform extract of its bark exhibiting cytotoxic activity when evaluated against a panel of human cancer cell lines (Ito et al. 2004).

3. Cyanocompounds

As stated, Sapindaceae are rich -in their seeds- in toxic cyanolipids (Figure 8), *e.g.* fatty acid esters of α- and γ- hydroxynitriles (Mikolajczak 1977; Seigler 1991). Although, some works report the occurrence of cyanolipids also in members of the Hippocastanaceae (Mikolajczak 1977; Bjarnholt & Møller 2008) and the Boraginaceae (Mikolajczak et al. 1969; Seigler et al. 1970), it appears that later investigations have confirmed that these metabolites are characteristic only of the Sapindaceae (Seigler 1976; Avato et al. 2005). The cyanolipids are usually extracted in the seed oils where the amounts vary broadly within the species (Dinesh & Hasan; Selmar et al. 1990; Hasan et al. 1994; Ucciani et al. 1994; Hasan & Roomi 1996; Sarita et al. 2002; Avato et al. 2005; Avato et al. 2006), ranging from only 3% in *Paullinia cupana var. sorbilis* (Guarana) (Avato et al. 2003), to 58% in *Schleichera trijuga* (Mikolajczak & Smith 1971).

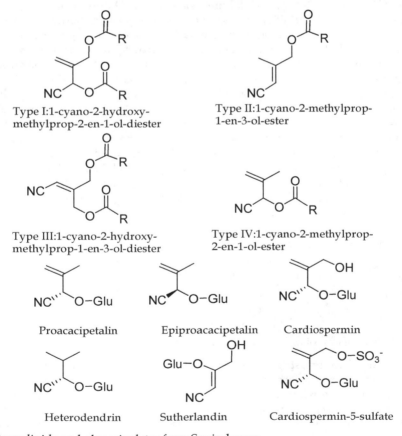

Type I:1-cyano-2-hydroxy-methylprop-2-en-1-ol-diester

Type II:1-cyano-2-methylprop-1-en-3-ol-ester

Type III:1-cyano-2-hydroxy-methylprop-1-en-3-ol-diester

Type IV:1-cyano-2-methylprop-2-en-1-ol-ester

Proacacipetalin

Epiproacacipetalin

Cardiospermin

Heterodendrin

Sutherlandin

Cardiospermin-5-sulfate

Fig. 8. Cyanolipids and glucosinolates from Sapindaceae.

Cyanolipids present in Sapindaceae belong to four types (Figure 8) which do not occur in all species. Although there is some controversy in the literature, it seems that each species has only one or two types of cyanolipids, but not all of them. The fatty acid moieties in cyanolipids vary within the species. In most of the studies, cyanolipids are esterified mostly by only one or two fatty acids. Eicosanoic acid (20:0) and eicosenoic acid (20:1 *n-9*) are present in high proportion in many species (Aichholz et al. 1997). In the genus *Paullinia*, the rare fatty acids paullinic acid (Z-13-eicosenoic acid) and cis-vaccenic acid (Z-11-octadecenoic acid) are ubiquitous constituents (Seigler 1974; Spitzer 1995; Spitzer 1996; Lago et al. 2000; Avato et al. 2003). Other fatty acids found go from dodecanoic (12:0) to docosanoic acids (22:0) (Mikolajczak et al. 1970a; 1970b; 1971; Aichholz et al. 1997; Avato et al. 2003).

Besides lipid derivatives, other cyanocompounds (the corresponding glycosides and salts) have also been isolated from aerial parts (Seigler et al. 1974; Bjarnholt et al. 2008) or roots of different species (Kumar et al. 2011). For example, cardiospermin (Figure 8) has been isolated from some species -and has been shown to be the responsible of the anxiolytic effects of ethanolic root extract from *Cardiospermum halicacabum* (Kumar et al. 2011); and in the particular case of *Cardiospermum grandiflorum* the corresponding sulphate-containing cyanogenic glucoside of cardiospermin was described (Hubel & Nahrstedt 1979).

Only cyanolipids of the type I and IV are cyanogenetic (Avato et al. 2005). Therefore cyanolipids do not work for all species as defensive compounds producing HCN. When HCN can be formed, its production works similarly to the one from glycosinolates, with a previous step of hydrolysis of the ester moieties catalyzed by estearases (Figure 9) (Wink et al. 1997). Cyanolipids can effectively work as plant defences, as it has been shown that their enzymatic breakdown produces the α-hydroxynitriles, from which HCN is released in a similar way than the one from glycosinolates (Wink et al. 1997). Moreover, it has been shown that *in vivo* HCN is produced upon wounding by herbivores (Selmar et al. 1990). Besides, as any plant defence, these defensive metabolites have been overcome in their original function by herbivores. In that sense it is well known that some Heteroptera (*Leptocorus* and *Jadera* spp.) not only are specialists on Sapindaceae, but are also able to sequester cyanolipids as such from their food plants, and biotransform them to the glycosylated derivatives (Braekman et al. 1982; Aldrich et al. 1990). Moreover, such acquisition of cyanocompounds from their host renders these gregarious, aposematic insects unpalatable to a variety of predators (Aldrich et al. 1990).

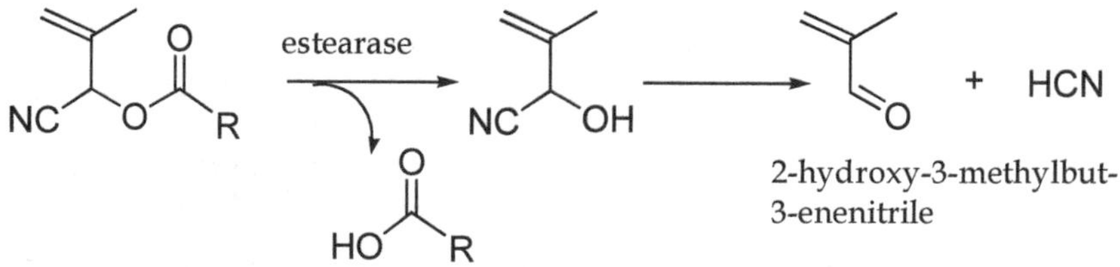

Fig. 9. Hydrolysis of cyanolipids to produce HCN (Wink et al. 1997).

Cyanolipids have been investigated in their potential as control agents for insects. In that sense, insect repellent and insecticidal properties of some of these products have been described. For instance, when adults of the red flour beetles, *Tribolium castaneum*, were exposed to seed oils from two species of *Cardiospermum* (*C. canescens* and *C. belicacabum*), the beetles preferred the arena zones where no oil was applied (Khan et al. 1983). However this repellency effect seems not to be a general pattern as in a pitfall trap bioassay, oils reach in cyanolipids stimulated aggregation in the cases of the saw toothed grain beetle, *Oryzaephilus surinamensis*, and of the rice weevil, *Sitophilus oryzae* (Mikolajczak et al. 1984). In this later work, when tested separately the four classes of cyanolipids (Figure 8) lost their repellent activity, suggesting synergism among them for this action. When the effect of cyanolipids was tested in a contact bioassay, only the ones belonging to classes I and IV showed a paralyzing effect against the saw toothed grain beetle, *Oryzaephilus surinamensis*. Eventhough one may be tempted to correlate this effect with the capacity of producing HCN by the pure compound tested, the conclusion is not again a general one since these cyanolipids did not have any effect on three other beetles (Mikolajczak et al. 1984). Finally, the European corn borer, *Ostrinia nubilalis*, was affected in its metamorphosis when cyanolipids of the classes II and IV were incorporated in its diet (Mikolajczak et al. 1984). On the whole, even though non- glycoside cyanogens were described in Sapindaceae as early as in the 1920` (cited by Mikolajczak et al. 1969), not many studies have been carried out in regard of either their ecological role as plant defences, or their potential as biopesticides. Although in the last case one can envisioned that cyanolipids of the classes I and IV will not be selective –because HCN is generated-, there is a chance that cyanolipids of the classes II and III may have some interest in this regard.

4. Conclusion

The family Sapindaceae includes many edible species, *e.g.* ackee, rambutan, longan and lychee (fruits from *Blighia sapida, Litchi chinensis, Nephelium lappaceum, Dimocarpus longan* respectively), which are widely consumed mainly in Asia and Australia (Diczbalis 2008; Vichitrananda & Somsri 2008; Diczbalis et al. 2010). Nevertheless, some species in this family produce in different phenological stages (including fruits in some cases) bioactive compounds with medicinal or toxicological properties. With reference to insect toxicity, up to now most of the studies carried out have found activity against species in the orders Lepidoptera and Diptera (mosquitoes). However, these results may be an artefact of the biodetection itself, as much of the research focuses on chewing armyworms and borers of economically significance, and on mosquitoes as important vectors of human diseases. In addition, products from Sapindaceae have revealed differential activity on insect targets from different orders, and even from the same order (Mikolajczak et al. 1984; Khan et al. 2002; Castillo et al. 2009); and have demonstrated selectivity when their activity was checked against beneficial insects while some other products have not (Castillo et al. 2009). These findings emphasise the need for widening the spectrum of biodetectors used in the tests.

The Sapindaceae are well characterized for the presence in their seeds of toxic cyanolipids, and the occurrence of this group of secondary metabolites seems to be restricted to this family. However, it is difficult to foresee that the Sapindaceae will find their way into the development of botanicals based on their unique cyanogens due to the intrinsic general toxicity of these compounds.

This family takes its name from the soapberry tree *Sapindus saponaria* (Emanuel & Benkeblia 2011), mostly known for being rich in saponins. Those chemical constituents provide its extracts with tensoactive properties, having been widely used not only as a source of soap but also for the application of its biological effects in medicine as well as in pest control. Indeed, the tensoactivity has been a property that has found application in pest control as shown by the fact that soaps are probably among the oldest insecticides in use (Silva et al. 2007).

Many other bioactive compounds belonging to different chemical groups have also been isolated from members of this family. For instance, from species studied for their anti insect activity, it has been reported the occurrence of clerodane diterpenoids and prenylated flavonoids (*Dodonaea* spp.); flavonols, phenols and triterpenes (*Paullinia* spp.); tannins (*M. pubescens*); lectins (*Talisia esculenta*); gallic acid, gallates and derivatives (*Acer rubrum*). Being confirmed in some cases that those compounds are the responsible for the activities found. Among these secondary metabolites, probably the ones with the strongest potential as anti-insect agents are the clerodanes. Eventhough the main sources of these diterpenes are species from Asteraceae and Lamiaceae, the Sapindaceae are showing to be also a good resource of them. So far, different studies have shown the antifeedant capability of clerodanes against many insects, including species from Lepidoptera, Coleoptera, and Orthoptera (Klein Gebbinck et al. 2002; Sosa et al. 2008).

All in all, the family Sapindaceae, which members are widely distributed in every continent and have been used since early days for different purposes –taking advantage of their medicinal and toxicological properties-, seems to be a promissory source of bioactive compounds to be used as biological control agents. However more extensive studies, not only on more species not yet prospected, but also concerning more diverse targets are still needed.

5. Acknowledgments

We acknowledge financial support from the following agencies: OPCW (research grant L/ICA/ICB/111817/06); TWAS (research grant N° 05-318); and LATU (Laboratorio Tecnológico de Uruguay), Facultad de Química (Universidad de la República) and Agencia Nacional de Investigación e Innovación (ANII, Uruguay) for doctoral fellowships to MD. Finally, we would like to thank the editors of this book for their kind invitation to participate in this endeavour.

6. References

Abdel, A., Shadia & Omer, E. A. (1995). Bio-evaluation of *Dodonaea viscosa* L. Jacq. extracts on the cotton leafworm, *Spodoptera littoralis* (Boisd.) as indicated by life table parameters. *Annals of Agricultural Science (Cairo)* Vol. 40, 2, pp. 891-900

Abou-Zaid, M. M., Helson, B. V., Nozzolillo, C. & Arnason, J. T. (2001). Ethyl m-digallate from red maple, *Acer rubrum* L., as the major resistance factor to forest tent caterpillar, *Malacosoma disstria* Hbn. *Journal of Chemical Ecology* Vol. 27, 12, pp. 2517-2527

Aichholz, R., Spitzer, V. & Lorbeer, E. (1997). Analysis of cyanolipids and triacylglycerols from sapindaceae seed oils with high-temperature gas chromatography and high-temperature gas chromatography chemical ionization mass spectrometry. *Journal of Chromatography A* Vol. 787, 1-2, pp. 181-194

Aldrich, J. R., Carroll, S. P., Lusby, W. R., Thompson, M. J., Kochansky, J. P. & Waters, R. M. (1990). Sapindaceae, cyanolipids, and bugs. *Journal of Chemical Ecology* Vol. 16, 1, pp. 199-210

Anonymous (2011). Germplasm Resources Information Network - (GRIN) National Germplasm Resources Laboratory, Beltsville, Maryland. In: *USDA, ARS, National*

Genetic Resources Program, 11 July 2011, Available from: http://www.ars-grin.gov/cgi-bin/npgs/html/gnlist.pl?999

Arruda, W., Cavasin Oliveira, G. M. & Garcia da Silva, I. (2003a). Morphological alterations of *Aedes aegypti* larvae exposed to *Magonia pubescens* extract. *Entomologia y Vectores* Vol. 10, 1, pp. 47-60

Arruda, W., Oliveira, G. M. C. & da Silva, I. G. (2003b). Toxicity of the ethanol extract of *Magonia pubescens* on larvae *Aedes aegypti*. *Revista da Sociedade Brasileira de Medicina Tropical* Vol. 36, 1, pp. 17-25

Arulmozhi, D. K., Veeranjaneyulu, A., Bodhankar, S. L. & Arora, S. K. (2005). Pharmacological studies of the aqueous extract of *Sapindus trifoliatus* on central nervous system: possible antimigraine mechanisms. *Journal of Ethnopharmacology* Vol. 97, 3, pp. 491-496

Avato, P., Pesante, M. A., Fanizzi, F. P. & Santos, C. A. D. (2003). Seed oil composition of *Paullinia cupana* var. *sorbilis* (Mart.) Ducke. *Lipids* Vol. 38, 7, pp. 773-780

Avato, P., Rosito, I., Papadia, P. & Fanizzi, F. P. (2005). Cyanolipid-rich seed oils from *Allophylus natalensis* and *A. dregeanus*. *Lipids* Vol. 40, 10, pp. 1051-1056

Avato, P., Rosito, I., Papadia, P. & Fanizzi, F. P. (2006). Characterization of seed oil components from *Nephelium lappaceum* L. *Natural Product Communications* Vol. 1, 9, pp. 751-755

Banerjee, H. N. (1936). Clerodin from *Clerodendron infortunatum*. *Science and Culture* Vol. 2, pp. 163

Benlekehal, H., Clotteau, M., Dornier, M. & Reynes, M. (2001). An Amazonian product particularly rich in caffeine: the seed of guarana [*Paullinia cupana* H.B.K. var. *sorbilis* (Mart.) Ducke]. *Fruits (Paris)* Vol. 56, 6, pp. 423-435

Bjarnholt, N. & Møller, B. L. (2008). Hydroxynitrile glucosides. *Phytochemistry* Vol. 69, 10, pp. 1947-1961

Boiça Junior, A. L., Medeiros, C. A. M., Torres, A. L. & Chagas Filho, N. R. (2005). Effect of plant aqueous extracts on the development of *Plutella xylostella* (L.) (Lepidoptera: Plutellidae), on collard greens. *Arquivos do Instituto Biologico Sao Paulo* Vol. 72, 1, pp. 45-50

Bousserouel, H., Litaudon, M., Morleo, B., Martin, M. T., Thoison, O., Nosjean, O., Boutin, J. A., Renard, P. & Sevenet, T. (2005). New biologically active linear triterpenes from the bark of three Newcaledonian *Cupaniopsis* species. *Tetrahedron* Vol. 61, 4, pp. 845-851

Braekman, J. C., Daloze, D. & Pasteels, J. M. (1982). Cyanogenic and other glucosides in a neo-guinean bug *Leptocoris isolata*: Possible precursors in its host-plant. *Biochemical Systematics and Ecology* Vol. 10, 4, pp. 355-364

Bruneton, J. (1995). Pharmacognosy, phytochemistry, medicinal plants. Paris, Lavoisier Publishing: 538-544

Bullangpoti, V., Visetson, S., Milne, J. & Pornbanlualap, S. (2004). Effects of Mangosteen's peels and rambutan's seeds on toxicity, esterase and glutathione-S-transferase in rice weevil (*Sitophilus oryzae* L.). *Kasetsart Journal, Natural Sciences* Vol. 38, 5, Supplement, pp. 84-89

Castillo, L., Gonzalez-Coloma, A., Gonzalez, A., Diaz, M., Alonso-Paz, E., Bassagoda, M. J. & Rossini, C. (2009). Screening of Uruguayan plants for deterrent activity against insects. *Industrial Crops and Products* Vol. 29, 1, pp. 235-240 da Silva, I. G., Guimaraes, V. P., Lima, C. G., da Silva, H. H. G., Elias, C. N., Mady, C. M., da Mota e Silva, V. V., de Paiva Nery, A., da Rocha, K. R., Rocha, C. & Isac, E. (2003). Larvicide and toxic effect of the ethanol crude extract of the stem coat of *Magonia pubescens* towards *Aedes aegypt* in laboratory. *Revista de Patologia Tropical* Vol. 32, 1, pp. 73-86

de Sousa Araújo, T. A., Alencar, N. L., de Amorim, E. L. C. & de Albuquerque, U. P. (2008). A new approach to study medicinal plants with tannins and flavonoids contents from the local knowledge. *Journal of Ethnopharmacology* Vol. 120, 1, pp. 72-80

Deepa, B. & Remadevi, O. K. (2007). Contact toxicity of the leaf extracts of *Dodonaea viscosa* (Sapindaceae) to *Spodoptera litura* (Fab.). *Hexapoda* Vol. 14, 1, pp. 66-67

Dharmani, P., Mishra, P. K., Maurya, R., Chauhan, V. S. & Palit, G. (2005). *Allophylus serratus*: A plant with potential anti-ulcerogenic activity. *Journal of Ethnopharmacology* Vol. 99, 3, pp. 361-366

Diczbalis, Y., Nicholls, B., Groves, I. & Lake, K. (2010). Sapindaceae production and research in Australia. *Acta Horticulturae* Vol. 863, pp. 49-58

Diczbalis, Y. A. (2008). Nutrition management of tropical fruits grown in north Queensland, Australia. *Acta Horticulturae* Vol. 772, pp. 375-379

Dinesh, K. & Hasan, S. Q. Cyanolipids in sapindaceous seed oils. *Asian Journal of Chemistry* Vol. 23, 6, pp. 2589-2591

dos Santos, W. L., Freire, M. D. M., Bogorni, P. C., Vendramim, J. D. & Macedo, M. L. R. (2008). Effect of the aqueous extracts of the seeds of *Talisia esculenta* and *Sapindus saponaria* on fall armyworm. *Brazilian Archives of Biology and Technology* Vol. 51, 2, pp. 373-383

El-Din, M. M. & El-Gengaihi, S. E. (2000). Joint action of some botanical extracts against the Egyptian cotton leafworm *Spodoptera littoralis* (Boisd.) (Lepidoptera: Noctuidae). *Egyptian Journal of Biological Pest Control* Vol. 10, 1/2, pp. 51-56

Emanuel, M. A. & Benkeblia, N. (2011). Processing of ackee fruit (*Blighia sapida* L.): present and future perspectives. *Acta Horticulturae* Vol. 894, pp. 211-213

Fernandes, F. F., D'Alessandro, W. B. & Freitas, E. P. S. (2008). Toxicity of extract of *Magonia pubescens* (Sapindales : Sapindaceae) St. Hil. to control the Brown Dog tick, *Rhipicephalus sanguineus* (Latreille) (Acari : Ixodidae). *Neotropical Entomology* Vol. 37, 2, pp. 205-208

Ferreira Barreto, C., Cavasin, G. M., Garcia da Silva, H. H. & Garcia da Silva, I. (2006). Study of the morphohistological modifications in larvae of *Aedes aegypti* (Diptera, Culicidae) submitted to the pure ethanolic extract of *Sapindus saponaria* Lin. (Sapindaceae). *Revista de Patologia Tropical* Vol. 35, 1, pp. 37-57

Gaillard, Y., Carlier, J., Berscht, M., Mazoyer, C., Bevalot, F., Guitton, J. & Fanton, L. (2011). Fatal intoxication due to ackee (*Blighia sapida*) in Suriname and French Guyana. GC-MS detection and quantification of hypoglycin-A. *Forensic Science International* Vol. 206, 1-3, pp. e103-e107

Getie, M., Gebre-Mariam, T., Rietz, R., Hohne, C., Huschka, C., Schmidtke, M., Abate, A. & Neubert, R. H. H. (2003). Evaluation of the anti-microbial and anti-inflammatory activities of the medicinal plants *Dodonaea viscosa*, *Rumex nervosus* and *Rumex abyssinicus*. *Fitoterapia* Vol. 74, 1-2, pp. 139-143

Getie, M., Gebre-Mariam, T., Rietz, R. & Neubert, R. H. H. (2002). Evaluation of the release profiles of flavonoids from topical formulations of the crude extract of the leaves of *Dodonaea viscosa* (Sapindaceae). *Pharmazie* Vol. 57, 5, pp. 320-322

Hasan, S. Q. & Roomi, Y. A. (1996). Cyanolipids in *Sapindus trifoliatus* seed oil. *Journal of the Oil Technologists' Association of India* Vol. 28, 1, pp. 23, 25

Hasan, S. Q., Roomi, Y. A. & Chitra, N. (1994). Reinvestigation of seed oil of *Sapindus saponaria* for its cyanolipid content. *Journal of the Oil Technologists' Association of India* Vol. 26, 3, pp. 77-79

Hassall, C. H. & Reyle, K. (1955). Hypoglycin-A and Hypoglycin-B, two biologically active polypeptides from *Blighia sapida*. *Biochemical Journal* Vol. 60, 1-4, pp. 334-339

Hassall, C. H., Reyle, K. & Feng, P. (1954). Hypoglycin A,B: Biologically active polypeptides from *Blighia sapida*. *Nature* Vol. 173, 4399, pp. 356-357

Hubel, W. & Nahrstedt, A. (1979). Cardiosperminsulfate - sulfur-containing cyanogenic glucoside from *Cardiospermum grandiflorum*. *Tetrahedron Letters* Vol. 20, 45, pp. 4395-4396

Isman, M., Matsuura, H., MacKinnon, S., Durst, T., Towers, G. & Arnason, J. (1996). Phytochemistry of the Meliaceae. *Recent Advances on Phytochemistry* Vol. 30, pp. 155-178

Ito, A., Chai, H. B., Kardono, L. B. S., Setowati, F. M., Afriastini, J. J., Riswan, S., Farnsworth, N. R., Cordell, G. A., Pezzuto, J. M., Swanson, S. M. & Kinghorn, A. D. (2004). Saponins from the bark of *Nephelium maingayi*. *Journal of Natural Products* Vol. 67, 2, pp. 201-205

Jain, S. C. (1976). Isolation of flavonoids from soapnut, *Sapindus emarginatus* Vahl. *Indian Journal of Pharmacy* Vol. 38, 6, pp. 141-142

Jefferie, P. R., Knox, J. R. & Scaf, B. (1973). Structure elucidation of some ent-clerodane diterpenes from *Dodonaea boroniaefolia* and *Cyanostegia angustifolia*. *Australian Journal of Chemistry* Vol. 26, 10, pp. 2199-2211

Kamala-Kannan, S., Park, S. M., Oh, B. T., Kim, H. M. & Lee, K. J. (2009). First report of aster yellows phytoplasma in Goldenrain tree (*Koelreuteria paniculata*) in Korea. *Journal of Phytopathology* Vol. 158, 3, pp. 197-199

Kanchanapoom, T., Kasai, R. & Yamasaki, K. (2001). Acetylated triterpene saponins from the Thai medicinal plant, *Sapindus emarginatus*. *Chemical & Pharmaceutical Bulletin* Vol. 49, 9, pp. 1195-1197

Kayode, J. (2006). Conservation of indigenous medicinal botanicals in Ekiti State, Nigeria. *Journal of Zhejiang University SCIENCE B* Vol. 7, 9, pp. 713-718

Khan, A. & Gumbs, F. A. (2003). Repellent effect of ackee (*Blighia sapida* Koenig) component fruit parts against stored-product insect pests. *Tropical Agriculture* Vol. 80, 1, pp. 19-27

Khan, A., Gumbs, F. A. & Persad, A. (2002). Pesticidal bioactivity of ackee (*Blighia sapida* Koenig) against three stored-product insect pests. *Tropical Agriculture* Vol. 79, 4, pp. 217-223

Khan, M. W. Y., Ahmad, F., Ahmad, I. & Osman, S. M. (1983). Nonedible Seed Oils as Insect Repellent. *Journal of the American Oil Chemists Society* Vol. 60, 5, pp. 949-950

Klein Gebbinck, E. A., Jansen, B. J. M. & de Groot, A. (2002). Insect antifeedant activity of clerodane diterpenes and related model compounds. *Phytochemistry* Vol. 61, 7, pp. 737-770

Koodalingam, A., Mullainadhan, P. & Arumugam, M. (2009). Antimosquito activity of aqueous kernel extract of soapnut *Sapindus emarginatus*: impact on various developmental stages of three vector mosquito species and nontarget aquatic insects. *Parasitology Research* Vol. 105, 5, pp. 1425-1434

Koodalingam, A., Mullainadhan, P. & Arumugam, M. (2011). Effects of extract of soapnut *Sapindus emarginatus* on esterases and phosphatases of the vector mosquito, *Aedes aegypti* (Diptera: Culicidae). *Acta Tropica* Vol. 118, 1, pp. 27-36

Kumar, R., Murugananthan, G., Nandakumar, K. & Talwar, S. (2011). Isolation of anxiolytic principle from ethanolic root extract of *Cardiospermum halicacabum*. *Phytomedicine* Vol. 18, 2-3, pp. 219-223

Lago, R. C. A., Simoni, M. d. L. P. S. C. & Pinto, A. d. C. (2000). On the occurrence of cyanolipids in *Paullinia carpopodea* Cambess and *P. cupana* Kunth seed oils. *Acta Amazonica* Vol. 30, 1, pp. 101-105

Lemos, T. L. G., Mendes, A. L., Sousa, M. P. & Braz-Filho, R. (1992). New saponin from *Sapindus saponaria*. *Fitoterapia* Vol. 63, 6, pp. 515-517

Lorenzi, H. (2004). Arvores brasileiras. Manual de identificaçao e cultivo de plantas arvoreas nativas do Brasil. Piracicaba: Plantarum

Macedo, M. L. R., Damico, D. C. S., Freire, M. D., Toyama, M. H., Marangoni, S. & Novello, J. C. (2003). Purification and characterization of an N-acetylglucosamine-binding lectin from *Koelreuteria paniculata* seeds and its effect on the larval development of *Callosobruchus maculatus* (Coleoptera: Bruchidae) and *Anagasta kuehniella* (Lepidoptera: Pyralidae). *Journal of Agricultural and Food Chemistry* Vol. 51, 10, pp. 2980-2986

Macedo, M. L. R., Freire, M. D. M., Novello, J. C. & Marangoni, S. (2002). *Talisia esculenta* lectin and larval development of *Callosobruchus maculatus* and *Zabrotes subfasciatus* (Coleoptera : Bruchidae). *Biochimica et Biophysica Acta* Vol. 1571, 2, pp. 83-88

Mahmoud, I., Moharram, F. A., Marzouk, M. S., Soliman, H. S. M. & El-Dib, R. A. (2001). Two new flavonol glycosides from leaves of *Koelreuteria paniculata*. *Pharmazie* Vol. 56, 7, pp. 580-582

Malarvannan, S., Giridharan, R., Sekar, S., Prabavathy, V. R. & Nair, S. (2009). Ovicidal activity of crude extracts of few traditional plants against *Helicoverpa armigera* (Hubner) (Noctuidae: Lepidoptera). *Journal of Biopesticides* Vol. 2, 1, pp. 64-71

Malarvannan, S., Kumar, S. S., Prabavathy, V. R. & Sudha, N. (2008). Individual and synergistic effects of leaf powder of few medicinal plants against American bollworm, *Helicoverpa armigera* (Hubner) (Noctuidae: Lepidoptera). *Asian Journal of Experimental Sciences* Vol. 22, 1, pp. 79-88

Malarvannan, S., Sekar, S., Prabavathy, V. R. & Sudha, N. (2008). Individual and synergistic efficacy of leaf powder of *Argemone mexicana* against tobacco caterpillar, *Spodoptera litura* Fab, (Noctuidae: Lepidoptera). *Emerging trends of researches in insect pest management and environmental safety, Volume I* Vol., pp. 155-164

Malarvannan, S. & Subashini, H. D. (2007). Efficacy of *Dodonaea angustifolia* crude extracts against spotted bollworm, *Earias vitella* (Fab.) (Lepidoptera: Noctuidae). *Journal of Entomology* Vol. 4, 3, pp. 243-247

Mazzola, E. P., Parkinson, A., Kennelly, E. J., Coxon, B., Einbond, L. S. & Freedberg, D. I. (2011). Utility of coupled-HSQC experiments in the intact structural elucidation of three complex saponins from *Blighia sapida*. *Carbohydrate Research* Vol. 346, 6, pp. 759-768

Medeiros, C. A. M., Boica Junior, A. L. & Angelini, M. R. (2007). Sub-lethal effect of aqueous vegetals extracts of *Azadirachta indica* A. Juss. and *Sapindus saponaria* L. against the biology of *Ascia monuste orseis* at kale. *Boletin de Sanidad Vegetal, Plagas* Vol. 33, 1, pp. 27-34

Meyer Albiero, A. L., Aboin Sertié, J. A. & Bacchi, E. M. (2002). Antiulcer activity of *Sapindus saponaria* L. in the rat. *Journal of Ethnopharmacology* Vol. 82, 1, pp. 41-44

Mikolajczak, K. L. (1977). Cyanolipids. *Progress in the chemistry of fats and other lipids* Vol. 15, 2, pp. 97-130

Mikolajczak, K. L., Madrigal, R. V., Smith, C. R. & Reed, D. K. (1984). Insecticidal effects of cyanolipids on three species of stored product insects, European Corn Borer (Lepidoptera: Pyralidae) Larvae, and Striped Cucumber Beetle (Coleoptera: Chrysomelidae) *Journal of Economic Entomology* Vol. 77, 5, pp. 1144-1148

Mikolajczak, K. L., Seigler, D. S., Smith Jr, C. R., Wolff, I. A. & Bates, R. B. (1969). A cyanogenetic lipid from *Cordia verbenacea* DC. Seed oil. *4* Vol. 6, 617-619,

Mikolajczak, K. L. & Smith, C. R. (1971). Cyanolipids of kusum (*Schleichera trijuga*) seed oil. *Lipids* Vol. 6, 5, pp. 349-&

Mikolajczak, K. L., Smith, C. R. & Tjarks, L. W. (1970). Cyanolipids of *Cardiospermum halicacabum* L and other sapindaceous seed oils. *Lipids* Vol. 5, 10, pp. 812-&

Mikolajczak, K. L., Smith, C. R. & Tjarks, L. W. (1970). Cyanolipids of *Koelreuteria paniculata* Laxm. seed oil. *Lipids* Vol. 5, 8, pp. 672-&

Mukherjee, S. N. & Joseph, M. (2000). Medicinal plant extracts influence insect growth and reproduction: A case study. *Journal of Medicinal and Aromatic Plant Sciences* Vol. 22-23, 4A-1A, pp. 154-158

Mukherjee, S. N. & Joseph, M. (2001). Medicinal plant extracts influence insect growth and reproduction: a case study. *Journal of Medicinal and Aromatic Plant Sciences* Vol. 22/23, 4A/1A, pp. 154-158

Murdock, L. L. & Shade, R. E. (2002). Lectins and protease inhibitors as plant defenses against insects. *Journal of Agriculture and Food Chemistry* Vol. 50, pp. 6605–6611

Muthukumran, P., Begumand, V. H. & Kalaiarasan, P. (2011). Antidiabetic activity of *Dodonaea viscosa* (L) leaf extracts. *International Journal of PharmTech Research* Vol. 3, 1, pp. 136-139

Nielsen, N. J., Nielsen, J. & Staerk, D. (2010). New resistance-correlated saponins from the insect-resistant crucifer *Barbarea vulgaris*. *Journal of Agrulture and Food Chemistry* Vol. 58, pp. 5509-5514

Niu, H. M., Zeng, D. Q., Long, C. L., Peng, Y. H., Wang, Y. H., Luo, J. F., Wang, H. S., Shi, Y. N., Tang, G. H. & Zhao, F. W. (2010). Clerodane diterpenoids and prenylated flavonoids from *Dodonaea viscosa*. *Journal of Asian Natural Products Research* Vol. 12, 1, pp. 7-14

Omosa, L. K., Midiwo, J. O., Derese, S., Yenesew, A., Peter, M. G. & Heydenreich, M. (2010). neo-Clerodane diterpenoids from the leaf exudate of *Dodonaea angustifolia*. *Phytochemistry Letters* Vol. 3, 4, pp. 217-220

Palanisamy, U., Cheng, H. M., Masilamani, T., Subramaniam, T., Ling, L. T. & Radhakrishnan, A. K. (2008). Rind of the rambutan, *Nephelium lappaceum*, a potential source of natural antioxidants. *Food Chemistry* Vol. 109, 1, pp. 54-63

Pelegrini, D. D., Tsuzuki, J. K., Amado, C. A. B., Cortez, D. A. G. & Ferreira, I. C. P. (2008). Biological activity and isolated compounds in *Sapindus saponaria* L. and other plants of the genus *Sapindus*. *Latin American Journal of Pharmacy* Vol. 27, 6, pp. 922-927

Pérez, D. & Iannacone, J. (2004). Insecticidal effect of *Paullinia clavigera* var. bullata simpson (Sapindaceae) and *Tradescantia zebrina* Hort. ex Bosse (Commelinaceae) in the control of *Anopheles benarrochi* Gabaldon, main vector of malaria in ucayali, Peru *Ecología Aplicada* Vol. 3, pp. 64-72

Pérez, D., Iannacone, J. & Pinedo, H. (2010). Toxicological effect from the stem cortex of the amazonic plant soapberry *Paullinia clavigera* (Sapindaceae) upon three arthropods. *Ciencia E Investigacion Agraria* Vol. 37, 3, pp. 133-143

Pérez, D., Iannacone, J. & Tueros, A. (2008). Toxicity of *Paullinia clavigera* schltdl. (Sapindaceae) and *Chondrodendron tomentosum* ruiz et pav. (Menispermaceae) on jumping lice of camu camu *Tuthillia cognata* (Hemiptera: Psyllidae). *Gayana Botanica* Vol. 65, 2, pp. 145-152

Pérez, D. D. & Iannacone, O. J. (2006). Effectiveness of botanical extracts from ten plants on mortality and larval repellency of *Rhynchophorus palmarum* L., an insect pest of the Peach palm *Bactris gasipaes* Kunth in Amazonian Peru. *Agricultura Tecnica* Vol. 66, 1, pp. 21-30

Peumans, W. J. & Van Damme, E. J. M. (1995). Lectins as plant defense proteins. *Plant Physiology* Vol. 109, pp. 347-352

Plasman, V., Plehiers, M., Braekman, J., Daloze, D., De Biseau, J. & Pasteels, J. (2001). Chemical defense in *Platyphora kollari* Baly and *Leptinotarsa behrensi* Harold (Coleoptera: Chrysomelidae). Hypotheses on the origin and evolution of leaf beetles toxins. *Chemoecology* Vol. 11, pp. 107-112

Porras, M. F. & Lopez-Avila, A. (2009). Effect of extracts from *Sapindus saponaria* on the glasshouse whitefly *Trialeurodes vaporariorum* (Hemiptera: Aleyrodidae). *Revista Colombiana De Entomologia* Vol. 35, 1, pp. 7-11

Prieto, J. M., Schaffner, U., Barker, A., Braca, A., Siciliano, T. & Boevé, J. L. (2007). Sequestration of furostanol saponins by *Monophadnus* sawfly larvae. *Journal of Chemical Ecology* Vol. 33, 513-524,

Rahman, S. S., Rahman, M. M., Begum, S. A., Khan, M. M. R. & Bhuiyan, M. M. H. (2007). Investigation of *Sapindus mukorossi* extracts for repellency, insecticidal activity and plant growth regulatory effect. *Journal of Applied Sciences Research* Vol., February, pp. 95-101

Rey, D., Pautou, M.-P. & Meyran, J.-C. (1999). Histopathological effects of tannic acid on the midgut epithelium of some aquatic Diptera larvae. *Journal of Invertebrate Pathology* Vol. 73, 2, pp. 173-181

Ribeiro, A., Zani, C. L., Alves, T. M. D., Mendes, N. M., Hamburger, M. & Hostettmann, K. (1995). Molluscicidal saponins from the pericarp of *Sapindus saponaria*. *International Journal of Pharmacognosy* Vol. 33, 3, pp. 177-180

Rodrigues, A. M. S., De Paula, J. E., Degallier, N., Molez, J. F. & Espindola, L. S. (2006). Larvicidal activity of some Cerrado plant extracts against *Aedes aegypti*. *Journal of the American Mosquito Control Association* Vol. 22, 2, pp. 314-317

Rodriguez, E. M. (1958). The wood of Argentine Sapindaceae. Structure, characteristics, and applications. *Rev Fac Agron Y Vet [Buenos Aires]* Vol. 14, (2), pp. 271-305

Saha, S., Walia, S., Kumar, J., Dhingra, S. & Parmar, B. S. (2009). Screening for feeding deterrent and insect growth regulatory activity of triterpenic saponins from *Diploknema butyracea* and *Sapindus mukorossi*. *Journal of Agricultural and Food Chemistry* Vol. 58, 1, pp. 434-440

Sarita, R., Geeta, N., Srivastava, V. K., Hasan, S. N. & Hasan, S. Q. (2002). Cyanolipids in *Sapindus obovatus* seed oil and reinvestigation of the seed oil of *Heliotropium indicum*, *H. eichwaldi* and *D. viscosa*. *Journal of the Oil Technologists' Association of India* Vol. 34, 2, pp. 69-71

Seigler, D. (1974). Determination of cyanolipids in seed oils of the Sapindaceae by means of their NMR spectra. *Phytochemistry* Vol. 13, 5, pp. 841-843

Seigler, D. S. (1976). Cyanolipids in *Cordia verbenacea*: A correction. *Biochemical Systematics and Ecology* Vol. 4, 4, pp. 235-236

Seigler, D. S. (1991). Cyanide and cyanogentic glycosides. In: *Herbivores: Their Interactions with Secondary Plant Metabolites*, Rosenthal, G. A. & Berembaum, M. R., pp. Academic Press San Diego, California

Seigler, D. S., Eggerding, C. & Butterfield, C. (1974). A new cyanogenic glycoside from *Cardiospermum hirsutum*. *Phytochemistry* Vol. 13, 10, pp. 2330-2332

Seigler, D. S., Mikolajczak, K. L., Smith Jr, C. R. & Wolff, I. A. (1970). Structure and reactions of a cyanogenetic lipid from *Cordia verbenacea* DC. seed oil. *Chemistry and Physics of Lipids* Vol. 4, pp. 147-161

Selmar, D., Grocholewski, S. & Seigler, D. S. (1990). Cyanogenic lipids: Utilization during seedling development of *Ungnadia speciosa*. *Plant Physiology* Vol. 93, 2, pp. 631-636

Sharaby, A., Abdel-Rahman, H. & Moawad, S. (2009). Biological effects of some natural and chemical compounds on the potato tuber moth, *Phthorimaea operculella* Zell. (Lepidoptera: Gelechiidae). *Saudi Journal of Biological Sciences* Vol. 16, 1, pp. 1-9

Sharma, A., Sati, S. C., Sati, O. P., Sati, D. M. & Kothiyal, S. K. (2011). Chemical constituents and Bioactivities of genus *Sapindus*. *International Journal of Research in Ayurveda & Pharmacy* Vol. 2, 2, pp. 403-409

Silva, H. H. G. d., Silva, I. G. d., dos Santos, R. M. G., Rodrigues Filho, E. & Elias, C. N. (2004). Larvicidal activity of tannins isolated of *Magonia pubescens* St. Hil. (Sapindaceae) against *Aedes aegypti* (Diptera, Culicidae). *Revista da Sociedade Brasileira de Medicina Tropical* Vol. 37, 5, pp. 396-9

Silva, T. M. S., Da Silva, T. G., Martins, R. M., Maia, G. L. A., Cabral, A. G. S., Camara, C. A., Agra, M. F. & Barbosa, J. M. (2007). Molluscicidal activities of six species of Bignoniaceae from north-eastern Brazil, as measured against *Biomphalaria glabrata* under laboratory conditions. *Annals of Tropical Medicine and Parasitology* Vol. 101, 4, pp. 359-365

Simpson, B., Claudie, D., Smith, N., Wang, J. P., McKinnon, R. & Semple, S. (2010). Evaluation of the anti-inflammatory properties of *Dodonaea polyandra*, a Kaanju traditional medicine. *Journal of Ethnopharmacology* Vol. 132, 1, pp. 340-343

Sofidiya, M. O., Jimoh, F. O., Aliero, A. A., Afolayan, A. J., Odukoya, O. A. & Familoni, O. B. (2008). Antioxidant and antibacterial properties of *Lecaniodiscus cupanioides*. *Research Journal of Microbiology* Vol. 3, 2, pp. 91-98

Sosa, M. E. & Tonn, C. E. (2008). Plant secondary metabolites from Argentinean semiarid lands: bioactivity against insects. *Phytchemical reviews* Vol. 7, 1, pp. 3-24

Sousa, S. A., Alves, S. F., de Paula, J. A. M., Fiuza, T. S., Paula, J. R. & Bara, M. T. F. (2009). Determination of tannins and methylxanthines in powdered guarana (*Paullinia cupana* Kunth, Sapindaceae) by high performance liquid chromatography. *Brazilian Journal of Pharmacognosy* Vol. 20, 6, pp. 866-870

Spitzer, V. (1995). GLC-MS analysis of the fatty-acids of the seed oil, triglycerides, and cyanolipid of *Paullinia elegans* (Sapindaceae): A rich source of cis-13-eicosenoic acid (paullinic acid). *Journal of High Resolution Chromatography* Vol. 18, 7, pp. 413-416

Spitzer, V. (1996). Fatty acid composition of some seed oils of the Sapindaceae. *Phytochemistry* Vol. 42, 5, pp. 1357-1360

Teffo, L. S., Aderogba, M. A. & Eloff, J. N. (2010). Antibacterial and antioxidant activities of four kaempferol methyl ethers isolated from *Dodonaea viscosa* Jacq. var. *angustifolia* leaf extracts. *South African Journal of Botany* Vol. 76, 1, pp. 25-29

Teixeira, J. R. M., Lapa, A. J., Souccar, C. & Valle, J. R. (1984). Timbós: ichthyotoxic plants used by brazilian indians. *Journal of Ethnopharmacology* Vol. 10, 3, pp. 311-318

Ucciani, E., Mallet, J. F. & Zahra, J. P. (1994). Cyanolipids and fattyacids of *Sapindus trifoliatus* (Sapindaceae) seed oil. *Fat Science Technology* Vol. 96, 2, pp. 69-71

Van Damme, E. J. M., Lannoo, N., Peumans, W. J. & Jean-Claude Kader and Michel, D. (2008). Plant Lectins. In: *Advances in Botanical Research*, pp. 107-209 Academic Press

Veeramani, C., Pushpavalli, G. & Pugalendi, K. V. (2010). In vivo antioxidant and hypolipidemic effect of *Cardiospermum halicacabum* leaf extract in streptozotocin-induced diabetic rats. *Journal of Basic and Clinical Physiology and Pharmacology* Vol. 21, 2, pp. 107-125

Vichitrananda, S. & Somsri, S. (2008). Tropical fruit production in Thailand. *Acta Horticulturae* Vol. 787, pp. 33-46

Waako, P. J., Gumede, B., Smith, P. & Folb, P. I. (2005). The in vitro and in vivo antimalarial activity of *Cardiospermum halicacabum* L. and *Momordica foetida* Schumch. et Thonn. *Journal of Ethnopharmacology* Vol. 99, 1, pp. 137-143

Weckerle, C. S., Stutz, M. A. & Baumann, T. W. (2003). Purine alkaloids in *Paullinia*. *Phytochemistry* Vol. 64, 3, pp. 735-742

Wilbert, W. & Haiek, G. (1991). Phytochemical screening of a Warao pharmacopoeia employed to treat gastrointestinal disorders. *Journal of Ethnopharmacology* Vol. 34, 1, pp. 7-11

Wink, M., Dey, P. M. & Harborne, J. B. (1997). Special nitrogen metabolism. In: *Plant Biochemistry*, pp. 439-486 Academic Press London

Wollenweber, E. & Roitman, J. N. (2007). New reports on surface flavonoids from *Chamaebatiaria* (Rosaceae), *Dodonaea* (Sapindaceae), *Elsholtzia* (Lamiaceae), and *Silphium* (Asteraceae). *Natural Product Communications* Vol. 2, 4, pp. 385-389

Zhao, H., Qin, X. & Yang, M. (2006). Preliminary study contact toxicity of extracts from *Dodonaea viscosa* seed aganist *Sitophilus oryzae*. *Journal of Yunnan Agricultural University* Vol. 22, 6, pp. 942-944

Zidan, Z. H., Gomaa, A. A., Afifi, F. A., Fam, E. Z. & Ahmed, S. M. S. (1993). Bioresidual activity of certain plant extracts on some stored grain insects, in relation to seed viability. *Arab Universities Journal of Agricultural Sciences* Vol. 1, 1, pp. 113-123

4

The Past and Present of Pear Protection Against the Pear Psylla, *Cacopsylla pyri* L.

Stefano Civolani

*Department of Biology and Evolution, University of Ferrara, Ferrara,
Agricultural Foundation "Fratelli Navarra", Malborghetto di Boara (Ferrara),
Italy*

1. Introduction

The pear psylla, *Cacopsylla pyri* L. (Hemiptera Psyllidae), is known in Europe for its extended infestations which may cause heavy economical losses to most pear growing regions. Pear (*Pyrus sp.* L.) is the second most relevant fruit species of temperate regions: the first one is apple (*Malus domestica* L.) and the third one is peach (*Prunus persica* L.). The top three world regions for pear production are China, Europe and North America. Total European pear production is presently stable around 2.6 million tones, according to 2010 data from the Fruit and Vegetable Service Center (CSO, Ferrara, Italy). In Europe, Italy and Spain are the largest producers, respectively with 35% and 20% of the total production. Pear production in France is decreasing (8%), mainly as a result of fireblight on "Passe Crassane", a highly susceptible variety. Until 2007, in the Netherlands and Belgium there was an increase in production with extensive planting of the "Conference" cultivar, but the first signs of saturation of the "Conference" market started to appear in 2007. The total pear production of the Netherlands and Belgium presently amounts to 9%.

All commercial varieties in Europe belong to the species *Pyrus communis* L. and they are all susceptible to *C. pyri*. The susceptibility increases when orchard techniques are aimed to maximize fruit production, such as the high density of plants per hectare, the large use of fertilizers and the intense irrigation (Fig. 1).

2. Damages caused by *C. pyri*

The damages that *C. pyri* may induce to pear trees are classified in two main types: 1) direct damages, weakening the plant by subtraction of nutrients; when the pest attack is intense, the plant wastes away with reduced production; 2) indirect damages, due to the production of a large amount of honeydew on which sooty molds develop (russetting fruits, Fig. 2), and also to the possible transmission of phytoplasms (Fig. 3).

In the first case the most damaging stages are the nymphs of all instars because of the high amount of honeydew (produced especially in spring and summer) dripping on everything including fruits. Besides lowering the fruit market value, honeydew favours the growth of sooty molds caused by saprophytic fungi, in turn causing indirect injury to the plant. Sooty molds actually induce alterations of photosynthesis, disruption of metabolism, leaf curling and premature loss, together with lower production. Prolonged attacks and intense weakening by *C. pyri* may lead to plant death.

Fig. 1. Pear orchard with high plant density.

Unlike nymphs, adults of *C. pyri* (similarly to those of the congener species *C. pyricola*) are responsible for the transmission of the phytoplasma of "pear decline". The phytoplasma is acquired by the psylla upon feeding on an infected plant and transmitted to an healthy one by salivation during feeding on phloem (Carraro *et al.*, 1998). The transmission mechanism is persistent-propagative because the phytoplasm reproduces in the insect body. The acquisition and inoculation of the pathogen requires for the vector insect to feed for 1-2 hours on phloem of infected plants. A latency period in the vector (about 1-2 weeks) follows the acquisition of phytoplasma in which the pathogen circulates and multiplies until it reaches the salivary glands. The first symptom of the disease appears during the summer-autumn period, when the leaves exhibit a red-purple hue contrasting with the yellow-green hue of senescent leaves of healthy plants. The leaves also have a stiff lamina with the edges curled upside and the apex folded downwards, and they fall prematurely beginning from the apical ones. In the next spring the infected trees show smaller, light green leaves with upward edges ("transparent canopy"). In some cases a sudden wilting is observed: although still on shoots, leaves become brown and dry. The tree may die within few days or weeks. Some authors (Giunchedi & Refatti, 1997; Davies *et al.*, 1998) showed that some plants undergo recovering during the winter dormancy because of degradation of aboveground phloem elements (Schaper & Seemüller, 1982). Moreover, the quinces *Cydonia oblonga* Mill used as rootstocks rarely allow the phyoplasma survival between vegetative cycles, unlike *P. communis* rootstocks which allow phytoplasma reproduction within roots in winter and in the next spring, and the following spreading of the pathogen in aboveground phloem elements. Therefore in pear trees grafted to quince rootstocks the phytoplasma population, in absence of reinfection from overwintering psylla adults, may progressively decrease, disappearing within some years.

Fig. 2. Damages to fruits due to sooty molds.

Fig. 3. Damages caused by phytoplasmas.

3. The key to improve *C. pyri* control is the detailed knowledge of its life cycle

In Europe *C. pyri* shows 4-5 generations per year and overwinters as adult in reproductive diapause. The adult winter forms appear at the beginning of September (Fig. 4) (Civolani & Pasqualini, 2003) and overwinter individually or in small groups sheltered in cracks of the tree bark, at branch crossing or at the base of buds. As soon as the weather conditions improve, the winter forms leave their shelter and reach the new apical leaves. Here they puncture the buds with their stylets, sucking and excreting fecal droplets which accumulate near the proctiger. In winter the adult male produces active sperms stored in the spermatheca and is therefore ready to mate and inseminate the female. However, the eggs can be fertilized only when oocytes reach maturity (Bonnemaison *et al.*, 1956). In January all females are mature, ready to mate and lay eggs. The main factor affecting the physiology of ovopository apparatus is temperature, which must be higher than 10° C for two consecutive days (thermal quiescence) (Nguyen, 1975).

Fig. 4. *C. pyri* adult winter form.

Fig. 5. *C. pyri* adult summer form.

From eggs laid in winter months, simultaneously to bud opening, the first instar nymphs emerge, infesting the new vegetation. Later, in April-May, the adult summer forms appear (Fig. 5) whose females lay a large number of eggs (Fig. 6) hatching in the second half of May. The first instar (Fig. 7) and late instar nymphs (Fig. 8) live on the shoots, excreting a large amount of honeydew responsible for heavy damages to the plant. The next generations overlap with all developmental stages until autumn. Aestivation is also observed during the summer months.

Fig. 6. *C. pyri* eggs.

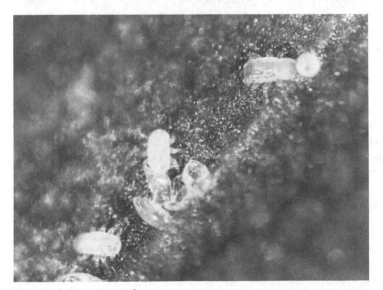

Fig. 7. Newly hatched *C. pyri* nymphs

Fig. 8. *C. pyri* late instar nymphs.

4. *C. pyri* Integrated pest management

In Europe this pest was considered of secondary importance until 1960, but its population rapidly increased simultaneously to heavy use of pesticides in pear orchards and intensive agriculture. The need to control pear infestations required the repeated use of different pesticides and several control strategies, but in most cases the efficiency of control progressively decreased, first because of the great adaptability and survival of the insect to the active ingredients, and second because of the negative effects on beneficial insects caused by excessive use of non-selective toxic ingredients. In this way the psylla population freely increased, leading to conditions in which the pest was hard or impossible to control.

Presently the defence against *C. pyri* is mainly based on integrated pest management (IPM), supported by natural control aimed to equilibrate the complex biological relationships of the field community. First of all, *C. pyri* control must follow IPM guidelines to optimize the activity of natural control agents and to reduce the chances of developing pesticide resistance by insect populations.

Among the basic strategies there are the "good agricultural practice" (GAP) techniques, such as reducing excessive plant growth and avoiding overstimulation by fertilizers or incorrect pruning. Also the general strategies to control other pest species may play a relevant role in psylla development and population increase. For example, the technique of mating disruption and the use of granulosis virus (CpGV) employed to control *Cydia pomonella* L. (Lepidoptera Tortricidae), favours the activity of entomophagous fauna by reducing the impact of synthetic products on the orchard, and may also help to reduce the treatments against *C. pyri*. However, in the last decade the chemical control has been the key defence strategy against the pear psylla in intensive pear orchards.

The main strategies of chemical control against *C. pyri* performed in the last 20 years in integrated and conventional farms are listed below. Each strategy has advantages and disadvantages, therefore its efficiency depends both on the active ingredients employed and the weather conditions at the time of treatment.

4.1 Autumn treatment

This treatment, performed at leaf fall, requires pesticides active against adult winter forms such as those belonging to the pyrethroid family (the same principles are also used in late winter against the same generation). These pesticides are completely non-selective, therefore dangerous for the beneficial insects: for this reason the treatment must be performed not too early (in October) when most individuals of *Anthocoris nemoralis* F. (Hemiptera Anthocoridae) are still active on the pear trees, but only at complete leaf fall (late November or early December), when *A. nemoralis* populations have already found shelter in bark crevices while *C. pyri* adult winter forms are still active on plants (Fig. 4) (Civolani, 2000; Civolani & Pasqualini, 2003).

Synthetic pyrethroids exhibit a very high abatement-contacticide action on psylla wintering adults. Table 1 shows the results of tests with the active ingredient ciflutrin: the abatement activity on adults is such that no egg deposition occurs in the next spring. Notwithstanding the good results that may be obtained by pyrethroid treatment in autumn (and also in late winter, as shown below) these strategies, common in France on *C. pyri* and in North America on *C. pyricola*, in the Italian pear growing regions are discouraged: here the pest population, after an initial sharp decline, soon recovers and increases again in May, reaching the economic threshold for spring-summer treatments. This event could be explained by the faster recovery of the pest in spring because the natural control by its predator *A. nemoralis* is limited. Moreover, considering only the pesticide efficiency, the results are strictly dependent on seasonal conditions. Indeed, if an early frost occurs at the beginning of autumn, most psylla adults take shelter early and survive to the late autumn treatments. Given the need to preserve auxiliary insect populations, an untimely autumn treatment with broad spectrum pyrethroids could be very dangerous for them and therefore is strongly discouraged in relevant pear growing areas such as Emilia-Romagna (Northern Italy).

Experimental tests have been performed to evaluate other active principles such as mineral oils and neonicotinoids. According to tests performed in 2001-2002 in the province of Ferrara (Emilia-Romagna), good results were obtained with mineral oil alone or associated to imidacloprid (Table 1) (Civolani, 2000), although for auxiliary insect preservation the use of neonicotinoids is not suggested against *C. pyri* in European pear growing areas, unlike North America where is found in field control guidelines against *C. pyricola*.

	Treatment date	Eggs per 100 bubs
untreated control	-	170
mineral oil	29 November 1999	30
mineral oil + imidacloprid	29 November 1999	12
ciflutrin	29 November 1999	0
mineral oil	3 March 2000	53
mineral oil + imidacloprid	3 March 2000	65
ciflutrin	3 March 2000	69

Table 1. Results of a field test with a pyrethroid and mineral oil with or without a neonicotinoid (Civolani, 2000).

4.2 Late winter treatment

This treatment is presently based on broad spectrum pyrethroids whose purpose is to break down the population of females emerging from winter shelters and about to lay eggs.

Formerly the highly toxic 4,6-dinitro-*o*-cresol (DNOC) was used, until finally banned in 1999. With a biological activity extended up to 15 days, DNOC was characterized by a strong contact activity especially towards eggs newly laid or about to be laid (Barbieri et al., 1986; Pollini et al., 1992). On pear psylla, the pyrethroids are exclusively active on adults through contact associated to an anti-feeding effect. The pyrethroid treatment is generally performed at pear bud swelling stage or at the latest when they break. The eggs are laid by *C. pyri* at the end of the thermal quiescent period, corresponding to a maximum temperature above 10° C for two consecutive days. The late winter strategy considerably lowers the amount of initial psylla population on pear trees, therefore the first generation is extremely reduced.

Concerning the pesticide activity, the best results of this strategy are obtained after a mild winter, because in these conditions almost all adults leave their shelters at the time of treatment. On the contrary, frost waves at the end of winter interrupt and delay the emergence of adults, reducing the efficiency of the treatment.

The control strategy based on late winter treatments with pyrethroids was recently disputed first because of the toxicity of the active ingredients involved, but mostly because the sharp reduction of the psylla first generation could starve the anthocorids, interfering with their settlement during early plant growth in spring. To avoid the side effects of pyrethroids, in the last ten years alternative solutions to synthetic pesticides have been repeatedly tested against the overwintering generations, and among them kaolin and some oily compounds.

Kaolin, a white, non-abrasive, fine-grained allumosilicate mineral that is purified and sized so that it can be easily dispersed in water, creates a mineral barrier on plants that prevents oviposition and insect feeding (Fig. 9 and Fig. 10) (Puterka et al., 2000). The efficiency of kaolin is shown by the results obtained in two experimental trials in 2001 and 2002 (Pasqualini et al., 2003). These authors found that a double treatment with kaolin affected egg laying of overwintering *C. pyri* by hindering their anchorage on the leaf surface and inhibiting host-plant acceptance (Table 2). It was also found that the body and wings of some adults became soiled, making insects less mobile and preventing them from reaching the laying site (host location) on plants. Indeed, the kaolin-treated plants were almost completely free of nymphs (Table 2). Due to the high mobility of *C. pyri* adults, the effects of kaolin treatment on the summer population were not assessed in this small plot trial. A larger plot trial was therefore performed in 2004 and again kaolin showed a very high control efficiency (Daniel et al., 2006). At the end of June 2004 some *C. pyri* individuals were observed in the kaolin treated plot, but the population density tended to be lower than that in the plot treated by a standard spirodiclofen strategy. In kaolin treated plants *C. pyri* was kept under the economic threshold until harvest (Daniel et al., 2006), therefore it could be an interesting alternative control strategy for this pest in organic and IPM orchards.

Some oily compounds could also be used in this period to interfere with egg deposition by *C. pyri* adults. It has been known from a long time that mineral oils and oily compounds could have negative effects on psylla egg deposition (Zwick & Westigard, 1978). A Turkish researcher (Erler, 2004) tested four types of oils, namely cotton seed oil, neem oil, fish-liver oil and summer oil, observing a delay in egg laying of about four weeks for fish-liver oil and summer oil, but of only one or two weeks for cotton seed oil and neem oil, probably depending on the stability of the oily material in open field. Other field trials aimed to interfere with egg laying by *C. pyri* adult winter forms were performed in 2001 and 2002 in Italy with pure mineral oil ("dormant oil"), obtaining a good reduction of the number of eggs laid (Pasqualini et al., 2003) (Table 2).

Fig. 9. Pear trees treated with kaolin in late winter.

Fig. 10. *C. pyri* adult with body soiled by kaolin.

	Treatment date	Eggs per 100 bubs	Nimphs per 100 flowers
First trial 2001			
Untreated control	-	136.75	6
kaolin	18 February and 10 March	1	0.25
mineral oil (dormant oil)	18 February.	30	2
Second trial 2002			
Untreated control	-	77.75	7.5
kaolin	11 and 19 February	0	0
mineral oil	11 February	12	3.5

Table 2. Results of two late winter field trials with kaolin and mineral oil (Pasqualini et al., 2003).

4.3 Spring-summer treatments

The treatments against summer generations can be performed towards eggs or nymphs. In the first case chitin inhibitors such as hexaflumuron (banned in 2004), triflumuron (banned in 2009), diflubenzuron and teflubenzuron have been employed. These active ingredients are used against second generation eggs, usually laid in the first decade of May. The treatment is usually performed against *C. pomonella* but shows a secondary effect on psylla. The chitin inhibitors provide the best results on *C. pyri* when they are applied on newly laid eggs (white eggs) or on eggs laid in a short time after the treatment. However, no activity is clearly exerted on eggs laid on the new shoots unreached by the treatment. Some authors observed that chitin-inhibitors show control effects similar to those of specific psyllicides described below: this could be explained by the absence of side effects on *C. pyri* natural predator, *A. nemoralis* (Souliotis & Moschos, 2008). However, the most relevant treatment employed against *C. pyri* in past and present times is the one against juvenile stages. This control strategy is base on specific synthetic active ingredients which are often acaricides, such as amitraz (commercially released in 1975 and banned in 2005), abamectin (commercially released in Italy in 1996) and spirodiclofen (commercially released in Italy in 2007), although in the past generic organophosphorates have been used such as monocrotofos and azinphos methyl.

Abamectin is presently the basic pesticide employed against *C. pyri*. Is should be briefly recalled that abamectin belongs to the chemical family of avermectins, compounds produced by the soil bacterium *Streptomyces avermitilis* (Lasota & Dybas, 1991). The activity of abamectin is mainly directed against young nymphs and secondarily against adults. The best results are therefore obtained when yellow eggs are mostly present and when the hatching peak, that could interfere with the pesticide activity because of honeydew, has not yet achieved (Pasqualini & Civolani, 2006). The product is not systemic but translaminary: the addition of mineral oil improves its penetration and after 24 hours no traces of the compound are found on leaf surface. One treatment timely performed against the second generation often represents the final solution, considering the high activity of the principle in comparison to amitraz (Table 3).

Spirodiclofen, commercially available since 2007, is the first member of a new chemical family, that of the tetronic acids (Nauen et al., 2000; Nauen, 2005), characterized by a new and original mechanism of action which interferes with biosynthesis of lipids in the target arthropods. The original mechanism of action of spirodiclofen plays a key practical role by allowing successful rotation strategies with abamectin to limit the risks of occurrence of resistance in *C. pyri* control. The best activity of spirodiclofen occurs when it is targeted on yellow eggs some days before the hatching of first instar nymphs: the active ingredient shows instead a decreasing activity with the advancing of the psylla developmental stages (Table 4). At the stage of yellow eggs the activity of spirodiclofen is improved by addition of mineral oil (Table 4) and the use of spirodiclofen followed by treatment with mineral oil (1000 ml/hl) may represent a good alternative to the treatment with abamectin in presence of high and prolonged infestations. In conditions of low-medium pressure by *C. pyri*, one treatment with spirodiclofen may be enough: the few individuals escaping the treatment may be easily captured by anthocorids, given the good selectivity of the product towards these valuable auxiliary insects (Pasqualini & Civolani, 2007). However, the efficiency of spirodiclofen is often lower than that of abamectin (Table 4) (Pasqualini & Civolani 2007; Boselli & Cristiani 2008; Marčić et al., 2009).

Other active ingredients have been employed on both *C. pyri* in Europe and *C. pyricola* in North America. For example, in *C. pyricola* the differences of abamectin efficiency observed in the field suggested to employ and recommend neonicotinoids in pear IPM programs, among which imidacloprid, introduced in 1995, thiametoxan in 2001, acetamiprid in 2002 and thiacloprid in 2004. Besides *C. pyricola*, the last ingredient is used mainly for codling moth *C. pomonella*.

Presently a new active ingredient, spirotetramat, a lipid biosynthesis inhibitor similar to the tetronic acid derivate spirodiclofen, is under investigation in Europe (Nauen et al., 2008) but already commercially available in North America. Due to its mode of action spirotetramat is especially effective against juvenile stages of sucking pests, psyllid included. In the case of female adults the compound significantly reduces fertility and consequently insect populations. Spirotetramat also exhibits unique translocation properties: after foliar uptake the insecticidal activity is translocated within the entire vascular system. This property allows the protection of new shoots or leaves appearing after foliar application (Nauen et al., 2008): given the high efficiency on *C. pyri* (unpublished data) this active principle could represent a future valuable alternative to abamectin in order to manage the risks of occurrence of resistance in *C. pyri* control.

It is also possible to control nymphs by simply washing the trees with high amounts of water to which insecticidal soaps (fatty acids salts) are added to remove the honeydew (Briolini et al., 1989). Recently some other products have been used, similar to liquid glue and able to trap by a physical mechanism small and scarcely active insects such as almost all juvenile instars of *C. pyri*. These products are synthetic sugar esters (sucrose octanoate) and represent a relatively new class of insecticidal compounds that are produced by the reaction of sugars with fatty acids. (Puterka et al., 2003).

After discussing the active ingredients that could be used against *C. pyri* and the different strategies that could be employed, once again it must be emphasized that an efficient control of *C. pyri* infestations could be obtained by an integrated pest management of the pear orchard which allows a balanced growth of plants and simultaneously favours the growth of populations of natural psylla antagonists.

During the spring growth period a key point is to protect the useful psylla predators, first of all the most important one, *A. nemoralis*. The populations of this anthocorid are low in early spring but increase in the second half of June, insuring the protection of pear trees until harvesting and providing the most relevant contribution against *C. pyri*.

Another aspect that should not be overlooked in *C. pyri* control is the relevant effect of weather conditions on pest populations. A late winter season with mild temperatures favours an early and fast emergence of adults from their winter shelters and a regular egg laying, while the late winter frosts interrupt the adult emergence and egg laying, producing a gradual development of first and second pest generation which interferes with the precise timing of treatments. Cold and rainy periods during blossoming and petal fall interfere with nymph spreading on plants: in this case the nymphs often crowded in the flower calyx, sometimes causing with their feeding activity russet blotches or young fruit drop. On the contrary, high summer temperatures tend to block psylla development because of high egg mortality and slowing of juvenile growth for a long period.

	Treatment date	Nymphs per shoot
First trial (2000, on the variety "Conference")		
untreated control	-	16.68
abamectin	6 May	0.15
amitraz	10 May	1.08
Second trial (2000, on the variety "William")		
untreated control	-	46.88
abamectin	6 May	2.23
amitraz	10 May	8.18
Third trial (2004, on the variety "Conference")		
untreated control	-	8.29
abamectin	6 May	0.98
amitraz	10 May	1.39

Table 3. Results of three field trials with abamectin and amitraz on two pear varieties (Pasqualini & Civolani, 2006).

5. Evolution of resistance of *C. pyri* to pesticides

As for all phytophagous pests, also for *C. pyri* the repeated use of chemical active ingredients causes the development of resistance. However, in Europe there are less resistance cases documented for *C. pyri* in comparison to those known since 1960 for *C. pyricola* in North America (Harries & Burts, 1965). Among the *C. pyri* resistance events in Europe the best known involve organophosphorates, pyrethroids and chitin inhibitors: in all documented cases a sharp decrease in pesticide activity was observed even after a few years of use. The active ingredient monocrotofos represents the best known case (Berrada *et al.*, 1995). The selection induced by this pesticide around the end of 1980 on some *C. pyri* populations near Toulouse (France) caused an increased resistance up to 140 fold in comparison to the susceptible strain in 30 generations, as shown by laboratory tests. Further tests showed that the mechanisms involved in the onset of resistance to this active

	Treatment date	Nymphs per shoot
First trial (2005, on the variety "Conference")		
untreated control	-	83.25
spirodiclofen	19 May (yellow eggs)	36.25
spirodiclofen + mineral oil	19 May (yellow eggs)	15.50
abamectin + mineral oil	19 May (yellow eggs)	0.75
Second trial (2002, on the variety "Abbé Fétel")		
untreated control	-	25
spirodiclofen	30 April (white eggs)	6
spirodiclofen	14 May (yellow eggs)	11
spirodiclofen	17 May (first hatching)	19
spirodiclofen	22 May (20-30 % hatching)	18
amitraz	14 May (yellow eggs)	2
Third trial (2007, on the variety "Beurré Bosc")		
untreated control	-	66.5
spirodiclofen	27 April	1.5
spirodiclofen and abamectin	27 April and 9 May	0
spirodiclofen and mineral oil	27 April and 9 May	0.6
abamectin	27 April	0.5
Abamectin and spirodiclofen	27 April and 9 May	0

Table 4. Results of three field trials with spirodiclofen on three pear tree varieties (Pasqualini & Civolani, 2007; Boselli & Cristiani, 2008).

ingredient were the enhanced activity of cytochrome P450 monooxygenase (MFO) and also the changes in acetyl cholinesterase, since the susceptibility to the pesticide could not be fully recovered by pretreating *in vivo* the adults with piperonylbutoxide (PBO), a specific inhibitor of MFO (Berrada *et al.*, 1994).

In 1994 in the Avignon region (France) the survey for *C. pyri* resistance was extended to 16 active ingredients belonging to five pesticide families, by topical laboratory tests on adults (Fig. 11). The tests showed that the resistance rates (RR) were extremely low for the family of carbamates (one- to 2.4-fold), relatively low for the family of pyrethroids (4.7- to 6.2-fold). For the family of organophosphorates insecticides, the resistance rates among the active ingredients were very different: lower than one (0.2-fold) for parathion-methyl, low for mevinphos and malathion (3.5- and 2.5-fold, respectively), and higher for chlorpyriphos-ethyl (10.2-fold), monocrotophos (26-fold), azinphos-methyl (62.2-fold) and phosmet (179.7-fold). In the same Avignon area, tests on resistance selection were performed in laboratory with the organophosphorate azinphos-methyl, which was frequently used in high amounts against *C. pomonella* and could indirectly cause selection also in *C. pyri*. The RR observed ranged from 10 to 40-fold in comparison to wild populations: these values were considerably lower than those reached by monocrotofos (Bues *et al.*, 2000). Further tests showed that the selection by azinphos-methyl produced a strong cross resistance with phosmet and monocrotofos (155 fold). By the same tests no cross resistance was shown for azinphos-methyl with amitraz, pyrethroids, carbamates and other organophosphorate pesticides such as chlorpyriphos and mevinphos (Bues *et al.*, 2000). The same authors started genetic studies by crossing the resistant and susceptible strains and showed that the resistance was autosomically inherited and semi-dominant in expression (Bues *et al.*, 2000). They also advanced by backcross the hypothesis that the resistance factor was monogenic. However, the result that the resistance to azinphos-methyl is semi-dominant is different from what previously obtained in Oregon by Van de Baan (1988) on pyrethroids: this author crossed two populations of *C. pyricola*, one susceptible and the other 240-fold resistant to the active ingredient fenvalerate (a pyrethroid), showing that the resistance was in this case semi-recessive (Van de Baan, 1988). Probably there are different mechanisms involved in the resistance to different pesticide families.

As mentioned before, the pyrethroids are another pesticide family largely employed in France against *C. pyri* and in North America against *C. pyricola* during leaf fall and late winter, before egg laying by overwintering adults. In laboratory topical tests on overwintering adults, some of these active ingredients showed very variable RR: for example, in tests performed in Avignon in 1994 the observed RR were 4 or 6-fold higher in comparison to the susceptible laboratory population (Buès *et al.*, 1999), but further studies in 1996 in the same southern Rhone valley, on a field population collected in Pont Saint Esprit, showed that the RR was 42.9-fold higher (Buès *et al.*, 1999). Later, the same authors confirmed similarly high resistance values to the active ingredient deltametrin in some Southern France populations in which the RR was 30-fold, always with the adult winter forms less susceptible in comparison to the same stage of the adult summer forms (Buès et al., 2003).

Concerning the mechanism of action producing resistance to pyrethroids, laboratory tests showed that the addition of PBO caused an recovery of pyrethroid efficiency almost complete: this shows that other mechanisms of induction of resistance could be secondarily involved.

Therefore the detoxifying enzyme MFO is involved in the mechanism of action of both organophosphorate and pyrethroid active ingredients.

Around 1995, in a region of West Switzerland a lower susceptibility was observed to teflubenzuron, active ingredient belonging to the family of chitin inhibitor (Schaub *et al.*, 1996). More recently this resistance to teflubenzuron was also observed in the Czech Republic, according to tests performed in 2004 and 2005 (Kocourek & Stará, 2006). The mechanisms of resistance to teflubenzuron have not yet been completely investigated.

As previously mentioned, abamectin is the most efficient and most used active ingredient against *C. pyri*: it was used against this pest for the first time in 1996, with only one spring treatment and only in some orchards, also because amitraz was an alternative until 2005, the year in which the ingredient was banned. Probably after the ban on this active ingredient the number of treatments per orchard and the area of employment were increased: this trend was also observed in the fruit growing area of Lleida, Girona and Huesca in Spain (Miarnau et al., 2010), and Ferrara and Modena in Italy. The current high selection pressure with this active ingredient, repeatedly applied over both geographical areas could induce selection for resistance.

Following the sudden outbreak of *C. pyri* populations in some fruit growing area in Emilia-Romagna in 2005, abamectin tests for resistance were performed for the first time only on overwintering adults (Fig. 11), but no resistance was detected, although LC_{50} and LC_{90} values appeared related to the time of adult field collection (Civolani et al., 2007).

Besides adults (Table 5) (Fig. 11), in 2007 and 2008 the abamectin tests in Emilia-Romagna were extended also to eggs and nymphs (Fig. 12 and Fig. 13), the stage target in *C. pyri* field treatments (Civolani et al., 2010). On adults, the abamectin topical tests performed in autumn 2007 and 2008 did not show significant differences among all populations tested, but the LC_{50} values were apparently related to the adult collection dates, as previously reported (Civolani et al., 2010).

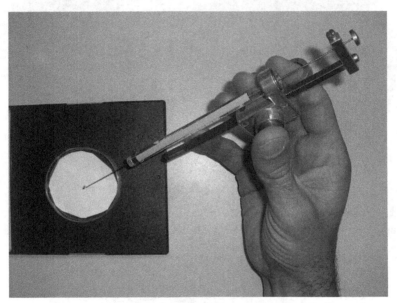

Fig. 11. Topical application of the insecticide solution on *C. pyri* adult winter form with hand-held manual micro-applicator.

The egg spray test (LC_{50} and LC_{90}) did not show relevant differences, although LC_{50} and LC_{90} values were always lower in the organic farm than in all others (Table 6). The results of leaf dip tests performed on young and old larvae were generally similar to the egg spray ones (Civolani et al., 2010).

Overall, the tests data indicate that no apparent resistance to abamectin has been developed up to now in *C. pyri* populations of Emilia-Romagna.

In 2009 and 2010 new cases of loss of efficiency of abamectin reappeared in some orchards in the province of Modena (Emilia-Romagna). In 2010 other tests were again performed on a *C. pyri* population which underwent 7 abamectin field treatment in 2009. However, in this case also the results of LC_{50} and LC_{90} did not show significant differences and the RR was just above 2 (unpublished data).

From 2004 to 2006 tests were carried out also in Spain to monitor the evolution of susceptibility to abamectin in orchards where the number of treatments changed from less than two until 2005 to an average of three after the ban on amitraz, with some orchards undergoing 6-7 treatments. The tests were performed both on adults and nymphs and no evidence of a high RR was found (Miarnau et al., 2010). However, there are some Spanish populations of *C. pyri* that show low susceptibility in adults as well as in nymphs: these populations come from the fields with the highest number of abamectin treatments per year. As in Emilia-Romagna, these cases also indicate a high risk of selection for resistance to abamectin, especially if the number of treatments per year is high and there are no alternative products to use in a resistance management program.

Fig. 12. Cages for potted pear plants to keep adults of different population separated.

Fig. 13. Potted young pear plants on which psylla eggs were laid and then treated with abamectin.

Population (locality)	Field strategy	2007		2008	
		LC$_{50}$	RR	LC$_{50}$	RR
Buondì-Vezzani (Diamantina)	organic	3.13		5.27	
Minotti (S. Martino)	integrated	4.98	1.59	3.24	0.61
Scagnolaro (Francolino)	integrated	2.80	0.89	4.16	0.78
Bonora (Boara)	integrated	-	-	5.95	1.12
Celati (Francolino)	integrated	-	-	4.23	0.80
Marchetti (Diamantina)	conventional	-	-	5.31	1

Table 5. Response of C. pyri overwintering adult populations to topical applications of abamectin (2007 and 2008) in the province of Ferrara (Civolani et al., 2010). The values of LC$_{50}$ are expressed in mg l^{-1} of abamectin.

Population (locality)	Field strategy	2007		2008	
		LC$_{50}$	RR	LC$_{50}$	RR
Buondì-Vezzani (Diamantina)	organic	0.20		0.15	
Minotti (S. Martino)	integrated	0.43	2.15	0.29	1.93
Scagnolaro (Francolino)	integrated	-	-	0.15	1
Bonora (Boara)	integrated	-	-	0.27	1.80
Celati (Francolino)	integrated	-	-	0.34	2.26
Marchetti (Diamantina)	conventional	-	-	0.44	2.93

Table 6. Response of C. pyri to egg spray applications (2007 and 2008) in the province of Ferrara (Civolani et al., 2010). The values of LC$_{50}$ are expressed in mg l^{-1} of abamectin.

Spirodiclofen is a recently introduced active ingredient in C. pyri control. Different LC$_{50}$ values were observed for spirodiclofen in tests again performed in Emilia-Romagna, regardless of its limited use as an alternative to abamectin. This active ingredient was tested on three populations of the province of Ferrara in 2006 and 2007 by laboratory assays on adults and eggs.
The topical tests on overwintering adults showed high susceptibility differences for this active ingredient between the adult population collected in the organic farm and those collected in the integrated (Minotti) and conventional (Marchetti) farms (Table 7).

Population (locality)	Field strategy	2006		2007	
		LC$_{50}$	RR	LC$_{50}$	RR
Buondì-Vezzani (Diamantina)	organic	19.09	-	89.01	-
Minotti (S. Martino)	integrated	2582.60	135.28	1931.70	21.,69
Marchetti (Diamantina)	conventional	25.93	1.358	943.72	10.60

Table 7. Results obtained in the topical test with spirodiclofen on C. pyri overwintering adults (2006 and 2007). The values of LC$_{50}$ are expressed in mg l^{-1} of spirodiclofen.

The results obtained by the spray test on yellow eggs before hatching, using the field doses of spirodiclofen, show low activity on the population collected in the Minotti integrated farm (both on first and second psylla generation). The activity levels in the Minotti farm, expressed as percentage of nymph mortality, are very different from those detected in the other two farms. However, the values detected in the conventional Marchetti farm are unexpectedly similar to those obtained in the organic farm (Table 8): these results were surprising because spirodiclofen was never employed before in none of these farms (unpublished data).

Population (locality)	Field strategy	Nymph mortality % on eggs laid by overwintering adults	Nymph mortality % on eggs laid by adult summer forms
Buondì-Vezzani (Diamantina)	organic	100	94,91
Minotti (S. Martino)	integrated	30.62	77.95
Marchetti (Diamantina)	conventional	-	94.81

Table 8. Results obtained in the spray test with spirodiclofen on *C. pyri* eggs laid by overwintering and summer adults in 2007. The results are expressed as percentage of nymph mortality.

6. Natural and biological control of *C. pyri*

In open field and especially in the pear orchard the techniques of biological control are not common because the fruit growers have always aimed to favour and exploit the development of the wild auxiliary insects, thus performing strategies of natural control. *Antochoris nemoralis*, common in all Europe, is known as the main predatory species of pear psylla. This species overwinters as an adult (Fig. 14) and starts to lay eggs in spring, inserting them under the leaf epidermis. The development of the juvenile forms occurs in 5 stages. *A. nemoralis* preys on both eggs and nymphs of psylla and in Emilia-Romagna typically shows three generations. Although generally preferring psylla, this anthocorid may feed on other insects and on the pear trees its activity against aphids and the pear sawfly *Hoplocampa brevis* Klug (Hymenoptera Tenthredinae) is very interesting. Laboratory tests showed an average predation of about 300 psylla nymphs during the entire life of an adult, which lasts about 60 days. The presence of *A. nemoralis* in pear orchards mainly depends on the type of control strategy applied in the farm. As previously mentioned, by limiting the use of pesticides to the minimum required and preferring the selective ones (see next chapter), it is favoured the development of the wild *A. nemoralis* populations which become a relevant factor to control the pest. Indeed, a dynamic equilibrium develops between the predator and *C. pyri* populations, often leading to the solution of the problem without the need for specific chemical treatments, or with treatments only reduced to tree washing. The main problem of this strategy is nevertheless the slow initial development of the predator population, which must have the prey available (in this case *C. pyri*) to rapidly increase in number within the pear orchard. Therefore, in the initial part of the season (around May-June) it is necessary to tolerate some amounts of the pest in the orchard to obtain later a good number of predator anthocorids on the trees. In other words, this means that when the presence of *C. pyri* rapidly increases in the third decade of May, usually we can expect a rapid increase of the predator population about two-three weeks later. This requires for the fruit grower to tolerate a relatively high presence of the pest and "resist" to the temptation to perform specific treatments. Since 1990, other field conditions added to these problems of the natural control, such as the increased chemical treatments against the codling moth, *C. pomonella*. These treatments caused a general weakening of the wild *A. nemoralis* populations, disrupting a very fluctuating natural equilibrium. Another problem recently emerged is that even in equilibrated pear orchards, not undergoing any heavy chemical treatment, for some reason the presence of predators remains low or undergoes high fluctuations over the years or even according to the seasons.

The previously mentioned limits of natural control led to artificially introduce anthocorids in the pear orchard, buying them from biofactories and thus performing a true biological control technique. The aim is to obtain a more numerous presence of the predator in the critical periods, anticipating the reproduction mechanism of the population which could occur naturally but with some delay. The introduction of predators is performed at the end of winter, between the end of March and the beginning of April. About 1000 adult individuals of *A. nemoralis* are placed per hectare of the pear orchard, in three consecutive weekly introductions. This biological control technique was common around 2000 in integrated pear orchards where the active ingredient amitraz was employed for the spring-summer control of *C. pyri*, then it was slowly neglected because of the limited results and the relatively high costs.

Fig. 14. Adult of *A. nemoralis*.

7. Selectivity of pesticides against *A. nemoralis*

In the last 20 years new families of pesticides have been developed with generally lower toxicity towards beneficial species in comparison to the previous ones. These pesticides are more suitable for the new techniques of integrated pest control (IPM) which preserve the contribution of beneficial insects to the natural control against pests, especially against *C. pyri*. Therefore it was necessary to perform tests on the new active ingredients, to verify their different toxicity degree and obtain the data required to improve their use.

Several studies have been recently performed to evaluate the selectivity of the new pesticides towards auxiliaries found in the pear orchards, first of all *A. nemoralis* (Fig. 14), the most relevant in the natural control against *C. pyri*. The active ingredients recently investigated include those directly employed against *C. pyri* (psyllicides) and also those largely employed on other key pests (non-psyllicides).

Concerning the psyllicides, in different toxicity tests performed in Emilia-Romagna since 1997 abamectin showed a medium degree of toxicity against the *A. nemoralis* population, mostly against first and second instar nymphs in comparison to adults. In some cases the mortality of nymphs reached 50% and, according to field data, the total population appears limited in comparison to untreated controls for about two weeks after the treatment. This result must nevertheless take into account the lower presence of prey as food for *A. nemoralis* (Pasqualini & Civolani 2007). Other products specific for *C. pyri*, such as amitraz (now banned), mineral oil, or insecticide soaps did not show relevant effects on the predator population (Civolani & Pasqualini, 1999; Pasqualini et al., 1999). Concerning spirodiclofen, the tests on toxicity towards *A. nemoralis* were performed again in Emilia-Romagna in years 2004-2006. The results show that the anthocorid populations undergoing treatment with spirodiclofen have similar development to the untreated ones, unlike abamectin for which the population first sharply decreases, then recovering according to the amount of prey available (Pasqualini & Civolani, 2007).

Concerning the non-psyllicides, those belonging to the chitin inhibitors, usually characterized by a long period of action, generally show a low toxicity on *A. nemoralis*, even if some of them, such as flufenoxuron, induce a heavy reduction of the anthocorid populations (Girolami et al., 2001, Pasqualini & Civolani 2002).

Among other non-psyllicides employed in pear orchards against the most relevant pests, there is spinosad, a pesticide of natural origin, whose activity derives from a toxin produced by *Saccharopolyspora spinosa* (Bacteria Actinomycetales), the indoxacarb belonging to the family of the oxadiazines, then methoxyfenozide and tebufenozide, synthetic molecules belonging to the family of moulting accelerator compounds (MAC), and the organophosphorates azinphos-methyl (banned in 2007), chlorpyrifos, chlorpyrifos-methyl and phosmet, all employed against relevant lepidopteran pests such as *C. pomonella*, *Cydia molesta* Busck and *Pandemis cerasana* Hübner. All these active ingredients have been more or less tested for toxicity against *A. nemoralis* and the results did not show relevant toxic effects on this predator (Civolani & Pasqualini, 1999).

As previously mentioned, the neonicotinoids are not employed in Italy and Europe as specific psyllicides, on the contrary of what happens in United States where they are employed as an alternative to abamectin against *C. pyricola*. However, a large amount of the above active ingredients are used against other pests, such as aphids and the pear sawfly *H. brevis*: the most relevant are imidacloprid, acetamiprid and tiametoxan, while against the codling moth *C. pomonella* the most frequently used is thiacloprid. All those active ingredients have significant toxic effects on *A. nemoralis*, thus they must be employed only for very few treatments along the year.

8. Conclusions

Concerning the strategies and methods to control *Cacopsylla pyri* (Hemiptera Psyllidae), and also their effects on the beneficial insects and the development of resistance, during the past decade Italian and European populations of *C. pyri*, in significant decline, have been apparently less capable to induce damage: this could be due to the success of defence programs based on integrated pest management (IPM). For pear, the IPM involves pest population control and auxiliary insect protection, associated to availability of new chemical or microbiological agents specifically targeted to pear key pests (mainly *Cydia pomonella*).

As far as known about *C. pyri* infestation, the biological control alone is not successful in preventing damage, especially when caused by second generation nymphs that feed on shoots and leaves in late spring and summer; thus chemical pest control strategies are also employed. For example, in Northern Italy the most common defense strategy against *C. pyri* in pear orchards involves chemical treatments on second-generation eggs or nymphs. Traditional treatments with chitin inhibitors, mainly aimed against *C. pomonella*, have also some secondary effects on *C. pyri*.

Specific treatments against *C. pyri* nymphs involve amitraz (now banned), abamectin and spirodiclofen. During spring and summer the treatment with the last two active ingredients, in addition to non-chemical treatments such as tree washing, is usually successful in limiting the honeydew damages caused by *C. pyri*.

During autumn and winter, the *C. pyri* management strategies involve treatments with synthetic pyretroids after leaf fall, to limit adult overwintering population, or at the end of winter, when females are ready to lay eggs. However, these treatments during autumn and winter, common in France and North America, are rarely employed in Italy because of their high toxicity against populations of auxiliary insects (such as Anthocoridae) which could still be present in the pear orchards. In late winter it is also possible to perform a non-chemical treatment against *C. pyri* by distribution of an aqueous suspension of kaolin on trees, in order to obtain a physical barrier to egg laying.

Each one of the above strategies shows favorable and unfavorable aspects in terms of efficacy, side effects on beneficial insects, timing of application and environmental conditions.

As for all phytophagous pests, also for *C. pyri* the repeated use of chemical active ingredients causes the development of resistance. Indeed, several insecticides employed in the past to control the pear psylla showed a sharp decline in activity because of resistance development. After the sudden outbreak of *C. pyri* populations in some fruit growing area of Ferrara and Modena (Italy) and Lleida, Girona and Huesca (Spain), abamectin tests were performed on winter form adults and nymphs. The results did not show relevant resistance effects, although LC_{50} and LC_{90} values were always higher in populations where abamectin treatment was repeated several times in the year. Overall, the results indicate that no apparent resistance to abamectin has been yet developed in *C. pyri* populations of the most important European areas of pear growth: nevertheless, the pear orchards in which *C. pyri* outbreaks recently occurred are presently under close investigation and careful survey.

9. Acknowledgments

The author wish to thank Dr Milvia Chicca (Department of Biology and Evolution, University of Ferrara) for kindly revising the manuscript and the English style.

10. References

Barbieri, R.; Becchi, R. & Pozza, M. (1986). Difesa fitosanitaria del pero, un quadriennio di prove e osservazioni anche con l'impiego di un piretrinoide in alternativa al D.N.O.C.. *Informatore Fitopatologico*, 36(12): 36-42

Berrada, S.; Fourner, D.; Cauny, A. & Nguyen T. X. (1994). Identifications of resistence meccanism in selected laboratory strain of *Cacopsylla pyri*, L. (Hom.: Psyllidae). *Pesticide Biochemistry and Physiology*, 48: 41-47

Berrada, S.; Nguyen, T. X.; Merzoug, D. & Fourner, D. (1995). Selection for monocrotophos resistance in Pear Psylla, *Cacopsylla pyri*, L. (Hom.: Psyllidae). *Journal of Applied Entomology*, 119: 507-510

Bonnemaison, L. & Missonnier, J., (1956). Le Psylle du poirier (*Psylla pyri* L.). Morfologie et biologie. Méthodes de lotte. *Annales des Epiphyties*, 7(2): 263-331

Boselli, M. & Cristiani, C. (2008). Diversificare le strategie di lotta contro la psilla del pero. *Terra e Vita*, 18: 64-67

Briolini, G.; Faccioli, G. & Pasqualini, E. (1989). A seven-year research on alternative methods to control Pear Psylla. *SROP/WPRS Bulletin*, 13(2): 89-92

Bues, R.; Boudinhon, L.; Toubon, J. F. & Faivre D'arcier, F. (1999). Geografic and seasonal variability of resistance to insecticides in *Cacopsylla pyri*, L. (Hom., Psyllidae). *Journal of Applied Entomology*, 123: 289-297

Bues, R.; Toubon, J. F. & Boudinhon, L. (2000). Genetic analysis of resistance to azinphosmethyl in the pear psylla *Cacopsylla pyri*. *Entomologia Experimentalis et Applicata*, 96: 159-166

Bues, R. ; Toubon, J. F. & Boudinhon, L. (2003). Resistance of pear psylla *Cacopsylla pyri* L.; Hom., Psyllidae to deltametrin and synergism with piperonil butoxide. *Journal of Applied Entomology*, 127: 305-312

Carraro, L.; Loi N.; Ermacora, N.; Gregoris, A.; Osler, R. & Hadidi, A. (1998). Transmission of pear decline by using naturally infected *Cacopsylla pyri* L.. *Acta Horticulturae*, 472: 665-668

Civolani, S. & Pasqualini, E. (1999). Tossicità nel breve periodo di diverse formulazioni di insetticidi su alcuni gruppi di entomofagi. *L'informatore Agrario*, 20: 87-90

Civolani, S. (2000). Elementi di studio sulla dinamica delle popolazioni svernanti di *Cacopsylla pyri* L. e del suo predatore *Anthocoris nemoralis* in relazione alla diffusione della moria del pero *Pyrus communis* L. in Emilia-Romagna. *Informatore Fitopatologico*, 50(11): 27-34

Civolani, S., Pasqualini, E., (2003). *Cacopsylla pyri* L. (Hom., Psyllidae) and its predators relationship in Italy's Emilia-Romagna region. *Journal of Applied Entomology.*, 127: 214-220

Civolani, S.; Peretto, R.; Caroli, L.; Pasqualini, E.; Chicca, M. & Leis, M., (2007). Preliminary resistance screening on abamectin in pear psylla (Hemiptera: Psyllidae) in Northern Italy. *Journal of Economic Entomology*, 100(5): 1637-1641

Civolani, S.; Cassanelli, S.; Rivi, M.; Manicardi, G. C.; Peretto, R.; Chicca, M.; Pasqualini, E. & Leis, M. (2010). Survey of susceptibility to abamectin of pear psylla *Cacopsylla pyri* L. (Hemiptera: Psyllidae) in northern Italy. *Journal of Economic Entomology*, 103(3): 816-822

Daniel, C. & Wyss, E., (2006). Pre-flowering kaolin treatments against the European pear sucker, *Cacopsylla pyri* (L.). *Deutsche Gesellschaft für allgemeine und angewandte Entomologie*, 15: 263-268

Davies, D. L.; Clark, M. F.; Adams, A. N. & Hadidi, A. (1998). The epidemiology of pear decline in the UK. *Acta Horticulturae*, 472: 669-672

Erler, E. (2004). Oviposition deterrency and deterrent stability of some oily substances against the Pear Psylla *Cacopsylla pyri*. *Phytoparasitica*, 32(5): 479-485

Girolami, V.; Berti, M. & Coiutti, C. (2001). Tossicità di nuovi insetticidi IGR su antocoridi predatori della psilla *L'Informatore Agrario*, 30: 63-66

Giunchedi, L. & Refatti, E. (1997). Il problema della moria del pero. *Frutticoltura*, 10: 59-61

Kocourek, F. & Stará, J. (2006). Managment and control of insecticide resistant pear psylla (*Cacopsylla pyri*). *Journal of Fruit and Ornamental Plant Research*, 14 (3): 167-174

Harries, F. H. & Burts, E. C. (1965). Insecticide resistance in the pear Psylla. *Journal of Economic Entomology*, 58 (1): 172-173

Lasota, J. A. & Dybas, R. A. (1991). Avermectins, a novel class of compounds: Implications for use in arthropod pest control. *Annual Review of Entomology*, 36: 91-117

Marčić, D.; Perić, P.; Prijović, M. & Ogurlić, I. (2009). Field and greenhouse evaluation of rapeseed spray oil against spider mites, green peach aphid and pear psylla in Serbia. *Bulletin of Insectology*, 62(2): 159-167

Miarnau, X.; Artigues, M. & Sarasua, M. J. (2010). Susceptibility to abamectin of pear psylla, *Cacopsylla pyri* (L.) (Hemiptera: Psyllidae) in pear orchards of North-East Spain. *IOBC/wprs Bulletin* 54: 593

Nauen, R.; Stumpf, N. & Elbert, A. (2000). Efficacy of BAJ 2740, a new acaricidal tetronic acid derivative, against tetranychid spider mite species resistant to conventional acaricides. In: *Proceedings of the Brighton Crop Protection Conference on Pests and Diseases*, 453-458

Nauen, R. (2005). Spirodiclofen : Mode of action and resistance risk assesment in Tetranychid pest mites. *Journal of Pesticide Science*, 30 : 272-274

Nauen, R.; Reckmann, U.; Thomzik, J. & Thielert, W. (2008). Biological profile of spirotetramat (Movento®)- a new two-way systemic (ambimobile) insecticide against sucking pest species. *Bayer CropScience Journal*, 61(2): 245-278

Nguyen, T. X. (1975). Évolution de la diapause ovarienne de *Psylla pyri* (Homoptera Psyllidae) dans les conditions naturelles de la région Toulousaine. *Bulletin de la Societe Zoologique de France*, 100(2): 241-245

Pasqualini, E.; Civolani, S.; Vergnani, S.; Cavazza, C. & Ardizzoni, M. (1999). Selettività di alcuni insetticidi su *Anthocoris nemoralis* F. (Heteroptera Anthocoridae). *L'Informatore Agrario*, 46: 71-74

Pasqualini, E. & Civolani, S. (2002). Mimic-Confirm (a. i. Tebufenozide): a tool for a soft and ecologically sound pest control in pear orchards. *IOBC/wprs Bulletin*, Vol. 25 (11): 97-106

Pasqualini, E.; Civolani, S. & Corelli Grappadelli, L. (2003). Particle film technology: approach for a biorational control of *Cacopsylla pyri* (Rhynchota: Psyllidae) in Northern Italy. *Bulletin of Insectology*, 55: 39-42

Pasqualini, E. & Civolani, S. (2006). Difesa della psilla del pero con abamectina. *L'Informatore Agrario*, 12: 50-54.

Pasqualini, E. & Civolani, S. (2007). Spirodiclofen, nuovo insetticida efficace contro la psilla del pero. *L'Informatore Agrario*, 11: 89 93

Pollini, A.; Bariselli, M. & Boselli, M. (1992). La difesa del pero dalla Psilla. *Informatore fitopatologico*, 5: 19-24

Puterka, G.; Glenn, D. M.; Sekutowski, D. G.; Unruh, T. R. & Jones S.K. (2000). Progress toward liquid formulations of particle films for insect and disease control in pear. *Environmental Entomology*, 29: 329-339

Puterka, G.; Farone, W,; Palmer, T. & Barrington, A. (2003). Structure-function relationships affecting the insecticidal and miticidal activity of sugar esters. *Journal of Economic Entomology*, 96: 636-644

Scaub, L.; Bloesch, B.; Bencheikh, M. & Pigeaud, A. (1996). Spatial distribution of teflubenzuron resistance by pear psylla in Western Switzerland. *OILB/SROP Bulletin*, 19 (4): 311-314

Schaper, U. & Seemmuller, E. (1982). Condition of the phloem and the persistence of mycoplasma-like organism associated with apple proliferation and pear decline. *Phytopathology*, 72: 736-742

Souliotis, C. & Moschos, T. (2008). Effectiveness of some pesticides against *Cacopsylla pyri* and impact on its predator *Anthocoris nemoralis* in pear-orchards. *Bulletin of Insectology*, 61: 25-30

Van De Baan H. E., (1988). Factors influencing pesticide resistance in *Psylla pyricola* Foerster and susceptibility in its Mirids predator, *Deraeocoris brevis* Knight. *Thesis, Oregon State University*

Zwick, R. W. & Westigard, P. H., (1978). Prebloom petroleum oil applications for delaying pear psylla (Homptera: Psyllidae) oviposition. *Canandian Entomologist*, 110: 225-236

5

Management of *Tuta absoluta* (Lepidoptera, Gelechiidae) with Insecticides on Tomatoes

Mohamed Braham and Lobna Hajji

*Centre régional de recherche en Horticulture et Agriculture Biologique;
Laboratoire d'Entomologie – Ecologie; Chott-Mariem,
Tunisia*

1. Introduction

Tomato, *Lycopersicon esculentum* Mill is a vegetable crop of large importance throughout the world. Its annual production accounts for 107 million metric tons with fresh market tomato representing 72 % of the total (FAO, 2002). It is the first horticultural crop in Tunisia with a production area of 25,000 hectares and a total harvest of 1.1 million metric tons (DGPA, 2009) of which nearly 70 % are processed (Tomatonews, 2011). Tomatoes are grown both under plastic covered greenhouses and in open field.

The tomato leafminer, *Tuta absoluta* Meyrick, (Lepidoptera : Gelechiidae) is a serious pest of both outdoor and greenhouse tomatoes. The insect deposits eggs usually on the underside of leaves, stems and to a lesser extent on fruits (photo 1). After hatching, young larvae penetrate into tomato fruits (photo 2), leaves (photo 3) on which they feed and develop creating mines and galleries. On leaves, larvae feed only on mesophyll leaving the epidermis intact (OEPP, 2005). Tomato plants can be attacked at any developmental stage, from seedlings to mature stage.

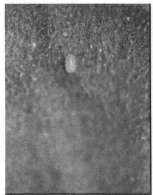  

Photo 1. *T. absoluta* egg   Photo 2. Larvae on fruit   Photo 3. Larva of *T. absoluta*

Originated from South America, *T. absoluta* was reported since the early 1980s from Argentina, Brazil and Bolivia (Estay, 2000); the insect rapidly invaded many European and

Mediterranean countries. It was first recorded from eastern Spain in late 2006 (Urbaneja, 2007), then Morocco, Algeria, France, Greece, Malta, Egypt and other countries (for a complete list see www.tutaabsoluta.com; Roditakis *et al.*, 2010, Mohammed, 2010).

Chemical control using synthetic insecticides is the primary method to manage the pest, but it has serious drawbacks, including reduced profits from high insecticide costs, destruction of natural enemy populations (Campbell *et al.*, 1991), build-up of insecticide residues on tomato fruits (Walgenbach *et al.*, 1991) and in the environment and fundamentally the rapid development of insecticide resistance. For example, resistance development has been reported against abamectin, cartap, methamidophos and permethrin in Brazil (Siqueira *et al.*, 2000a, Siqueira *et al.*, 2000b) and against deltamethrin and abamectin in Argentina (Lietti *et al.*, 2005). Thus, in order to avoid selection of resistant biotypes, a careful management with frequent changes of active ingredients is desirable. Furthermore, modern integrated pest management recommends effective pesticides that have low mammalian toxicity, low persistence in the environment and high degree of selectivity. Since insecticide control currently remains an indispensable tool, the goal is to minimize the amount and impact of pesticides through the diversification of active ingredients used.

In this paper, we present the data from insecticides trials conducted in 2009 and 2010 under laboratory and field conditions, in which the efficacy of several hitherto untested insecticides and natural products was compared with that the widely used insecticides to manage *T. absoluta* in Tunisia such as spinosad, indoxacarb and pyrethroids compounds.

2. Material and methods

2.1 Laboratory trials
2.1.1 Laboratory assays in 2009
Tomato seeds (cv Topsun) were sown on 30 January 2009. Seeds were deposited in 110 cm3 cells in a rectangular polyester tray of 60 cm x 40 cm x 5 cm filled with peat (Potgrond H, Germany). On March 3, 2009, seedlings were transplanted into 1 liter plastic flowerpot (bottom diameter =8 cm, top diameter = 12 cm and height = 12 cm) filled with peat without fertilization and watered as required. The tomato plants were maintained in the laboratory until use. Three days before the assay, plants (having four to six true leaves) were deposited in a tomato crop situated in the vicinity of the laboratory to permit *T. absoluta* egg-laying then transferred to the laboratory. Leaves were examined under binocular microscope and *T. absoluta* larvae were counted. Insecticides were sprayed using a hand sprayer (1 liter of capacity). After drying, the treated plants were kept in an unsealed empty greenhouse bordering the laboratory. There were four replications (plants) for each product and an untreated plant was used as a check. The efficacies of the products were tested twice: 48 hours following sprays and 12 days later. The Insecticides and natural plant extracts used are given in table 1.

2.1.2 Laboratory assays in 2010
A colony of *T. absoluta* was established from larvae and pupae collected from tomato infested field in the Chott-Mariem region. The insect was reared and maintained in a small greenhouse (10*6 m). From time to time, tomato leaves harboring *T. absoluta* pre-imaginal stages collected in the field were introduced in the rearing greenhouse.

Tomato seeds (cultivar Riogrande) were sown on February 13, 2010 in a rectangular polyester tray as mentioned before. Plants having four to six true leaves were transferred to the rearing greenhouse and remained there for 2 to 3 days to allow egg-laying. Thereafter

Active ingredients	Trade name	Companies	Dose cc/ hl water
deltamethrin	Decis EC25	Bayer Crop Science	100 cc/hl
bifenthrin	Talstar	FMC Corporation	100 cc/hl
acetamiprid	Mospilan 200 SL	Basf	50 cc/hl
methomyl	Lannate 25	Dupont de Nemours	150 cc/hl
metamidophos	Tamaron 40	Bayer Crop Science	150 cc/hl
abamectin	Vertimec	Syngenta	30 cc/hl
Spinosad	Tracer	Dow-Agroscience	60 cc/hl
Rotenone	Rotargan	Atlantica Agricola (Spain)	300 cc/hl
Neem extract	Oleargan	Atlantica Agricola	100 cc/hl

Table 1. Insecticides and natural plant extracts used in the laboratory trial in 2009.

returned to the laboratory and put in wooden cages for insecticide trials. Leaves were examined under binocular microscope and T. *absoluta* larvae were counted just before insecticide spray (April 3, 2010) and regularly after 2 to 3 days post-treatment. Dead larvae following trial were recorded. The second insecticide spray was done on April 19, 2010 (two weeks later). The Insecticides and natural plant extracts used are given in table 2.

Active ingredients	Trade name	Companies	Dose cc/ hl water
diafenthiuron	Pegasus	Syngenta	125 cc/hl
triflumuron	Alystin SC 480	Bayer Crop science	50cc/hl
emamectin benzoate	Proclaim®	Syngenta	2500 grams/hl
Plant extracts	Tutafort	AltincoAgro (Spain)	125 cc/hl

Table 2. Insecticides and natural plant extracts used in the laboratory in 2010.

2.2 Field trials
2.2.1 Trials using natural products
Field experiments using botanical extracts, Spinosad and Kaolin Clay were conducted from March 2010 to May 2010 in a half commercial tomato greenhouse (34 meters long x 8 meters width) in Saheline region, Tunisia (35°42′ North, 10°40′East). Tomato seeds (cv Sahel) were sown on 27 October 2009 in an expanded polyester tray under plastic protected nursery bed. Four double rows of tomato were transplanted on 23 November 2009. The plot (greenhouse) was prepared according to usual cropping practices in the region. Ploughing, tillage and second tillage to incorporate manure, bed formation, irrigation device establishment and drip irrigation.

Active ingredients	Trade name	Companies	Dose cc/ hl water
Spinosad	Tracer 240	Dow- Agroscience	60 cc/hl
Neem extract	Oleargan	Atlantica Agricola (Spain)	100 cc/hl
Kaolin Clay	Surround WP™	Engelhard Corporation (NJ.U.S.A)	5 kg/hl
Orange extract	Prev-am™	ORO Agri International Ltd	300 cc/hl
Botanical extracts	Deffort	AltincoAgro (Spain)	350 cc/hl
Botanical extracts	Armorex	Soil Technologies Corp (U.S.A)	60 cc/hl
Botanical extracts (*Quassia amara* and Neem)	Conflic	Atlantica Agricola (Spain)	250 cc/hl

Table 3. Natural products experimented in 2010.

Plots measured 4 m2 each (10 plants) arranged in a randomized block design with four replications. The active ingredients, the trade name and doses of the natural products are given in table 3. The products were diluted with tap water and applied at field rates based on the recommended label dilutions without surfactants.

2.2.2 Trials using insecticides

Trials using insecticides were undertaken during the same period in the second half greenhouse. Plot measured 8 square meters each (20 plants) arranged in a randomized block design with four replications. Three chemical compounds were used (table 4).

Active ingredients	Trade name	Companies	Dose cc/ hl water
indoxacarb	Avaunt 150EC	Dupont	50 cc/hl
triflumuron	Alsystin SC 480	Bayer Crop science	50 cc/hl
diafenthiuron	Pegasus 500SC	Syngenta	125 cc/hl

Table 4. Insecticides compounds experimented under tomato greenhouse in 2010.

Insect monitoring

To assess the *T. absoluta* infestation prior to the trial, thirty leaf samples, taken from about 30 different plants were weekly collected (from January to March 2010) at random from the entire greenhouse. The sample was placed in a plastic bag and taken to the laboratory. Leaves were examined under binocular microscope (Leica MZ12.5); eggs, larvae pupae, of *T. absoluta* live or dead as well as mines were recorded. However, only larvae (live or dead) were presented in this study.

2.3 Statistical analysis

Data on the effectiveness of various insecticides were analyzed using the Minitab Software for Windows (Minitab 13.0). The mean number of live larvae per plant or per leaf was tested for Normality assumption by Kolmogorov-Smirnov test then the data were square root transformed. General linear model procedures were used to perform the analysis of variance. Wherever significant difference occurred, Tukey's multiple comparison test was applied for mean separation.

In the laboratory trial of 2010, due to the low number of live larvae in the control, a one way-ANOVA percentage of mortality was used instead of corrected mortality.

The percentages of efficacies of insecticides were evaluated either:

i.   Abbott formula : the percentage of efficacy = (Ca-Ta)/Ca*100   where Ca is the average live larvae in the control and Ta is the mean survival score in the treatment.

ii. The percentage of larval mortality = mean number of dead larvae/(mean number of dead larvae + mean number of live larvae)*100.

3. Results

3.1 Laboratory trials
3.1.1 Assays in 2009

One day before the assay, the mean number of total live larvae (L1 to L4 instars) per plant varied from 0.75 to 3. There is no significant difference between treatments (GLM-ANOVA. F= 0.99, df= 9,30; P = 0.47, table 5). Three days after the first application, the mean number of live larvae per plant decreases in all treatments except in the control (Table 5). All insecticides significantly reduced *T. absoluta* larvae when compared with non treated control (F= 4.24, df = 9,30; P= 0.001, Table 5). However, the level of suppression by acetamiprid and bifenthrin did not differ significantly from the control (Table 5).

Mean number of larvae/plant on indicated days before treatment (DBF) and days after treatment (DAT)					
Insecticides !	1DBT1!!	3DAT1	5DAT1	8DAT1	12DAT1
spinosad(1)	1.75a	0.5a(86.66)*	0.50a(85.71)*	0.5a (87.5)*	0.25a(93.75)*
neem extract(2)	1.5a	0.75a(80)	0.75a(78.50)	0.5a(87.5)	0.5a(87.5)
rotenone(3)	0.75a	0.25a(93.33)	0.25a(92.90)	0.5a(87.5)	0.75a(81.25)
deltamethrin(4)	0.5a	0a(100)	1a(71.42)	0.75a(81.25)	1.5ab(62.5)
acetamiprid(5)	2a	1.25ab(66.66)	1.25ab(64.28)	1.25ab(68.75)	0.50a(87.5)
methomyl(6)	3a	0.5a(87)	0.5a(86)	0.50a(88)	0.75a(81)
metamidophos(7)	2a	0.75a(80)	0.75a(79)	0.75a(81)	1.00a(75)
abamectin(8)	2.25a	0.75a(80)	0.75a(79)	0.5a(88)	0.25a(94)
bifenthrin(9)	2a	1.25ab(67)	2ab(43)	1.25ab(69)	1.00a(75)
Control	2.5a	3.75b	3.5b	4b	4b
Statistical analysis	F= 0.99	F= 4.24	F= 3.69	F= 4.20	F= 4.66
ANOVA-	df = 9,30	df = 9,30	df = 9,30	df = 9,30	df = 9,30
GLM	P = 0.47	P= 0.001	P= 0.003	P= 0.001	P=0.003

! denote commercial compounds: (1): Tracer, (2): Oleargan, (3): Rotargan, (4): Decis, (5): Mospilan, (6): Lannate (7): Tamaran, (8): Vertimec, (9): Talstar

!! Means followed by the same letter within a column are not significantly different at P= 0.05 (ANOVA-GLM procedure) followed by Tukey multiple comparison

* Data in brackets denote percent Abbott mortality (Abbott, 1925)

Table 5. Mean number of *T. absoluta* total live larvae/plant on indicated days before treatment (DBF) and days after treatment (DAT) (the first treatment was done on April 1, 2009).

Five days following the first application, all the products performed well except acetamiprid and bifenthrin which show no significant difference compared with the control (Table 5). Eight days after the first application, the mean number of total live larvae per plant varied from 0.5 to 4. All the tested products reduced significantly the density of live larvae per plant compared with the control (F= 4.20; df = 9,30; P= 0.001). Still, acetamiprid and bifenthrin showed mild efficacy (table 5). At 12 days following treatments, all the products performed well (F= 4.66, df = 9,30 ; P= 0.003), yet the plants treated with deltamethrin show increasing mean live larvae per plant (table 5).

Regarding the corrected mortality according to Abbott formula, Spinosad and rotenone gave satisfactory results post-treatment (88.4 % and 88.7% respectively) followed by Lannate (85%), Vertimec (85%), neem extract (83. 22), and Tamaran (79%). However, Decis (78.8%), Mospilan (71.8) and Talstar (63%) showed mild efficacy. Though, Decis performed well till 8 days following the first application (84.2%).

Mean number of total live larvae/plant on indicated days after the second treatment (DAT)				
Insecticides	0DBT2!!	2DAT2	4DAT2	8DAT2
spinosad	0a	0a(100)*	0a(100)*	0.75a(83.33)*
neem extract	0.5a	0.5a(92.85)	0.75a(78.60)	1.75a(61.11)
rotenone	0.25a	0.5a(85.71)	0.5a(85.71)	1.25a(72.22)
deltamethrin	0.75a	0.75a(78.60)	1a(71.42)	1.25a(72.22)
acetamiprid	0.5a	0.5a(85.71)	0.75a(78.60)	1.5a(66.66)
methomyl	1.25a	0.75a(78.60)	0.75a(78.60)	2a(83.33)
metamidophos	0.75a	0.75a(78.60)	0.5a(85.71)	1a(77.71)
abamectin	0.5a	0.5a(85.71)	0.5a(85.71)	1.5a(66.66)
bifenthrin	1a	1a(64.28)	1.5ab(57.14)	2a(55.55)
Control	3.5b	3.5b	3.5b	4.5b
Statistical	F= 6.07	F= 7.24	F=5.84	F= 4.39
analysis	df = 9,30	df = 9,30	df = 9,30	df = 9,30
	P = 0.00	P= 0.00	P= 0.00	P= 0.001

* Data in brackets denote percent Abbott mortality (Abbott, 1925)
!! : Means followed by the same letter within a column are not significantly different at P= 0.05 (ANOVA-GLM procedure) followed by Tukey multiple comparison

Table 6. Mean number of total *T. absoluta* live larvae/plant the day of the second treatment and thereafter (DAT2) (the treatment was undertaken on April 21)

Just before the second application, the mean number of live larvae in treated plants remained low compared with the control. It varied between zero (Tracer) and 3.5 (control) (table 6). Two days following the second insecticide application, all tested compounds show good efficacy compared with control (F=4.24; df = 9,30; P<0.001). Spinosad (Tracer) performed well (100 % efficacy according to Abbott corrected mortality formula). However, bifenthrin (Talstar) shows mild efficacy (table 6). The same conclusion can be formulated four days following treatments (table 6). At eight days after trial, the insecticide spinosad remains active and performed well (83.33 % efficacy) (table 6).

The overall efficacy according to Abbott formula (1925) shows the good performance of spinosad (Tracer), rotenone (Rotargan), methomyl (Lannate), abamectin (Vertimec) (Fig. 1.).

However, the percentage of larval mortality (number of dead larvae/sum of dead and live larvae) following the first and second insecticide application shows the best performance of spinosad (91 %), neem extract (71 %) and abamectin (71%).

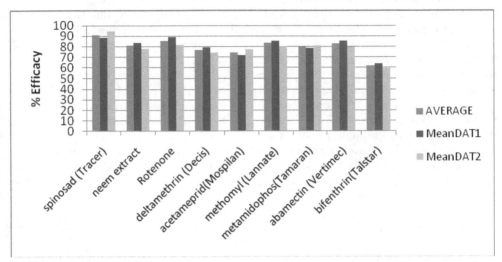

Fig. 1. Overall percentage of efficacy according to Abbott formula (1925). DAT1 = days after the first treatment, DAT2 = days after the second treatment (laboratory trial, 2009).

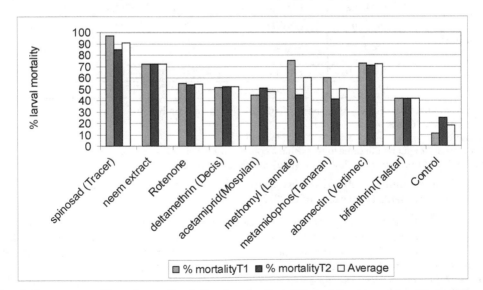

Fig. 2. Percentage of larval morality following the first (T1) and the second treatment (T2) (mean number of four dates after the fist treatment and 3 dates after the second treatment).

3.1.2 Assays in 2010

Just before the first spray (April 3, 2010), the mean number of live larvae (first to fourth instars) per leaf varied from 0.12 (Control) to 0.52 (Proclaim®). Although there is no significant difference between treatments (ANOVA-GLM F= 1.37, df = 4, 116; P=0.24), the control plants harboured less live larvae (table 7). There is no larval mortality.

Two days following the first spray (April 5), there is no significant difference between treatments regarding live larvae (GLM; F= 0.93, df = 4, 116; P= 0.46. Table 7). However, the percentage of larval mortality did vary (ANOVA, 1 factor, F = 4.17; df = 4, 120; P= 0.003) showing the best performance of Proclaim® (57.14 %; Table 7).

Nine days after the first insecticide application (April 12), the mean number of live larvae per leaf did not significantly vary between treatments (ANOVA-GLM procedure Table 7). However, the percentage of mortality significantly varies between treated and untreated plants (ANOVA 1 factor, F= 3.07; df = 4, 120; P= 0.021). The maximum percentage of mortality is given by Proclaim® (45.70%, table 7).

At 11 days after the first insecticide application (on April 14), the mean number of live larvae did not significantly vary among treated and untreated plants (ANOVA - GLM procedures Table 7). However, the percentage of mortality did vary according to treatments (F = 3.16, df = 4, 120; P= 0.017) showing the good efficacy of Proclaim® (52.93 % Table 7).

Mean number of live larvae/leaf on indicated days before treatment (DBF) and days after treatment (DAT)µ					
Insecticides!	0DBT!!	2DAT1	9DAT1	11DAT1	13DAT1
(1)	0.36(0)a	0.36(10)a	0.37(13.61)a	0.44(12.66)a	0.34(12.82)a
(2)	0.32(0)a	0.2(37.5)a	0.24(29.47)a	0.34(23.52)a	0.34(20.05)a
(3)	0.52(0)a	0.24(57.14)a	0.29(45.70)a	0.25(52.93)a	0.23(51.51)a
(4)	0.44(0)a	0.48(0)a	0.26(17.91)a	0.20(21.91)a	0.18(27.39)a
(5)	0.12(0)a	0.24(0)a	0.22(0)a	0.25(0)a	0.20(0)a
Statistical analysis	F= 1.37	F=0.90	F= 0.57	F=0.63	F=0.27
ANOVA	df =4,116	df =4,116	df =4,116	df =4,116	df =4,116
-GLM	P =0.24	P =0.46	P = 0.67	P=0.64	P= 0.89

! :(1):triflumuron(Alystin), (2) plant extract (Tutafort), (3) emamectin benzoate (Proclaim®) (4) diafenthiuron (Pegasus) and (5) Control.
µ: Data under brackets denote percentage of mortality
! ! : Means followed by the same letter within a column are not significantly different at P= 0.05 (ANOVA-GLM procedure) followed by Tukey multiple comparison

Table 7. Mean number of live *T. absoluta* larvae on indicated days before treatments and days after treatments (laboratory trial, 2010)

At 13 days after the first application, the mean number of live larvae did not significantly vary between treatments and control (Table 7). However, the percentage of mortality significantly varies between treated and control plants (F = 3.53 df = 4, 120; P= 0.009) showing the good efficacy of the compound Proclaim® (51.51 %, table 7).

At 16DAT1 and just before the second spray, the mean number of live larvae shows no significant difference between treated and control plants (table 7. continued). However, the percentage of mortality did significantly vary between treated and control plants (One way

ANOVA F= 4.95 df = 4, 120; P= 0.001). The compound Proclaim® shows the highest mortality percentage (54.83 % table 7.Cont.).

At three days after the second insecticide application, there is no significant difference regarding the mean number of live larvae per leaf (GLM-ANOVA). Nevertheless, plants treated with the product Proclaim® harbour zero live larvae per leaf suggesting the good efficacy of this insecticide. This is confirmed by the high percentage of mortality (100 %) as well as the significant difference between treated and control plants (One way ANOVA, F= 4.51 df = 4, 120; P= 0.002).

Mean number of live larvae/leaf on indicated days before treatment (DBF) and days after treatment (DAT)(μ)					
Insecticides	16DAT1!	3DAT2	5DAT2	8DAT2	10DAT2
(1)	0.33(12.55)a	0.19(51.71)a	0.16(59.91)ac	0.15(59.4)ac	0.15(59.14)ac
(2)	0.31(21.47)a	0.06(80.15)a	0.06(83.64)ac	0.06(83.35)ac	0.06(83.35)ac
(3)	0.20(54.83)a	0(100)a	0(100)bc	0(100)bc	0(100)bc
(4)	0.20(19.60)a	0.16(0)a	0.1(59.55)ac	0.09(59)ac	0.09(59)ac
(5)	0.20(0)a	0.16(0)a	0.16(0)a	0.16(0)a	0.06(0)a
Statistical	F= 0.27	F= 2.02	F= 1.85	F= 1.85	F= 1.56
analysis	df =4, 116	df =4, 116	df =4, 116	df =4, 116	df =4, 116
GLM-	P= 0.89	P= 0.096	P= 0.123	P= 0.096	P=0.189
ANOVA					

μ : Data under brackets denote percentage of mortality
! : Means followed by the same letter within a column are not significantly different at P= 0.05
(ANOVA-GLM procedure) followed by Tukey multiple comparison

Table 7. (continued). Mean number of live *T. absoluta* larvae on indicated days before treatments and days after treatments (laboratory trial, 2010)

Five days after the second spray, the mean number of live larvae did not vary among treated and untreated plants (table 7. Cont.). But the percentage of mortality significantly varies (ANOVA one factor F= 3.98 df = 4, 120; P= 0.03) showing again the good performance of Proclaim® (table 7.Cont.).

At eight days after the second spray, there is no significant difference between treated plants and control regarding the mean number of live larvae (table 7.Cont.). However, the percentage of mortality varies (ANOVA, one factor, F= 3.88 df = 4, 120; P= 0.005). The compounds Proclaim® and Tutafort are the best (100 % and 83.35 respectively, table 7.Cont.).

At 10 days after the second insecticide application, there is no significant difference between treated plants and control (GLM-ANOVA, Table 7.Cont.). Concerning the percentage of mortality, there is a significant difference between treated and control plants (ANOVA, one factor, F= 3.99 df = 4, 120; P= 0.006). Proclaim® followed by Tutafort performed well (100 % and 83.35 respectively, table 7.Cont.).

3.2 Field trials
3.2.1 Natural products experimented in 2010 under greenhouse
The first spray was undertaken on March 26, 2010, then on April 8 and on April 19, 2010.
At three days following the first application, the mean live larvae (small and old larvae) per leaf did not significantly vary between treated and control plots (GLM-ANOVA Procedure,

P= 0.09). Although, plots treated with spinosad show the minimum live larvae as demonstrated by 70% efficacy according to Abbott formula (Table 8). The details of larval instars (small larvae: first and second instars and old larvae: three and fourth instars) show a significant difference between insecticides tested. The compounds Tracer, Armorex and Deffort performed well (table 9).

Mean number of total larvae/leaf on indicated days before treatment (DBF) and days after treatment (DAT)					
Insecticides	1DBT!!	3 DAT1*	10DAT1(μ)	2DAT2	6DAT2
Armorex(1)	0.30a	0.20(20)a	0.1(69.23)a	0(100)a	0.325(0)a
Deffort(1)	0.30a	0.25(0)a	0.45(0)a	0.475(0)b	0.3(0)a
Oleargan (1)	0.20a	0.32(0)a	0.225(30.76)a	0.05(33.33)a	0.25(0)a
Konflic(1)	050a	0.57(0)a	0.2(38.46)a	0.125(0)a	0.075(0)a
Prev-am™ (2)	0.32a	0.37(0)a	0.45(0)a	0.1(0)a	0.2(0)a
Surround WP™ (3)	0.30a	0.32(0)a	0.25(23.07)a	0.075(0)a	0.15(0)a
Tracer(4)	0.1a	0.075(70)a	0.05(84.61)a	0(100)a	0.025(0)a
Control	0.20a	0.25a	0.325a	0.075a	0.025a
Statistical	F= 1.42	F= 1.94	F= 1.61	F= 1.61	F= 1.92
Analysis	df =3, 309	df =3, 309	df =3, 309	df =3, 309	df =3, 309
GLM- ANOVA	P= 0.120	P= 0.09	P= 0.131	P=0.008	P=0.066

(1): Botanical extracts
(2): Orange extract
(3) : Kaolin
* Corrected mortality according to Abbott formula
μ = second spray
!! : Means followed by the same letter within a column are not significantly different at P= 0.05
(ANOVA-GLM procedure) followed by Tukey multiple comparison

Table 8. Mean number of total live larvae following natural products applications under tomato greenhouse (Saheline, Tunisia, 2010).

At 10 days after the first natural products applications, the ANOVA-GLM procedure shows no significant difference between treatments regarding the mean number live larvae (Table 8). The Abbott's percentages of efficacy show the performance of spinosad (84.61 %) and the plant extract (Armorex; 69.23%).

At two days after the second spray, (April 10) there is a significant difference between treated plots (AVOVA-GLM procedure, P= 0.008, table 8). The plots treated with Deffort show the maximum density of mean live larvae per leaf (table 8). However, there is no significant difference between the other products and control. The details of larval stages confirm the low efficacy of Deffort compared with the other products and control (small larvae : P= 0.026; Old larvae P= 0.019; table 9).

Six days following the second application (April 14), the mean number of live larvae shows no significant difference between treated and untreated plots (Table 8).

At eleven days after the second spray, the mean number of live larvae per leaf is relatively similar among treatments and did not significantly vary (ANOVA-GLM procedure P= 0.211) varying from 0.1 to 0.9. Plots treated with Kaolin (Surround) harbour the minimum density.

Four days after the third spray (April 23, 2010), the treated plot differed significantly showing the good performance of the compounds neem extract, Tracer and Konflic (table 8). This is confirmed by the analysis of detailed larval instars (table 9).

At nine days after the third spray, the mean number of total larvae varied between 0.2 and 2.05. The ANOVA-GLM procedure showed a significant difference between treatments. The products Tracer, Armorex and Deffort were effective in reducing *T. absoluta* larval densities (table 8).

Mean number of total larvae/plant on indicated days before treatment (DBF) and days after treatment (DAT)				
Insecticides	11 DAT2!	4DAT3!!	9DAT3	18DAT3
Armorex(1)	0.525(0)a	0.1(85.18)a	0.3(85.36)b	0.9(12.2)a
Deffort(1)	0 .925(0)a	0.3(55.55)a	0.2(90.24)b	0.65(36.85)a
Oleargan (1)	0.325(13.33)a	0.075(88.88)ab	0.55(73.17)b	0.375(63.4)a
Konflic(1)	0.475(0)a	0(100)ab	0.825(59.75)a	0.325(68.3)a
Prev-am™ (2)	0.225(40) a	0.175(74.07)a	1.675(18.30)a	0.7(31.7)a
Surround(3)	(3)0.1(73.33)a	0.2(70.37)a	0.55(73.17)b	0.375(63.4)a
Tracer(4)	0.35(6.66)a	0.1(85.18)a	0.25(87.80)b	0.75(26.8)a
Control	0.375a	0.675a	2.05a	1.025a
Statistical	F=1.41	F=2.49	F=2.49	F=1.36
Analysis	df= 7,309	df= 7,309	df= 7,309	df= 7,309
GLM-ANOVA	P=0.201	P=0.017	P=0.000	P=0.220

! : third spray
!! : Means followed by the same letter within a column are not significantly different at P= 0.05
(ANOVA-GLM procedure) followed by Tukey multiple comparison

Table 8. (Continued) Mean number of total live larvae following natural products applications under tomato greenhouse (Saheline, Tunisia, 2010).

Mean number of live larvae/leaf on indicated days before treatment (DBF) and days after treatment (DAT)				
Insecticides	3DAT1!!		10DAT1	
	SL*	OL*	SL*	OL*
Armorex(1)	0.075(40)a	0.125(0)a	0.1(50)a	0(100)a
Deffort(1)	0.05(60)a	0.2(0)a	0.225(0)a	0,225(0)a
Oleargan (1)	0.2(0)a	0.125(0)a	0.175(12.5)a	0.05(60)a
Konflic(1)	0.425(0)b	0.15(0)a	0.1(50)a	0.1(20)a
Prev-am™ (2)	0.25(0)ab	0.125(0)a	0.275(0)a	0.175(0)a
Surround WP™ (3)	0.3(0)a	0.025(80)a	0(100)a	0.25(0)a
Tracer(4)	O(100)a	0.075(40)a	0.05(75)a	0(100)a
Control	0.125(0)ab	0.125(0)a	0.2(0)a	0.125(0)a
Statistical Analysis	F= 4.03	F= 0.77	F= 1.76	F= 1.53
	df= 3,309	df= 3,309	df= 3,309	df= 3,309
GLM-ANOVA	P= 0.00	P= 0.611	P= 0.096	P=0.157

*: SL : Small larvae (L1-L2), OL: Old larvae (L3-L4)
!! : Means followed by the same letter within a column are not significantly different at P= 0.05
(ANOVA-GLM procedure) followed by Tukey multiple comparison.

Table 9. Mean number of live small and old larvae following natural products applications under tomato greenhouse (Saheline, Tunisia, 2010).

Mean number of larvae/plant on indicated days before treatment (DBF) and days after treatment (DAT)				
Insecticides	2 DAT2!!		6DAT2	
	SL*	OL*	SL*µ	OL*
Armorex(1)	0(100)a	0(100)b	0.1b	0.225(0)a
Deffort(1)	0.175(0)b	0.3(0)a	0.025a	0.275(0)a
Oleargan (1)	0.025(0)a	0.025(50) b	0.175b	0.075(0)a
Konflic(1)	0.05(0)a	0.075(0)b	0a	0.075(0)a
Prev-am™ (2)	0(100)a	0.1(0)b	0.125b	0.075(0)a
Surround(3)	0.025(0)a	0.05(0)b	0.1b	0.05(0)a
Tracer(4)	0(100)a	0(100)b	0a	0.025(0)a
Control	0.025a	0.05b	0a	0.025a
Statistical analysis GLM-ANOVA	F= 2.31	F= 2.44	F= 2.18	F= 1.34
	df= 3,309	df= 3,309	df= 3,309	df= 3,309
	P= 0.026	P=0.019	P=0.036	P=0.069

µ: undetermined Abbott percentage of efficacy (zero Small larvae in the control plot)
SL: Small larvae (L1-L2), OL: Old larvae (L3-L4).
!!: Means followed by the same letter within a column are not significantly different at P= 0.05 (ANOVA-GLM procedure) followed by Tukey multiple comparison

Table 9. (Continued) Mean number of live small and old larvae following natural products applications under tomato greenhouse (Saheline, Tunisia, 2010).

Mean number of larvae/plant on indicated days before treatment (DBF) and days after treatment (DAT)				
Insecticides	11 DAT2!!		4DAT3	
	SL*	OL*	SL*	OL*
Armorex(1)	0.375(0)a	0.15(0)a	0(100)a	0.1(84.61)ab
Deffort(1)	0.7(0)a	0.225(0)a	0.15(0)b	0.15(76.9)ab
Oleargan (1)	0.25(0)a	0.075(50)a	0.025(0) a	0.05(92.30)b
Konflic(1)	0.425(0)a	0.05(66.66)a	0(100)a	0(100)b
Prev-am™ (2)	0.075(66.66)a	0.15(0)a	0(100)a	0.175(73.0)b
Surround WP™ (3)	0.07(66.66)a	0.02(83.33)a	0.125(0)b	0.075(88.4)b
Tracer(4)	0.275(0)a	0.075(50)a	0.075(0)ab	0.02 (96.15)b
Control	0.225a	0.15a	0.025a	0.65a
Statistical Analysis	F=1.31	F=1.10	F=2.75	F=2.82
	df=3,309	df=3,309	df=3,309	df=3,309
GLM-ANOVA	P=0.246	P=0.361	P=0.009	P=0.007

*: SL : Small larvae (L1-L2), OL: Old larvae (L3-L4)
!!: Means followed by the same letter within a column are not significantly different at P= 0.005 (ANOVA-GLM procedure) followed by Tukey multiple comparison

Table 9. (Continued) Mean number of live small and old larvae following natural products applications under tomato greenhouse (Saheline, Tunisia, 2010).

Mean number of larvae/plant on indicated days before treatment (DBF) and days after treatment (DAT)				
Insecticides	9 DAT3! !		18DAT3	
	SL*	OL*	SL*	OL*
Armorex(1)	0.125(72.22)a	0.175(89.06)b	0.75(0)a	0.15(75) a
Deffort(1)	0.025(94.44)b	0.175(89.06)b	0.4(5.88)a	0.25(58.33)a
Oleargan (1)	0.225(50)a	0.325(79.68)b	0.075(82.35)a	0.3(50)a
Konflic(1)	0.275(38.88)a	0.55(65.62)a	0.175(58.82)a	0.15(75) a
Prev-am™ (2)	0.375(16.78)a	1.3(18.75)a	0.45(0)a	0.25(58.33)a
Surround WP™ (3)	0.1(77.77)a	0.45(71.87)a	0.2a	0.175(70.83)a
Tracer(4)	0.125(72.22)a	0.125(92.18)b	0.45(0)a	0.3(50)a
Control	0.45a	1.6a	0.425a	0.6a
Statistical Analysis	F= 2.33	F=5.68	F=1.41	F= 1.97
	df = 3,309	df = 3,309	df = 3,309	df = 3,309
ANOVA-GLM	P= 0.00	P=0.000	P=0.201	P=0.06

* Data in brackets denote percent Abbott mortality (Abbott, 1925)
! !: Means followed by the same letter within a column are not significantly different at P= 0.05
(ANOVA-GLM procedure) followed by Tukey multiple comparison

Table 9. (Continued) Mean number of live small and old larvae following natural products applications under tomato greenhouse (Saheline, Tunisia, 2010).

Three days following the first insecticide application, the mean number of live larvae (small and large) did not vary significantly between treated and untreated plots (ANOVA-GLM Procedure F= 1.94, df = 3, 309 P= 0.063). However, the plants treated with spinosad (Tracer) harbor the minimal larval density (Table 9).

3.2.2 Insecticides compounds experimented under tomato greenhouse in 2010

Four days before the first insecticide application, the mean number of live larvae per leaf varied between 0.6 and 0.97 showing no significant difference between treatments and control (ANOVA. GLM, F= 0.82, df =3, 156; P=0.82).

Two days following the first treatment (March 24), the mean number of live larvae remains relatively low and did not significantly vary between treatment and control (F = 0.34; df = 3, 153; P= 0.79). The corrected mortality according to Abbott formula shows slight efficacy of tested products (Table 10).

At 12 days following the first application, the mean number of live larvae significantly differed between treatments (GLM, F=2.90, df = 3, 156; P= 0.037). The Tukey multiple comparisons showed the good performance of indoxacarb (Avaunt) (Table 10). There is no significant difference between plot treated with triflumuron (Alystin), diafenthiuron (Pegasus) and untreated plots.

Three days after the second treatment, there is a significant difference between treated plots and control (GLM, F= 16.45 df = 3, 153; P= 0.000). The three compounds performed well particularly Avaunt (92.30 % according to Abbott formula).

Nine days following the second spray, all insecticides performed well compared with the control (F= 46.7 df =3,153; P=0.000) with the best performance of indoxacarb (Avaunt) (96.87 % efficacy according to Abbott formula, Table 10).

Mean number of larvae/leaf on indicated days before treatment (DBF) and days after treatment (DAT)μ					
Insecticides	4DBT1!!	2 DAT1	12DAT1	3DAT2	9DAT2
indoxacarb	0.87a	0.7(15. 15)a!	0.2(71.42) a!	0.05(92 .30)a	0.075(96.87)a
triflumuron	0.97a	0.6(27.27)a	0.52 (25)ab	0.1(84.61)a	0.4(83.33)a
diafenthiuron	0.6a	0. 72 (12.12)a	0.4(42.85)ab	0.125(80.76)a	0.30(87.5)a
Control	0.87a	0.85a	0.7 b	0.65b	2.4b
Statistical	F=0.82	F= 0.43	F=2.90	F= 16.45	F= 46.7
analysis	df =3, 153	df = 3, 153	df = 3, 153	df =3, 153	df =3, 153
GLM-ANOVA	P=0.48	P=0.72	P=0.037	P=0.000	P=0.000

μ : the first treatment was undertaken on March 22, 2010.
! : data in brackets denote percentage of efficacy (Abbott Formula)
!! : Means followed by the same letter within a column are not significantly different at P= 0.05
(ANOVA-GLM procedure) followed by Tukey multiple comparison

Table 10. Mean number of *T. absoluta* larvae/leaf on indicated days before treatment (DBF) and days after treatment (DAT) (Saheline tomato greenhouse, 2010).

Mean number of larvae/leaf on indicated days before treatment (DBF) and days after treatment (DAT)			
Insecticides	18DAT2!!	3DAT3	12DAT3
indoxacarb (Avaunt)	0.05(95.83)a	0.075(95.45)a	0.35(78.12)a
triflumuron (Alystin)	0.05((95.83)a	0.5(69.69)a	0.7(56.25)a
diafenthiuron (Pegasus)	0.075(93.75)a	0.325(80.30)a	0.32(87.5)a
Control	1.2b	1.65b	1.6b
Statistical analysis	F= 40.88	F= 20.91	F=10.87
	df =3, 153	df =3, 153	df =3, 153
	P = 0.00	P= 0.00	P= 0.000

!! : Means followed by the same letter within a column are not significantly different at P= 0.05
(ANOVA-GLM procedure) followed by Tukey multiple comparison

Table 10 (continued). Mean number of *T. absoluta* larvae per leaf on indicated days before treatment (DBF) and days after treatment (DAT) (Saheline tomato greenhouse, 2010).

At 18 days following the second application, the mean number of live larvae significantly varies between treated and control plots (GLM   F= 40.88; df = 3, 153; P= 0.000). The efficacy of tested insecticide remains high compared with the control.

At 3 and 12 days following the third insecticide application all tested insecticides continue to be effective compared with the control (F= 20.91 df =3, 153; P= 0.00 ; F=10.87; df =3, 153; P= 0.00). Nevertheless, indoxacarb (Avaunt) tend to be a powerful suppressor of *T. absoluta* larvae (table 10).

4. Discussion

In Argentina, the primary *T. absoluta* management tactic was chemical sprays (Lietti *et al.*, 2005). Organophosphates were initially used for *T. absoluta* control then were gradually replaced by pyrethroids during the 1970s. During the early 1980s, cartap which alternates with pyrethroids and thiocyclam were sprayed showing the good effectiveness of the former. During the 1990s, insecticides with novel mode of actions were introduced such as abamectin, acylurea, insect growth regulators, tenbufenozide and chlorfenapyr (Lietti *et al.*, 2005).

Our laboratory results demonstrate the efficacy of spinosad (Tracer), rotenone (Rotargan), methomyl (Lannate) and abamectin (Vertimec). Methomyl was only tried due to its highly used frequency in tomato production against Noctuid larvae in Tunisia.

Spinosad, a mixture of spinosyns A and D, is derived from the naturally occurring actionomycete, *Saccharopolyspora spinosa* (Sparks *et al.*, 1998). Because of its unique mode of action, involving the postsynaptic nicotinic acetylcholine and Gamma-aminobutyric (GABA) receptors, spinosad has strong insecticidal activity against insects (Salgado, 1998) especially Lepidoptera (e.g. *Helicoverpa armigera* (Wang *et al.*, 2009), *Spodoptera frugiperda* (Méndez *et al.*, 2002), Diptera (King and Hennesey 1996; Collier and Vanstynwyk , 2003 ; Bond *et al.*, 2004), some Coleoptera (Elliott et al., 2007) as well as stored grains (Hertlein *et al*, 2011).

To date, spinosad is considered a good alternative control of Lepidopteran pests due to its high activity at low rates and its use in integrated pest management programs. The product possesses advantages in term of safety for farm workers and consumers due to its low mammalian toxicity and rapid breakdown in the environment (Sparks et al., 1998). The compound is considered as a standard product for the control of *T. absoluta* in Brazil (Maraus *et al.*, 2008) showing, however low efficacy compared with the insecticide novaluron.

Rotenone has been reported to be an excellent insecticide against a wide range of insect pests. Davidson (1930) found that rotenone was a toxic and effective contact insecticide against several species of whiteflies, aphids, caterpillars and mites. Also, Turner (1932) reported a high toxicity of rotenone to larvae of the Colorado potato beetle *Leptinotarsa decemlineata* (Say).

Azadirachtin, a tetranortriterpenoid isolated from the seeds of neem tree, *Azadirachta indica* (Meliaceae), and the fruit of chinaberry, *Melia azaderach* (Meliaceae) acts as an antifeedant and inhibits the growth and the development of several insects (Meisner *et al.*, 1981, Raffa, 1987; McMillian *et al.*, 1969). The antifeefant effects of azadirachtin are partly due to sensory detection and avoidance by insects (Simmonds and Blaney 1984).

Acetamiprid (Mospilan) is a neonicotinoid insecticide that is formulated for both soil and foliar application. It is a broad-spectrum insecticide effective against several groups of insects including Lepidopterans, Coleopterans, Hemipterans and Thysanopterans. The insecticide has an ingestion and stomach action and has a strong osmotic and systemic action (Takahashi *et al.*, 1998). The compounds interact with Acetylcholine receptors (AChRs) in a structure-activity relationship, resulting in excitation and paralysis followed by death (Ishaaya *et al.*, 2007).

Abamectin a mixture of avermectins is extracted by the fermentation of the soil bacterium *Streptomyces avermitilis* (Strong & Brown 1987). The insecticide acts on the GABA receptor activating the chloride channel (nerve and muscles) (Aliferis and Jabaji, 2011).

Throughout the assay, the product emamectin benzoate (Proclaim®) showed the best efficacy strongly suppressed *T. absoluta* larval populations. Indeed, several authors reported the performance of this product against several insects, for example, Seal (2005), reported the efficacy of emamectin benzoate at various rates in reducing the densities of the melon thrips, *Thrips palmi* adults and larvae. Stanley *et al.*, (2005) reported the high acute toxicity of emamectin benzoate to *Helicoverpa armigera* under laboratory conditions.

Cook et al., (2004) conducted field and laboratory trials on cotton and soybean for the control of the beet armyworm *Spodoptera exigua* (Hübner) and the fall armyworm *Spodoptera frugiperda* using indoxacarb, pyridalyl, spinosad methoxyfenozide and emamectin benzoate demonstrated the good efficacy of tested products compared with the control. Plots treated with indoxacarb, spinosad and emamectin benzoate had significantly fewer beet armyworm larvae.

Avermectins are a family of 16-membered macrocyclic lactone natural product homologues produced by the soil microorganisms, *Streptomyces avermitilis*. They act as agonists on GABA and glutamate gated chloride channels. The chloride ion flux produced by the direct opening of channels into neuronal cells results in loss cell function and disruption of nerve impulses. Consequently, arthropods are paralyzed irreversibly and stop feeding. Maximum mortality is achieved within four days (Jansson *et al.*, 1997).

Emamectin benzoate (Proclaim) is a novel semi-synthetic derivative of the natural product abamectin in the avermectin family. This insecticide has a high potency against a broad spectrum of lepidopterous pests with an efficacy of about 1,500-fold more potent against certain armyworm species (Jansson *et al.*, 1996)

Insect growth regulators like triflumuron, lufenuron are claimed to be safe and have little impact on beneficial arthropods compared with conventional insecticides and thus attracted considerable attention for their inclusion in IPM programs (Ishaaya *et al.*, 2007). In this study, triflumuron showed low efficacy against *T. absoluta* larvae. These results are in accordance with data reported by El-Sheikh and Aamir (2011) suggesting the greater efficiency of lufenuron in controlling *Spodoptera littoralis* Boisd compared with triflumuron or flufenoxuron. Similarly, low effectiveness of triflumuron (Alystin SC48) for the control of *Cactoblastis cactorum* (Lepidoptera: Pyralidae) was reported in Argentina by Labos et al., (2002). Yet the concentration used was lower (30 cc/ hl). Regarding the control of the Mediterranean fruitfly, *Ceratitis capitata*, triflumuron (Alystin 25) failed to give satisfactory results (a concentration of 150 ppm did not kill adults, Zapata *et al.*, (2006)).

Diafenthiuron (Pegasus) is a new type of thiourea derivative that affects respiration in insects. It disrupts oxidative phosphorylation by inhibition of the mitochondrial ATP synthase, an enzyme with essential role in cellular bioenergetics (Ishaaya, 2010). It is an insecticide and acaricide which kills larvae, nymphs and adults by contact and/or stomach action, showing also some ovicidal action (e-pesticide manual, 2005). In our laboratory trial, diafenthiuron (Pegasus) shows little efficacy in *T. absoluta* larval suppression (table 10).

Tutafort (plant extract) shows little efficacy after the first application but increases effectiveness after the second application engendering about 80 % of larval mortality (table 7.Cont.). Yet according to manufacturer, (Altinco, 2011), the product has a preventive action and should be applied against eggs and adults. The compound acts by contact penetrating the insect cuticle and dissolves the cell membranes causing the insect dehydrate and its death (Altinco, 2011).

Management of resistance to prevent or delay the development of resistance to an insecticide and cross resistance to additional insecticides is necessary for increasing the chance of chemical control of *T. absoluta*. Thus, the avoidance of resistance requires the development of pest management programs in which efforts are made to take advantages of natural enemies of pests, plant resistant cultivars, if available, appropriate cultural and physical methods.

Accordingly, diversification of control tactics should be implemented with the minimum use of chemicals. Insecticides should be applied only as needed basis and only used as the last form of control. When insecticides are applied, the way that they are used should be rationalized and optimized to exploit the full diversity of synthetic chemicals and natural products mostly used at rotational basis.

Development of resistance in *T. absoluta* is an important problem in regions where the insect is established. The expanding international trade of plant material not only spread the pest but also spreads the resistance genes associated with the pest (Denholm and Jespersen, 1998). It is possible that the Mediterranean populations of *T. absoluta* already carried gene resistance from South American counterpart populations and thus, may already express high level of resistance to one or multiple insecticide. Indeed, Cifuentes *et al.*, (2011), demonstrated high genetic homogeneity of *T. absoluta* populations came from Mediterranean basin and from South America countries using ribosomal and mithochondrial markers.

Our field results (tomato greenhouse) suggest the good performance of the tested compounds (indoxacarb, triflumuron and diafenthiuron). So far, the product indoxacarb tend to be a powerful suppressor of *T. absoluta* larvae.

Indoxacarb is reported by several authors as a powerful insecticide in managing many Lepidopteran pests. Wakil *et al.* (2009) in their study for the management of the pod borer, *Helicoverpa armigera* Hubner (Lepidoptera : Noctuidae) in Pakistan showed the integration of weeding, larvae hand picking and indoxacarb sprays was the most effective in reducing the larval population, pod infestation and maximum grain yield. Also, in Cameroon, Brévault et al., (2008) reported a good efficacy of indoxacarb as a larval insecticide of *H. armigera*.

In the United Kingdom, three insecticides were registered for the control of *T. absoluta* under protected tomato, pepper and aubergine: *Bacillus thuringiensis var. kurstaki*, indoxacarb and spinosad (FERA, 2009).

Indoxacarb belongs to a novel class of insecticides, the oxadiazines. It a broad spectrum non-systemic insecticide active especially against Lepidoptera. Indoxacarb affects insect primarily through ingestion but also by contact with treated plant surface. It kills by binding to a site of sodium channels and blocking the flow of sodium ions into nerve cells. The result is impaired nerve function, feeding cessation, paralysis and death (Wing *et al*, 2000).

5. Conclusions

T. absoluta has been a serious pest of tomatoes in Tunisia since the autumn 2008. Farmers have gradually come to understand that conventional insecticides such as organophosphates and carbamates are not effective against the insect. Even though more expensive compared with other insecticides, spinosad (Tracer) is now the widely used bio-insecticide to manage the insect.

It is not the intent in this study to advocate one insecticide over another but to enlarge the array of effective insecticide and bio-insecticides with different modes of action. These studies clearly demonstrated the efficacious of several chemicals such as spinosad, abamectin, emamectin benzoate, triflumuron and diafenthiuron. Although, plant extracts such as Armorex and Deffort show mild efficacy in controlling *T. absoluta* larvae, they can be used in conjunction with chemical products and integrated in a whole program of control.

The efficacies of sprayings using mixtures of natural products and synthetic chemicals for the control of the pest are planned in our laboratory studies. Indeed, insecticides that work in synergy when mixed together are an avenue to explore in *T. absoluta* control. It has been proposed that pesticides mixtures with different modes of action may delay the onset of resistance developing in pest populations (Bielza *et al.*, 2009). However, some problems need to be considered when two or more insecticides are mixed together especially phyto-toxicity.

The use of insecticides to control *T. absoluta* must not divert attention from the implementation of alternative pest management strategies including cultural, mass-trapping and biological control that can reduce reliance to chemical products.

Chemical pesticides continue to be an important component of insect pest management even with the development of other control methods (mass-trapping, plant resistance…). The use of insecticides based on different chemistries and with varying modes of action is an important component of an integrated pest management strategy. Hence, insecticides will continue to be an integral component of pest management programs due mainly to their effectiveness and simple use. However, the principal factor account for the possible reluctance to shift to the newer insecticides is the high cost.

6. Acknowledgements

The financial support for this work was provided by the Institution of Agricultural Research and Higher Education (IRESA. Ministry of Agriculture and Environment. Tunisia) through the research project "*Tuta absoluta*". We wish to thank the former President Pr Mougou A. and the current President Pr Amamou H. for their help.

We thank Bensalem A., Benmaâti S., Hajjeji F., Ammar A. and Rhouma J. for technical assistance in both laboratory and field.

7. References

Abbott W.S. 1925. A method for computing the effectiveness of an insecticide. J. Econ. Entomol. 18:265-267.

Aliferis K.A., & Jabaji, S., 2011. Metabolomics- A robust bioanalytical approach for the discovery of the modes of action of pesticides: A review. Pesticide Biochemistry and physiology. 100: 105-117.

Altinco, 2011. www.altinco.org. Accessed on July 7, 2011.

Bielza P., Fernández E., Graválos C., and Albellán J., 2009. Carbamates synergize the toxicity of acrinathrin in resistant western flower thrips (Thysanoptera: Thripidae). J. Econ. Entomol. 102: 393-397.

Bond J.G., Marina C.F., and Williams T., 2004. The naturally derived insecticide Spinosad is highly toxic to *Aedes* and *Anopheles* mosquito larvae. Medical and Veterinary Entomology 18: 50-56.

Brévault B., Oumarou Y., Achaleke J., Vaissayre M. & Nebouche S. 2008. Initial activity and persistance of insecticides for the control of bollworms (Lepidoptera : Noctuidae) in cotton crops. Crop Protection. 28: 401-406.

Campbell C.D., Walgenbach J.F., and Kennedy G.C., 1991. Effect of parasitoids on Lepidopterous pests in insecticide-treated and untreated tomatoes in western North Carolina. J. Econ. Entomol. 84: 1662-1667.

Cifuentes D., Chynoweth R., and Bielza P., 2011. Genetic study of Mediterranean and South American populations of tomato leafminer *Tuta absoluta* (Povolny, 1994) (Lepidoptera: Gelechiidae) using ribosomal and mitochondrial markers. Pest. Manag. Sci. Wileyonlinelibrary.com. DOI10.1002/ps.2166.

Cook D.R. Leonard B.R., and Gore J., 2004. Field and laboratory performance of novel insecticides against armyworms (Lepidoptera: Noctuidae). Florida Entomologist. 87(4): 433-439.

Davidson W.M. 1930. Rotenone as a contact insecticide. J. Econ. Entomol. 23 : 868-874.

Denholm I., and Jepersen J.B ., 1998. Insecticide resistance management in Europe: recent developments and prospects. Pesticide Science. 52: 193-195.

DGPA, Direction Générale de la Production Agricole 2009. Tunisian Ministry of Agriculture. Annual report.

Collier T.R., & Vanstynwyk R., 2003. Olive fruitfly in California: prospects for integrated control. California Agriculture, 57: 28-32.

EstayP. 2000. Polilla del tomate *Tuta absoluta* (Meyrick). http://alerce.inia.cl/docs/informativos/informativo09.pdf. Accessed, 12 May 2010.

Elliott R.H. Benjamin M.C., & Gillott G., 2007. Laboratory studies of the toxicity of spinosad and deltamethrin to *Phyllotreta cruciferae* (Coleoptera : Chrysomelidae). Can. Entomol. 139: 534-544.

El-sheikh E. A. and Aamir M. M., 2011. Comparative effectiveness and field persistence of insect growth regulators on a field strain of the cotton leafworm, *Spodoptera littoralis*, Boisd (Lepidoptera : Noctuidae). Crop Protection. 30 : 645-650.

E-pesticide manual. Version 3.3. 2005. Edited by British Crop Production Council (BCPC). ISBN. 1901396401.

FAO. 2002. Production year book 2000. Vol. 54. FAO. Italy.

FERA, 2009. The food and environment Research Agency. Plant Pest notice. South American tomato moth *Tuta absoluta*. N° 56: 1-4.

Hertlein M.B, Thompson G.D., Subramanyam B. & Athanassiou C.G. 2011. Spinosad : A new natural product for stored grain protection. Journal of stored products research. Doi : 10.1016/j.jspr.2011.01.004 : 1-15.

Ishaaya I., 2010. Biochemical sites of insecticide action and resistance. Isaac Ishaaya (Ed.) Springer-Verlag. Berlin, Heidelberg.

Ishaaya I., Barazani A., Kontsedalov S., & Horowitz A.R., 2007. Insecticides with novel modes of action : mechanism, selectivity and cross resistance. Entomological Research. 37 (Suppl.1) A2-A10.

Jansson R.K., Brown R., Cartwright B. Cox D., Dunbar D.B., Dybas R.A., Eckel C., Lasota J.A., Mookerjee P.K., Norton J.A., Peterson R.F., Starner V.R., & White S., 1996. Emamectin benzoate : a novel avermectin derivate for control of lepidopterous pests. Proceedings : the management of Diamonback Moths and other crucifers pests. Chemical control 171-177. Entomology. Cornell University.

Jansson R.K, Peterson R.F., Mookerjee P.K., Halliday W.R., Argentine J.A., & Dybas R.A., 1997. Development of a novel soluble granule formulation of emamectin benzoate for control of Lepidoptera pests. Florida Entomologist. 80(4): 425-443.

King J.R & Hennesey M.K.1996. Spinosad bait for the Caribbinean fruitfly (Diptera: Tephritidae). Florida Entomologist. 79(4):526- 531.

Labos E., Ochoa J., and Soulier C., 2002. *Cactoblastis cactorum* Berg (Lepidoptera: Pyralidae) preliminary studies for chemical control. Proc. 4th International Conference on Cactus pear and Cochineal. ISHS Acta Horticulturae. Eds. Nefzaoui & Inglese. N° 581: 247-254.

Lietti M.M., Botto E., & Alzogaray R.A., 2005. Insecticide Resistance in Argentine Populations of *Tuta absoluta* (Lepidoptera: Gelechiidae). Neotropical Entomology. 34(1): 113-119.

Maraus P.F., Catapan V., Faria D.S., Filho J.U.T.B., Santos H.S., & Hora R.C. 2008. Eficiência de insecticidas no controle da traça (*Tuta absoluta*) na cultura do tomateiro. Horticultura Brasileira 26: 2929-2933.

(McMillian W.W., Bowman M.C., Starks K.J., & Wiseman B.R., 1969. Extract of Chinaberry leaf as a feeding deterrent and growth retardant for larvae of the corn earworm and fall armyworm. J. Econ. Ent. 62: 708-710.

Meisner J., Ascher K.R.S., Aly R., & Warhben J.D.J. 1981. Response of *Spopdoptera littoralis* (Boisd.) and *Earias insulata* (Boisd.) larvae to azadirachtin and salannin. Phytoparasitica 9:27-32.

Méndez W.A., Valle J., Ibarra J.E., Cisneros J., Penagos D.I., & Williams T., 2002. Spinosad and nucleopolyhedrovirus mixtures for control of *Spodoptera frugiperda* (Lepidoptera: Noctuidae) in maize. Biological control 25:195-206.

Mohammed A.S. 2010. New record for leafminer, *Tuta absoluta* (Lepidoptera: Gelechiidae) infested tomato plantations in Kafer El-Sheikh region. J. Agric. Res. Kafer El-Sheikh. Uni. 36(2): 238-239.

OEPP/EPPO. European and Mediterranean Plant Protection Organization. 2005. *Tuta absoluta*. Bulletin OEPP/ EPPO Bulletin. 35:434-435.

Raffa, K.F., 1987. Influence of host plant on deterrence by azadirachtin of feeding by fall armyworm larvae (Lepidoptera : Noctuidae). J. Econ. Entomol. 80: 384-387.

Roditakis E., Papachristos D., & Roditakis N.E. 2010. Current status of the tomato leafminer *Tuta absoluta* in Greece. OEPP/EPPO Bulletin. 40: 163-166.

Salgado V.L., 1998. Studies on the mode of action of Spinosad: insect symptoms and physiological correlates. Pesticides Biochemistry and Physiology. 60: 91-102.

Seal D.R., 2005. Management of melon thrips, *Thrips palmi* Karny (Thysanoptera: Thripidae) using various chemicals. Proc. Fla. State. Hort. Soc. 118: 119-124.

Simmonds M.S.J., and Blaney W.M., 1984. Some neurophysiological effects of azadirachtin on lepidopterous larvae and their feeding response. Natural pesticides from the neem tree (*Azadirachta indica* A. Juss) and other tropical plants. Proc. 2nd. Int. Neem. Conf. Edited by Schmutter H. and Asher K.R.S. German Agency for technical cooperation. Germany.

Sparks T.C., Thompson G.D., Kirst H.A., Hertlein M.B., Larson L.L. Worden T.N & Thibault M.B 1998. Biological activity of Spinosyns, new fermentain derived insect control agents on tobacco budworm (Lepidoptera: Noctuidae) larvae. J. Econ. Ent. 91: 1277-1283.

Siqueira H.A.A, Alvaro A., Guedes R.N.C, & Picanço M.C., 2000a. Insecticide resistance in populations of *Tuta absoluta* (Lepidoptera : Gelechiidae). Agric. Forest. Entomol. 2:147-153.

Siqueira H.A.A, Guedes R.N., & Picanço M.C., 2000b. Cartap resistance and synergism in populations of *Tuta absoluta* (Lepidoptera: Gelechiidae). J. Appl. Ent. 124: 233-238.

Stanley J., Chandrasekaran & Regupathy A. 2005. Basesline toxicity of emamectin and spinosad to *Helicoverpa armigera* (Lepidoptera : Noctuidae) for resistance monitoring. Entomological Research. 39: 321-325.

Strong L., & Brown T.A., 1987. Avermectins in insect control and biology. Bull. Entomol. Res. 77: 357-389.

Takahashi H., Takakusa N., Suzuki J., & Kishimoto T., 1998. Development of a new insecticide, acetamiprid. J. Pestic. Sci. 23: 193-198.

Tomatonews, 2011. http://www.tomatonews.com/resources.html. Accessed March 10, 2011.

Turner N. 1932. Notes on rotenone as an insecticide. J. Econ. Entomol. 25: 1228-1237.

Tuta absoluta. 2011. www.tutaabsoluta.com/agrinewsfull.php?news=3. Accessed 15 March 2011.

Urbaneja A., Vercher R., Navarro V., Garcia-Mari F., & Porcuna J.L., 2007. La polilla del tomate *Tuta absoluta*. Phytoma España. 194:16-23.

Wang D., Cong P.Y., Li M., Qui X.H., & Wang K.Y. 2009. Sublethal effects of Spinosad on survival, growth and reproduction of *Helicoverpa armigera* (Lepidoptera : Noctuidae). Pest. Man. Sci. 65: 223-227.

Wakil W., Ashfaq M., & Ghazanfar M.U., 2009. Integrated management of in chickpea in rainfed areas of Punjab, Pakistan. Phytoparasitica. 37: 415-420.

Walgenbach J.F., Leidy R.B., & Sheets T.J., 1991. Persistence of insecticides on tomato foliage and implications for control of tomato fruitworm (Lepidoptera : Noctuidae). J. Econ. Entomol. 84: 978-986.

Wing K.D., Sacher M., Kagaya Y., Tsurubuchi   Y., Muldirig L.,& Connair M.,   2000.
    Bioactivation and mode of action of the oxidiazine indoxacarb in insects. Crop
    Protection 19: 537-545.
Zapata N., Budia F., Viñuela E., & Pilar M. 2006. Laboratory evaluation of natural pyrehrins,
    pymetrozine and triflumuron as alternatives to control Ceratitis capitata adults.
    Phytoparasitica 34(4): 420-427.

6

Use and Management of Pesticides in Small Fruit Production

Carlos García Salazar[1], Anamaría Gómez Rodas[2] and John C. Wise[2]
[1]Agriculture & Agribusiness Institute,
Michigan State University Extension, West Olive, MI
[2]Department of Entomology, Trevor Nichols Research Center, Fennville, MI
USA

1. Introduction

Historically, pesticide management has been an inefficient process for pest control in tree fruit production. The amount of active ingredient released into the pest habitat is thousands of times greater than the amount of pesticide required to kill it if the treated population were confined in a jar or a small space. To illustrate this, lets take the oral LD50 for methyl parathion in rats (18 to 50 mg/kg) (1, 2) and the field dose for controlling the Brown Stink Bug (BSB) in pecans (Penncap-M, 227 grams of active ingredient (AI) per acre) (3). If the weight of the population of BSB in one acre of pecans were the equivalent of 1 kg, taking the LD50 of 50 mg/kg for methyl parathion we may need only 100 mg to kill the entire population if treated in a jar or a small confined space. But since they are distributed all over the field it takes thousand times more product than needed to kill every insect present in the treated field.

In today's world, due to the increased cost of the pesticides a more efficient approach is needed, as well as more potent poisons requiring smaller application rates and new environmental legislations. In 1996, the enactment of the Food Quality Protection Act (FQPA) in the United States of America changed for good the way we controlled pests in agriculture in general (4). The enactment of the FQPA legislation demanded the re-registration of all pesticides used in agriculture until 1996. One of the major changes was the introduction of a 10 X safety margin for children for pesticides shown to have a deleterious effect on children and woman during pregnancy.

The most important changes occurring under FQPA were that pesticides had to be evaluated in relation to the aggregated exposure taking into consideration all potential sources of exposures to residues in food, water and residential use. Pesticides with a common mode of action potential exposure had to be assessed as a cumulative exposure. Finally, potential risk must be assessed without considering the benefits.

Organophosphate (OP) or conventional insecticides used extensively in fruit production were some of the most affected by FQPA. This created the need for new chemicals to substitute those products that were removed from the list of approved materials. Products that were not cancelled required mitigation measures that restricted their use. These substitute pesticides are less effective and require multiple applications to obtain similar results than the ones obtained with conventional OP's. This difference in the performance of the new pesticides increased the cost of pest control. New products are more pest-specific, a small amount of a.i. is required per application (< 1 oz /acre) and intensive site-specific IPM pest scouting is required to effectively use them.

Under this new paradigm, an effective use and management of pesticides require growers and IPM practitioners to know the chemistry of the pesticide being used, the biology of the pest and its behavior, the influence of weather, plant structure, and equipment used to release the pesticide into the pest habitat.

2. Problems associated with the use of pesticides in agriculture

Every year there are a number of claims to farm insurance companies that are the result of errors during the application of pesticides. The main causes for losses of crops due to misapplications are equipment failure (24%), improper tank mixes (33%) and drift (33%). Other causes are application in the wrong field and problems reading the label of the product (5). The most important main factors affecting the efficacy of pest control are, the prevailing weather conditions during the application, the type of nuzzles and conditions of the spray equipment, the characteristics of the crop and the behavior of the pest. Although the applicator can not control some of these factors there are always measures that may help reduce the pesticide environmental impact and increase their effectiveness.

3. Weather conditions during the application

Wind velocity. It is usually the most critical factor of all meteorological conditions affecting the efficacy of the pesticide application. The greater the wind speed, the farther off-target a droplet of a given size will be carried. The larger the droplet, the less it is affected by the wind and the faster it will fall. High winds however, can cause even larger droplets to move off-target. Therefore, spraying operations should be stopped if wind speeds are excessively high (> 10 miles/hour or 16 Km/hour). In 2001, the US Environmental Protection Agency (EPA) issued a recommendation that indicated that for pesticide applications in orchards and vineyards and other fruit crops the maximum wind speed during the application should be between 3 and 10 miles/hour (4.5-16 Km/hour) (EPA 730-N-01-006). To illustrate the effect of wind velocity on pesticide droplets we may take the data from Rose and Lambi (1985) (6). They showed that droplets 100 microns in diameter travel 4.6 m in a wind speed of 1.6 Km/hour. However, in a wind speed of 8 Km/hour they travel 23.1 m. A droplet of 400 microns under the same conditions travels 1 and 5 m, respectively.

Relative Humidity. As droplets fall through the air, they evaporate into the atmosphere. This evaporation reduces the size and mass of the particle enabling it to remain airborne longer and, under the right conditions, to drift farther from the application site. The rate at which water evaporates from the spray particles depends primarily on air temperature and relative humidity. At 70 % relative humidity and 78°F, a 100-micron droplet will fall 5 feet and hit the ground before evaporating to half its original diameter. However, at 30% relative humidity and 78°F, a 100-micron droplet quickly evaporates and becomes one-eighth of its original volume while falling only 2.5 feet. While evaporative loss of spray materials occurs under almost all atmospheric conditions, these losses are less pronounced under the environmental conditions that occur during the cooler parts of the day - early morning and late afternoon. The relative humidity is usually highest during these cooler periods (7).

Atmospheric stability. It is an important factor influencing drift. Under normal (stable) meteorological conditions, the air temperature decreases by 5.4°F per 1,000 feet of height (1°

C per 100 meters). Cool air tends to sink, displacing lower warm air and causing vertical mixing. As a warm air layer rises, suspended droplets rise with it and dissipate into the upper layers by normal air turbulence and vertical mixing. Vertical mixing is the result of the ADIABATIC effect. Adiabatic effect refers to the behavior of a parcel of air when it moves up and down through a gradient of pressure in the atmosphere—this being greatest at the surface and diminishes with height. This process occurs as follows: **Rising Air Parcel.** This movement causes *adiabatic cooling*. As the air parcel rises it enters a low pressure region and this causes the parcel to expand. This sudden expansion causes a decrease in the parcel's temperature and it starts cooling down. This effect is similar to opening a canister of propane gas, while it escapes the gas forms a frosty cap at the tip of the exit. **Falling Air Parcel.** This creates the *adiabatic warming*. As the parcel of air starts falling down, it enters a region of high pressure. This change in atmospheric pressure compresses the parcel of air and causes an increase in its temperature. The parcel of air continues descending until its pressure and temperature equals the surrounding air (Figure 1).

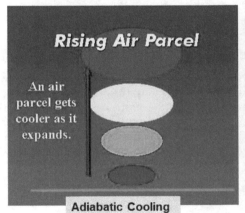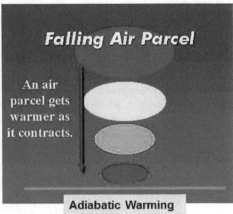

Fig. 1. Adiabatic effect.

When the rising parcel of air gets cooler than the surrounding air, it stops rising. This produces a situation in which the temperature at ground level and the temperature several hundred meters above the ground are the same. There is no vertical movement of the air and the atmosphere is in a condition of stability. Conversely, if the rising parcel gets warmer than the surrounding air, it continues rising creating the condition of atmospheric instability where parcels of air move up and down as they warm up and cool down.

Movement of pesticide droplets. When water-encapsulated pesticide droplets exit the nozzle their final deposition depends on the droplet size, the air and water temperature, and the environmental relative humidity. Big droplets (400-500 microns) will reach the foliage before losing substantial amount of volume. But all other droplets that do not reach the foliage are affected by the adiabatic effect. When those droplets are falling to the ground they are compressed by the atmosphere, their temperature rises beyond the surrounding air and they become airborne until their temperature equals the surrounding air. As those droplets cool down they fall back and the process is repeated again (Figure 2).

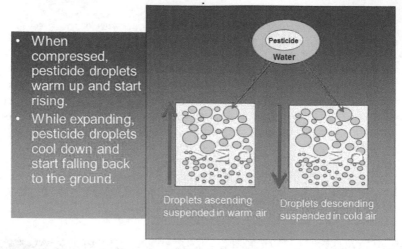

Fig. 2. Adiabatic effect on droplets containing the pesticides as they exit the nuzzle.

4. Spray application under unstable conditions

As warm air rises, suspended droplets rise with it and dissipate into the upper layers by normal air turbulence and vertical mixing. Under these conditions, the opportunity for crop injury at any specific off target site is very small because the pesticide is dispersed and diluted into the atmosphere. **Figure 3**, shows the application with a cannon sprayer in blueberries under unstable conditions. The picture shows that most of the drift remains confined to the first 5-6 rows facing the sprayer and the cloud of droplets that constituted drift moves up with little horizontal movement out of the target area

Fig. 3. Spray application under unstable conditions in a blueberry field.

5. Spray application under stable conditions

Under stable air conditions since the temperature at ground level is the same over the next few hundred meters above the ground, vertical air movement is hindered and there is little or no atmospheric mixing. Consequently, fine droplets will hang in the air and slowly diffuse superficially on relatively level terrain.

Fig. 4. Spray application early in the morning with no vertical mixing.

In Figure 4 the effect of atmospheric stability is presented. Fine spray droplets remain aloft traveling horizontally away from the target area. In this picture, pesticide drift goes over the road falling on passing by vehicles. Applications under these circumstances are the most problematic because they cause dangerous drifts that end up creating environmental pollution.

A very common type of atmospheric stability is the one created by an inversion of the gradient of temperature occurring in the atmosphere. Under conditions of instability the air temperature decreases with an increase in altitude, 5.4°F per 1,000 feet of height (1° C per 100 meters, approximately). Very often the air temperature near the ground is cooler than the air a few meters above ground; the surrounding air gets warmer with increasing altitude. This creates a layer of hot air that acts like a cap that does not allow the vertical movement of the air holding down cooler air underneath. This phenomenon is what we call thermal inversion and is very frequent very early in the morning and during the evening hours when the ground cools the air layer immediately above it.

Spray application under thermal inversion conditions. Pesticide applications conducted under conditions of thermal inversion are prone to produce large quantities of dangerous drifts. Pesticide particles suspended in the cool layer cannot move anywhere except laterally, possibly for several miles. **Do not spray** during thermal inversions, when air close to the ground is colder than the air above it. You should not spray early in the morning in still air if sensitive areas are nearby (especially down gradient if the terrain is sloped). Applicators concerned about these adverse spraying conditions should wait until late afternoon or early evening to spray, when there is less chance of the atmosphere being inverted and conditions are more favorable.

6. Spray equipment used for pesticide applications in small fruit crops

The most important factor in providing an efficient pesticide cover is type of equipment. The efficacy of your pest control depends on the right selection of your spraying equipment. Different types of spraying equipment are used in small fruit crops for pest control. Air-blast spraying equipment is the most common equipment employed for pesticide application (Figure 5).

Fig. 5. Air assisted sprayers: right, a cannon sprayer; left, conventional air-blast sprayer.

However, when the crop canopy makes it difficult to apply pesticides using ground equipment fixed-wing aircrafts are mainly used (Figure 6).

Fig. 6. Aerial application of insecticide in blueberries near harvest with fixed-wing aircrafts (Fennville, Michigan 2006).

Different air-assisted equipment produces different types of spray which in turn have different distribution on the target crop. Table 1 shows the properties of different air-assisted equipment on the deposition of pesticides on the target canopy.

Application Technology	Droplet size	Canopy Penetration	Field Distribution
Air-blast Sprayer	Fine – Very Coarse	High	Uniform
Cannon Sprayer	Very Fine – Medium	Medium	Horizontally-Variable
Aerial Application	Fine – Medium	Medium	Vertically Variable
Proptec Sprayer	Fine – Medium	High	Uniform

Table 1. Pesticide application equipment and characteristics of spray deposition.

Show (1996) gave a very good definition of every type of spray depending on the droplet size (5). Accordingly, a very fine spray (VF) is made of droplets with a diameter < 182 microns (μm); A fine spray (F) is made of droplets with a diameter ranging from 183 to 280 μm; Medium spray (M) is made of droplets ranging from 281 to 429 μm; A coarse spray (C) is made of droplets ranging from 430 to 531 μm in diameter, etc.

The type of spray produced by the equipment determines the volume of water needed to spray the recommended dose of pesticide. The following table indicates how much water is needed to cover one acre of blueberries using different types of sprays.

Spray type	Volume type	Droplet diameter (μm)	Application rate gal/acre
Coarse	High	500	108
Medium	Medium	200 - 500	32 -108
Fine	Low	100 - 200	6 - 32

Table 2. Type of spray and its effect on the amount of water needed to spray one acre of blueberries.

7. Plant structure and canopy: where do your chemicals end up?

After selecting the right environmental conditions and the equipment for the spray application the next factors determining the success of your applications will be, type of canopy and plant structure.

Plant structure is important in determining the final destination of your spray. It also influences the success of pest control. Canopy is the place where the pest is found scouting for food, mating, looking for oviposition sites or escaping from natural enemies and adverse environmental factors. This also includes escaping from the applications of pesticides. Therefore, in determining the amount of water and the equipment required to control a pest problem, type of target crop and plant structure are crucial.

In blueberries, for example, the structure of the bush is important in the efficacy of the application of insecticides and fungicides using air blast sprayers. A well pruned bush will allow for better penetration of the pesticide. Thus, you could achieve better pest control and savings in product. On the other hand, thick canopies prevent the placement of the product in the habitat of the pest resulting in poor and more expensive pest control. Van Ee et al.(2000) demonstrated how different types of pruning affected pesticide deposition in "highbush" blueberries (*Vaccinium coymbosnm* L) at different times during the plant growth season. Their goal was to determine how sprayer type, pruning severity, and canopy development interacted to affect spray deposition patterns. Deposition was measured as the percentage of the surface area of card targets that was covered following applications of black dye. Light measurements indicated that the canopy of blueberry bushes, regardless of pruning treatment, closed by the middle of June, and light levels within the canopy changed little from then until fruit harvest in August. They used a standard airblast sprayer that pushed spray up and white kromcoat cards clipped to each target. Targets were clipped to branches in a random orientation in the same two bushes used for the light measurements. Figures 7.1, 7.2, and 7.3 showed the conditions of pruning of treated bushed before treatments were applied.

Fig. 7.1. Lightly pruned bushes.

Fig. 7.2. Moderately pruned bushes.

Fig. 7.3. Heavily pruned bushes.

Van Ee, found that spray penetration and deposition in all types of canopies was greater in the bush's side facing the sprayer. However, spray penetration and coverage was greatly reduced at the far side of the bush when the field was lightly pruned. Conversely, medium or heavy pruning allowed more penetration and deposition than in light pruned bushes. Moderate or heavy pruning allowed similar deposition, except at the top of the bush. In the heavily pruned bushes pesticide deposition was greater than at any other canopy structure (7.3).

In 2006, we conducted a study to determine the spray deposition patterns in mature blueberries using different spray equipment. A mature Jersey field was sprayed with Surround WP (kaolin clay) utilizing a cannon sprayer, an airblast sprayer and a 20-gallon aerial application (Wise et al. unpublished results).

We found that a spray application of 50 gal/acre in light pruned bushes resulted in high deposition at the bottom and middle portion of the canopy and less deposition on the top of the bush. The same volume applied with a cannon sprayer resulted in higher spray deposition on top of the bush but less deposition on the middle section and much less at the bottom of the bush.

The application of a spray volume of 20 gallons to the same field with fixed-wing aircraft resulted in an even deposition of the spray from top to bottom of the bush.

These results were very important to establish what type of equipment we need to use to apply pesticides in blueberry fields in order to optimize the efficacy of the pesticide.

8. Pest habitat and behavior

Finally, the purpose of the spray application is to place the pesticide in contact with the pest. This objective can be achieved in two ways; 1) depositing the pesticide in the habitat of the pest were the pest may enter in contact with the pesticide or 2) placing the pesticide in contact with the pest by taking advantage of its behavior. In this case the pest "finds" the product while moving around the canopy of the crop.

For plant diseases and insects that do not move a lot and remain confined to a small area (a leaf, branch, bush or section of the canopy) we need to bring the pesticide to the pest by targeting their habitat. Spray cover has to be very good with enough number of droplets per square centimeter of foliage to maximize the probability of the pest getting in contact with the spray. In the case of blueberry aphids and fruitworms, pest control measures require placing the insecticide in the habitat of the pest. Blueberry aphids live at the bottom, inside of the bush. Thus, spray applications should reach the bottom of the bush. If the product applied is systemic, droplet size does not really matter, the product will be translocated into de plant and the insect will acquire the lethal dose while sucking up sap. With the Cranberry fruitworm (*Acrobasis vaccinii* Riley) and the Cherry fruitworm (*Grapholita packardi* Zeller), if pest control is directed against eggs and larvae it should be directed into the calyx of the fruit. The product needs to be placed also into the calyx where larvae will hatch. In this case a large volume of spray needs to be deposited on the top of the blueberry bush where the fruit is produced and the fruitworms mate and deposit their eggs. In both cases, a large number of small droplets of pesticide need to be applied into the habitat of the insect.

With an insect or pest that moves a lot or explores intensively the canopy of the crop, like the Blueberry Maggot (*Rhagoletis mendex*) or the Cranberry fruitworm first instar larvae. A few large droplets will be enough to have a good control of these pests. Adult blueberry maggot explores intensively the blueberry canopy in search of mates, food and sites to oviposit. Also, first instar Cranberry fruitworm larvae after hatching crawl out of the berry calix to search for a place to enter the berry at the site where the peduncle and the fruit meet. In both insects the wandering period is critical to put the pesticide in contact with the pest. Table 3 presents a summary of the habitat and behavior of several insects that attack blueberries in Michigan.

Pest	Targeted Life Stage	Pest Behavioral Activity Level	Location on the Plant	Exposure Period
Fruitworms	Egg/larva	Low	Fruit Cluster	Short
Leafrollers	Larva	Medium	Upper Canopy	Long
Blueberry Maggot	Adult	High	Upper Canopy	Short
Japanese Beetle	Adult	High	Upper Canopy	Long
Blueberry Aphid	Adult/Nymphs	Low	Lower Canopy	Long
Bud mites	Adult/Nymphs	Low	Buds	Short

Table 3. Habitat and behavior of several insect pest of "Highbush" blueberries.

9. Pesticide management summary

We can summarize the recommendations for optimizing the efficacy of the pesticides as follow:

The pesticide efficacy and efficiency will depend on:

Weather conditions that affect the movement of pesticides in the environment.

-       Avoid spraying under low relative humidity,
-       Avoid spraying when wind speed is greater than 10 mph,

-    Avoid spraying under stable or thermal inversion conditions

Equipment. This affects pesticide deposition in the target site (amount of product and percentage of surface area covered).
-    Use the appropriate spraying equipment,
-    Calibrate your sprayer,
-    Use the appropriate spray volume.

The conditions of the sprayer will limit the biological effect of the spray. Important factors to consider are droplet size, volume and distribution of the application

Crop conditions. This affects the penetration of chemicals and pesticide deposition. Proper pruning facilitates:
-    spray penetration and deposition,
-    and placement of the pesticide into the pest habitat,
-    removes refuge sites where the pest can find shelter.

In the case of crop conditions, the canopy density limits the spray's penetration and deposition. It will be more difficult to implement a successful pesticide management program in a slightly pruned field than in a field that is properly pruned. A thick canopy provides many places where the pest can be protected from natural enemies and from our pesticides! Under these circumstances, only a limited amount of our spray will reach the pest habitat limiting the effectiveness of our pest control program.

Pest behavior. Each pest species has its particular behavior. Therefore, the efficacy of your pest control will depend on:
-    your understanding of the pest biology and behavior,
-    the behavior of a pest that limits the critical window to control it

Knowing the pest is important because it is the objective of our pest control program. If we do not understand the pest biology and behavior, chances are we will miss the critical times when the pest is more susceptible to our pesticides.

Finally, the success of your **Pesticide management program will depend on** your ability in putting the pesticide in contact with the pest either by:
-    Placing the pesticide in the pest habitat and/or,
-    Taking advantage of the pest behavior

10. References

[1] Hayes, W.J. and E.R. Laws (ed.). 1990. Handbook of Pesticide Toxicology, Vol. 3, Classes of Pesticides. Academic Press, Inc., NY.

[2] Meister, R.T. (ed.). 1992. Farm Chemicals Handbook '92. Meister Publishing Company, Willoughby, OH.

[3] Crop Data Management Systems, Inc. 2007. PennCap-M.
       http://www.cdms.net/LDat/ld403009.pdf

[4] U.S. Food and Drug Administration. 2009. The Food Quality Protection Act of 1996.
       http://www.fda.gov/RegulatoryInformation/Legislation/FederalFoodDrugandC
       osmeticActFDCAct/SignificantAmendmentstotheFDCAct/ucm148008.htm.

[5] Show, B.W. 1996. Minimizing Spray Drift. Texas Agricultural Extension Service, The Texas A&M University System. College Station, Texas.
       agsafety.tamu.edu/Programs/Ag-Chemical/drift.ppt

[6] Ross, Merrill A. and Carole A. Lembi. 1985. Applied Weed Science. Burgess Publishing Company, Minneapolis, MN.

[7] H. Erdal Ozkan. 1998. Effect of Major Variables on Drift Distances of Spray Droplets. The Ohio State University Extension (AEX-525-98). http://ohioline.osu.edu/aex-fact/0525.html.

[8] Van Ee, G., Richard Ledebuhr, Eric Hanson, Jim Hancock, and Donald Ramsdell (2000) Canopy Development and Spray Deposition in Highbush Blueberry. HortTechnology 10(2) 354-357.

7

Effects of Kaolin Particle Film and Imidacloprid on Glassy-Winged Sharpshooter (*Homalodisca vitripennis*) (Hemiptera: Cicadellidae) Populations and the Prevention of Spread of *Xylella fastidiosa* in Grape

K.M. Tubajika[1], G.J. Puterka[2], N.C. Toscano[3], J. Chen[4] and E.L. Civerolo[4]
[1]*United States Department of Agriculture (USDA), Animal and Plant Inspection Service (APHIS), Plant Protection and Quarantine (PPQ), Center for Plant Health Science and Technology (CPHST), Raleigh, North Carolina,*
[2]*United States Department of Agriculture (USDA), Agricultural Research Service (ARS), Stillwater, Oklahoma;*
[3]*Department of Entomology, University of California, Riverside, California,*
[4]*United States Department of Agriculture (USDA), Agricultural Research Service (ARS), San Joaquin Valley Agricultural Sciences Center (SJVSC), Parlier, California, USA*

1. Introduction

The glassy-winged sharpshooter (GWSS), *Homalodisca vitripennis* (Germar), (Hemiptera: Cicadellidae), is a major pest of important agronomic, horticultural, landscape, ornamental crops and native trees in California (Blua et al., 1999; Purcell et al., 1999; Purcell & Saunders, 1999). This insect is an invasive species in California that was first detected in the state in 1989 (Sorensen & Gill, 1996). This sharpshooter is a key vector of *Xylella fastidiosa* and has changed the epidemiology of *X. fastidiosa* (*Xf*) diseases affecting important agronomic and horticultural crops as well as landscape, ornamental and native trees in California based on the infection of these crops by the bacteria (Blua et al., 1999; Purcell et al., 1999; Purcell & Saunders, 1999) (Fig. 1). It is not clear if management strategies for Pierce's Disease (PD) developed in GWSS-free regions of California are suitable to manage the disease in vineyards where GWSS has become established.

Data from earlier studies suggest that GWSS has continued to increase in number (population density) and geographical range in California (Purcell & Saunders, 1999; Tubajika et al., 2004). GWSS populations are widely distributed over a large number of hosts including perennial agronomic crops, ornamental plantings, and weedy plant species (Hill & Purcell, 1995; Purcell et al., 1999; Raju et al., 1980). Previous studies by Blua et al. (1999) and Perring et al. (2001) showed that GWSS populations utilize citrus plants as their primary over-wintering host (other plants are available) when grapes, stone fruits, ornamental hosts, and weedy species are dormant during the winter. This vector may have a great impact on grape-growing areas in the Coachella, Temecula and the lower San Joaquin Valley due to numerous citrus orchards that support overwintering populations. Effective insect and disease management strategies are dependent upon knowledge of inoculum sources, and biology and the ecology of insect vectors and their natural enemies (Blua et al., 1999; Hill & Purcell, 1995; Tubajika et al., 2004). The GWSS transmits Xf to grape (Fig. 1), as well as to oleander (Costa et al., 2000) with the transmission efficiency greater than that of other vectors such as the green sharpshooter (GSS) [*Draeculacephala mineroa* Ball] and red-headed sharpshooter (RHSS) [*Carneocephala fulgida* Nottingham], RHSS) (Purcell & Saunders, 1999), but much less than that of the blue-green sharpshooter (BGSS) [*Graphocephala atropunctata* (Signoret] (Purcell & Saunders, 1999).

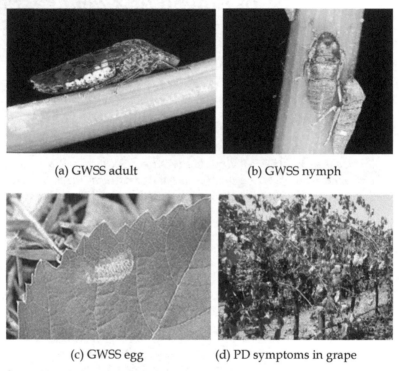

(a) GWSS adult   (b) GWSS nymph

(c) GWSS egg   (d) PD symptoms in grape

Fig. 1. Glassy-winged sharpshooter (GWSS) adult (a), nymph (b) (photos by B. Stone-Smith), and egg (c) found in grape in Bakersfield, CA. The GWSS transmits the bacterium *Xylella fastidiosa*, which causes Pierce's disease (PD) (d) in grape.

Imidacloprid, (Admire, Bayer Corp., Kansas City, MO, USA), is an neonicotinoid insecticide which interrupts the binding of nicotinergic acetylcholine in post-synaptic receptors of the insect (Hemingway & Ranson, 2005; Romoser & Stoffolano, 1998). In Georgia, treatment of grapevines with Admire slowed the rate of PD spread, but ultimately extended vineyard life by only a year when infestations were severe (Krewer et al., 2002). Imidacloprid is the most widely used neonicotinoid for the prevention of infestation of grapevines by GWSS in California. This product may be useful in reducing Xf transmission by the GWSS and in slowing the development of PD in grape. Contact insecticides offer short-term protection against GWSS infestations because of the continued movement of sharpshooter adults from citrus and non-treated hosts/areas which re-infest the grapevines. In a study on the transmission of *Xf* to grapevines by *H. vitripennis* (syn. *coagulata*), reported that *H. vitripenns* (syn. *coagulata*) transmitted *Xf* to grapevines in a persistent manner(Almeida & Purcell, 2003). This research also showed that the nymphal lost infectivity during molting, but there was no evidence of a latent period (delay after acquisition) of the pathogen by adult populations.

Likewise, kaolin (Surround®, NovaSource Tessenderlo Kerley, Inc., Phoenix, AZ.) is a potential alternative pest management product with improved safety to pesticide handlers and reduced environmental impact (Glenn et al., 1999)). It protects plants from insect feeding and oviposition by coating the plant surfaces with a protective mineral barrier (aka particle film) (Glenn et al., 1999; Puterka et al., 2000). Kaolin has been shown to suppress pear psylla, *Cacopsylla pyricola* (Forster); the spirea aphid, *Aphis spiraecola* Pagenstecher; the two-spotted spider mite, *Tetranychus urticae* Koch; and the potato leafhopper, *Empasca fabae* (Harris) in previous research (Glenn et al., 1999) (Fig. 2). The mechanism involved in reduction of pest density is reported to be reduced oviposition and feeding on treated plants (Glenn et al., 1999). Puterka et al. (2000) found Surround® reduced *C. pyricola* populations and controlled Fabraea leaf spot, fungal disease caused by *Fabraea maculata* Atk. Moreover, Glenn et al., (1999) reported that kaolin films contributed to control of fungal and bacterial plant pathogens by preventing the formation of a liquid film on the surface of pear leaves. In this chapter, the effects of kaolin particle film and imidacloprid on *H. vitripennis* (syn. *coagulata*) population levels and the prevention and spread of *Xf* in grape are described.

    (a) Field study using kaolin   (b) Close-up of kaolin-treated grape leaves

Fig. 2. Field study showing the application of kaolin (a) and a close-up of kaolin-treated grape used to control Glassy-winged sharpshooter in Kern County, Bakersfield in California.

2. Impacts of kaolin and imidacloprid on *Homalodisca vitripennis*

The use of insecticides is common for controlling pest populations (Hemingway & Ranson, 2005). However, the development of resistance and cross-resistance among widely used and newly applied systemic insecticides can affect the development of new strategies in integrated pest management. The GWSS has become a significant problem to California agriculture because it feeds readily on grape and, in doing so, transmits Xf, the causal agent of Pierce's Disease (PD) in grape (Fig.1d). Prior to the appearance of GWSS, California grape growers were able to manage PD in grape that is vectored by a number of other indigenous sharpshooters (Anon, 1992). Unfortunately, the GWSS more adapted to the citrus/grape agroecosystem than other sharpshooters, thus, making it a serious vector of PD that now threatens the grape industry in California (Purcell & Saunders, 1999). Although contact insecticides offer short-term protection against infestations of GWSS, the continued movement of GWSS adults from citrus and uninfested areas often re-infest grapes. Clearly, there is a need to investigate other approaches or technologies that could repel GWSS infestations or prevent them from feeding on grape vines and transmitting PD (Puterka et al., 2003).

In an effort to assess the efficacy of control using kaolin, experimental potted lemon trees in field cages at the Bakersfield location were used to test the efficacy kaolin on GWSS infestation and survival under no-choice conditions. 'Eureka' lemon trees were either treated with 6% kaolin (60 grams kaolin per liter of water) or left untreated as a control. Trees were sprayed with kaolin by a 4-liter hand-pump sprayer until all of the foliage was wetted and then allowed to dry. Either a treated or untreated tree was placed in screen-covered cages measuring 1 m² by 2 m high, and either 50 GWSS adults (Fig. 1a) or 10 instar nymphs (third to fourth stage) (Fig. 1b) were placed at the bases of the trees to disperse. Numbers of GWSS on trees and within the cages were recorded daily for four days to determine infestation rates and survival. GWSS numbers were converted to percentages and analyzed using analysis of variance (ANOVA) (SAS Institute Inc., Cary, NC).

2.1 Survival effect of kaolin and imidacloprid on GWSS in citrus

Experimental results showed that treatment of grapes with kaolin significantly affected (P < 0.01) the number of nymph and adult populations. GWSS nymphs and adults were greatly repelled by kaolin application in treatments where choice and no-choice experiments were conducted (Fig.3ab). Very few nymphs or no adults were able to colonize lemon treated with kaolin in experiments where either GWSS nymphs or adults were given no choice for colonization, feeding, and oviposition on plant leaves treated with kaolin and the untreated citrus. There were no significant day or treatment-by-day interactions for nymphs ($P = 0.4$) or adults ($P = 0.19$) in the no-choice tests suggesting that GWSS did not vary over the 4-day period after they had found a suitable site on which to settle (Fig.3ab). Survival of nymphs under no-choice conditions was significantly reduced by kaolin treatments ($P < 0.001$) after being exposed for 4 days, and averaged 8.3 ± 4.2% on kaolin versus 75.0 ± 15.2% on untreated trees. There was no Adult GWSS survival on kaolin-treated lemons and 18.8 ± 4.4% on the untreated lemon 1 day after infestation (Fig.3ab). This relationship did not change over the 4-day period. Adult survival was extremely poor in caged tests because many individuals did not settle on lemon trees under caged conditions and clung to the side of cages until death.

Likewise, the choice tests produced very similar results as the no-choice tests. Neither GWSS nymphs nor adults settled on kaolin-treated lemon, which resulted in no colonization over a 4-day period (Fig.3ab). Treatment preferences of nymphs or adults did not change significantly over time. Survival of nymphs significantly declined over time, from 90.0 ± 4.5% at 1 day to 53.3 ± 6.4% at 4 days after infestation. We hypothesize that this decline in population levels was due to the nymphs falling from PF treated trees or from e movement between treated and untreated foliage and being unable to return to the trees. The number of surviving adults did not decline in untreated trees in the no-choice test indicating they had favorable conditions to colonize and thrive. Some GWSS nymphs that were able to find the few untreated spots on kaolin-treated foliage probably remained at those sites during the study period. In contrast to the above findings, adult survival on kaolin-treated foliage did not change over time in either the choice or no-choice tests. We observed that kaolin treated plants are undesirable hosts to both GWSS adults and nymphs in choice and no-choice environment.

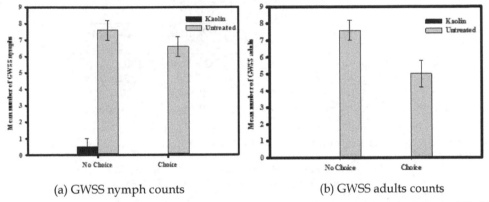

(a) GWSS nymph counts   (b) GWSS adults counts

Fig. 3. Mean number of glassy-winged sharpshooter nymph (n=20) (a) and adults (n=50) (b) per tree 4 days after being released on trees that were treated or not treated with kaolin films in choice versus no-choice experiments.

However, in no-choice experiments, we observed that few nymphs managed to find an untreated site on which to settle when presented with the kaolin-treated plants. This finding implies that good coverage is important when using this material for insect control. When given a choice between kaolin-treated and untreated plants, no GWSS nymphs or adults infested kaolin treated plants.

The choice study may be more representative of what occurs under field conditions because GWSS can utilize more than 75 species of plants (Turner & Pollard, 1959). Therefore, GWSS could easily move to other plants or plant species if they encountered kaolin treated- plants. Results from other studies indicate that kaolin particle films protect plants from insect feeding, oviposition, and by repelling GWSS nymphs and adults by coating the plant surfaces with a mineral barrier (Puterka et al., 2003). This finding is supported by previous reports using kaolin particle film treatments (Glenn et al., 1999; Puterka et al., 2000; Sisterson et al., 2003). There are numerous examples where kaolin particle films repel insect infestations, thus prevent feeding and oviposition although there are at least six other possible mechanisms of action that depend on specific insect-plant relationships (Glenn and Puterka 2005, Puterka and Glenn 2008).

2.2 Mortality effect of kaolin and imidacloprid on GWSS on grape

The mortality effects of kaolin and imidacloprid on GWSS were assessed on grape artificially infested by X. fastidiosa. The GWSS was caged for 48 h on grape vines that was previously infested by X. fastidiosa (Tubajika et al., 2007). After acquisition, GWSS, in groups of 10, were caged for 24 or 48 h in small sleeve cages (45 x 55 cm) containing plants either treated or not treated with kaolin and imidacloprid. After 24hr or 48 hr post-treatment, the number of affected GWSS was recorded. The criterion for mortality was insects without any movement (ataxic)[1]. The GWSS mortality was then expressed as percentage.

The GWSS mortality (%) was significantly (P < 0.05) impacted by particle film treatment (Tubajika et al., 2007). After 24 hr feeding time, GWSS mortality ranged from 0% (Negative and positive control plants) to 58% (kaolin-treated plants). For the 48 hr feeding time, the mortality ranged from 0% (negative control plants) to 100% (kaolin treated-plants) (Table 1).

Percent (±SE) Plant mortality[y]		
Treatment\Feeding time	24-hour	48-hour
Kaolin-treated plants	58 ± 14aB	100 ± 25aA
Imidacloprid-treated plants	34 ± 1bB	71 ± 21bA
Positive Control (Infested plants)[x]	0 ± 0cA	9 ± 1cA
Negative Control (Non infested plants)[x]	0 ± 0cA	0 ± 0cA

[x]Untreated plants control plants consisted of water-treated plants exposed or not exposed, respectively to infective GWSS.
[y]Calculated from the means for 5 replications. Within columns (small letters) and across rows (capital letters), means followed by the same letter did not differ significantly (P < 0.05) according to Fisher's least significant difference test.

Table 1. Effect of kaolin and imidacloprid treatments and length of glassy-winged sharpshooter (GWSS) feeding time on GWSS mortality in greenhouse experiments.

Overall, the GWSS mortality was greater on imidacloprid-, kaolin-, and the untreated plants when the GWSS fed for 48 h than when they were allowed to feed for 24 h.

However, mortality of the GWSS on untreated control plants and non-infested plants with X. fastidiosa was not impacted by the feeding time (negative controls) in these experiments (Table 1). The occurrence of insecticide resistance in GWSS depends on the insecticide used and duration of exposure (Tubajika et al., 2007). Therefore, effective insecticide management strategies and their implementation are necessary for the prevention of rapid resistance development. The mechanism of action of particle films do not rely on toxicity to insects thus resistance to particle films appears unlikely (Puterka & Glenn 2008).

2.3 Pierce's Disease (PD) incidence
Similarly, the impact of kaolin and Imidacloprid on incidence of PD was also assessed on the same grape vines. Each plant was visually assessed at 30-day intervals for PD symptom development which includes stunting, delayed growth, marginal leaf necrosis, and abscission between the petiole and leaf blade. On each assessment date, the total number of plants exhibiting PD symptoms was recorded. Ten leaf samples from each plant were bulked and assayed for the presence of X. fastidiosa using assays enzyme-linked immunosorbent assay (ELISA) and immunocapture polymerase chain reaction (IC-PCR)

---

[1] An inability to coordinate voluntary muscular movements that is symptomatic of some nervous disorders.

Effects of Kaolin Particle Film and Imidacloprid on Glassy-Winged Sharpshooter (Homalodisca vitripennis)
(Hemiptera: Cicadellidae) Populations and the Prevention of Spread of Xylella fastidiosa in Grape
143

(Minsavage et al., 1997)). In determining the incidence of PD, the number of plants with symptoms of PD was expressed as percentage of total plant assessed.

The treatment of plants with kaolin and imidacloprid, exposure time of plants to the infectious GWSS, and the interaction of treatment by exposure time had an effect on PD incidence (Tubajika et al., 2007). PD incidence was 4%, 7%, and 32% in kaolin-, imidacloprid-treated plants, and in untreated controls, respectively (Table 2).

Percent (±SE) Pierce's Disease incidence[yz]		
Treatment\Feeding time	24-hour	48-hour
Kaolin-treated plants	8 ± 2bA	0 ± 0bB
Imidacloprid-treated plants	9 ± 2bA	4 ± 1bB
Positive Control (Infested plants)[x]	48 ± 12aA	19 ± 3aB
Negative Control (Non infested plants)[x]	0 ± 0bA	0 ± 0bA

[x]Untreated plants control plants consisted of water-treated plants exposed or not exposed, respectively to infective GWSS.
[y]Calculated from the means for 5 replications. Within columns (small letters) and across rows (capital letters), means followed by the same letter did not differ significantly ($P < 0.05$) according to Fisher's least significant difference test.
[z]Leaf samples from each plant were assayed for the presence X. fastidiosa using ELISA and IC-PCR.

Table 2. Effect of kaolin and imidacloprid treatment and length of glassy winged sharpshooter feeding time on Pierce's Disease incidence.

The incidence of PD did not differ among plants treated with imidacloprid and kaolin. Xf detection by ELISA assays was 8%, 11%, and 34% in kaolin-, imidacloprid-treated plants, and the untreated controls, respectively (data not shown). However, incidence of PD was greater when the GWSS were allowed to feed for 24 h versus 48 h. There was no difference in PD incidence on plants treated with imidacloprid and kaolin regardless of the length of the GWSS feeding time. In the untreated control plants, disease incidence was higher in plants exposed to infectious GWSS for 24 h than for 48 h (Table 2). The anti-feedant property of imidacloprid may be important in reducing the acquisition of PD bacterium, decreasing the transmission efficiency of infected GWSS, and subsequently, reducing the spread of PD bacterium (Tubajika et al., 2007). This finding is consistent with results obtained by Krewer et al., (2002), who showed that imidacloprid treatment slowed the development of PD in the field but was not effective in preventing the infection of PD in areas with prevalent sources of inoculum and high vector abundance.

3. Kaolin treatment as a barrier to GWSS movement into vineyard

Previous studies by Blua et al., (1999) and Perring et al., (2001) have shown that GWSS population utilize citrus as their primary reproductive host, and citrus is the predominant overwintering host, when grapes are dormant during winter. The application of insecticide to control insects can decrease spread and minimize yield loss in certain pathosystems (Purcell & Finley, 1979). Contact insecticides only offer short-term protection against GWSS infestations because of the continued immigration of sharpshooter adults from citrus which re-infest the grapes (Purcell & Finley, 1979).

The prevention of movement of GWSS from citrus to adjacent vineyards by kaolin particle film treatments as barriers was examined. The vineyard, which was comprised of an assortment of Thompson Seedless, Flames Seedless, and Chenin Blanc grape cultivars, was divided into six blocks 164.6 m wide by 365.7 m long (6.5 ha) and assigned treatments of kaolin or conventional insecticides (Puterka et al., 2003). Kaolin barrier treatments only extended 247.5 m into each block; the remaining 152.4 m was left untreated. The insecticide treatment block s had applications the entire 365.7 m distance of the block. Grapes received three bi-weekly applications of kaolin (11.36 kg kaolin, 378.5 per liter water) and six weekly applications of Dimethoate (Dimethoate 400, Platte Chemical Co., Greeley, CO, USA) applied at 4.67 l per ha, methomyl (Lannate LV, E. I. DuPont de Nemours & Co., Wilmington, DE, USA) applied at 2.33 l per ha, and Naled (Dibrom 8E, AMVAC Chemical Corp., Los Angeles, CA, USA) applied at 0.74 l per ha (Puterka et al., 2003). GWSS adults and nymphs were collected from the yellow sticky traps spaced every 30.5 m along the 365.7 m transects per block. Because of edge effect on movement of GWSS from citrus into grapes, plots were partitioned into four distances (0 to 4.3 m (interface); 4.4 to 123.7 m; 123.8 to 247.5 and 247.6 to 365.7 m) to better estimate the number of GWSS. GWSS adults and nymphs will be assessed by weekly monitoring of GWSS adults and nymphs caught per yellow sticky trap (Trece, Salinas, California placed 1 m high in the grape vines in vineyard bordering citrus (interface) and placed every 30.5 m into 365.7 m transects extending into each treatment block (transect) in Kern County, California. GWSS egg masses were sampled by inspecting 25 leaves per vine every 30.5 m along the sticky trap transects in each block.

3.1 GWSS adult counts from traps

Traps in kaolin-tread plants at the edge of vineyard (0-4.3 m) caught fewer GWSS than traps in the insecticide-treated plants based on one experiment from 9 March to 6 April, except on 16 March (Fig. 4a). After 6 April, there were no differences in Interface trap catches between treatments. Transect traps in kaolin barrier treatment resulted in fewer GWSS on 22 March and 6 April. Data from all other trap dates showed that treatment effects were not significantly different. Counts of GWSS adults were significantly different between sample dates, treatments, and blocks (Fig. 4ab).

Visual counts of in the grape-orange grove interface (0-4.3 meters) revealed higher number of GWSS adult in the insecticide treatment than in the kaolin treatment on week 1 (22 March) (kaolin = 0.06 ± 0.06, insecticides = 0.6 ± 0.6), week 2 (29 March) (kaolin = 0.14 ± 0.02, insecticide = 0.56 ± 0.10), and week 3 (6 April) (kaolin = 0.0 ± 0.0, insecticides = 0.9 ± 0.5). Overall, the GWSS adult counts were much lower in traps beyond the particle film barrier in grape (Fig. 4b). The treatment comparison of three applications of kaolin to six insecticide applications over a 2-month period showed that the kaolin barrier was equally effective or better than numerous insecticide applications in reducing GWSS movement into grapes (Puterka et al., 2003). This may appear to be contrary given the fact that the insecticide used was systemic; however, it appears that the kaolin was a sufficient barrier to prevent GWSS movement into the grape vineyards.

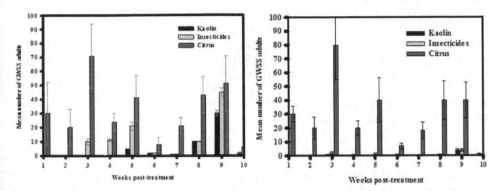

Fig. 4. Effects of kaolin and conventional insecticides on the mean number (±SE) of glassy-winged sharpshooter adults caught per yellow sticky trap (Trece, Salinas, California placed 1 m high in the grape vines in vineyard bordering citrus (interface) and placed every 30.5 m into 365.7 m transects extending into each treatment block (transect) in Kern County, California.

3.2 GWSS egg counts on grape

GWSS egg counts differed significantly (P = 0.0001) between the insecticide and barrier treatments in leaf samples taken by increasing distances from the grape-orange grove interface into the grape vineyard (P = 0.02). The insecticide treatment had significantly (P < 0.0001) larger numbers of GWSS eggs at the edge (0 meters) than at the distances away from the edge of grape leaves in vineyard bordering the citrus (Table 3). This suggest that oviposition on foliage can be affected by proximity to locales of insecticide application.

	GWSS egg counts (±SE)[y]	
Distance from edge of vineyard (m)[x]	Kaolin	Insecticides
0.0 - 4.3 (block 1)[y]	0.00 ± 0.02aA	14.1 ± 5.1aB
4.4 - 123.7 (block 2)	0.06 ± 0.02aA	5.3 ± 2bB
123.8 - 247.5 (block 3)	0.06 ± 0.02aA	1.6 ± 0.3cB
248.6 - 365.7 (block 4)	0.90 ± 0.03bA	1.9 ± 0.2cB

[x]Vineyard was partitioned to monitor the movement of glassy-winged sharpshooter into the grapes. Distance in meters from citrus.
[y]Values are means of 20 observations. Means followed by the same letter did not differ significantly (P < 0.05) according to Fisher's least significant difference test.
[z]Interface.

Table 3. Effects of kaolin barrier and conventional insecticides on the numbers ((±SE) of glassy-winged sharpshooter eggs sampled in Bakersfield vineyard bordering citrus from 0 – 365.7 meters from the edge. Yellow sticky traps were monitored from March 9 to May 9 on 25 grape leaves per vine.

In contrast, the oviposition averaged 0.0 to 0.9 GWSS eggs per block in the kaolin barrier treatment and did not differ (P = 0.50) from one another. Overall, the kaolin barrier resulted in undetectable levels of oviposition, whereas insecticides did not prevent oviposition (Puterka et al., 2003). The egg mass sampling has been shown as the best measure of how kaolin barrier treatments and insecticide treatments affected GWSS activity and host suitability (Puterka et al., 2003). This study suggests that unlike other vectors of Pierce's Disease, the GWSS disperse well into vineyard and is able to vector *Xf*, agent causal of PD in grape. This is consistent with the pattern of PD spread that we observed in Kern County, Bakersfield, CA (Tubajika et al., 2004).

4. Impact of imidacloprid on GWSS in citrus

The epidemic of PD in the vineyards of Temecula in Riverside County, California brought into focus the urgent need to control GWSS populations in CA (Castle et al., 2005) around vineyards. Prior to the first applications of imidacloprid which was made in the spring of 2000 for control of GWSS infestations in Temecula, CA; there had been very limited experience with imidacloprid in citrus or against GWSS in any crop (Catle et al., 2005)). Experiments with imidacloprid treatment were carried out at the University of California's Agricultural Operations at Riverside, CA during 2001 and 2002. The experiment was conducted in a block of 10 rows (6.4 m centers) which were split equally between 30 year-old orange trees (var Frost Valencia grafted on Troyer citrange) and lemons (var Lupe grafted on Cook) situated in the center of a 12-ha orchard. The GWSS nymphs and adults were collected in each bag and counted to determine the effect of imidacloprid on GWSS infestations between treated and untreated citrus trees. A sample consisted of four to six rapid thrusts at five locations around each tree. The contents of the collecting jar were then emptied into pre-labeled ziplock bags before moving on to the next tree

4.1 GWSS nymph counts in citrus

Based on the counts in the above experiments, sampling date proved to be a source of variation a two year study. In 2001, ifferences among the three sampling dates from April to June September. GWSS adults and nymphs were significantly ($P < 0.0001$) greater as observed among four dates in 2002 (P < 0.0001). Differences in GWSS counts between imidacloprid-treated and untreated trees were observed in 2001 (Fig. 5a) when populations were much larger than in 2002 (Fig. 5b). For the first four weeks following treatment, nymphal counts were high in both treated and untreated trees (Fig 4a). At week 6 post-treatment, a sharp decline in nymphal counts occurred in the imidacloprid-treated oranges coinciding with mean titers of imidacloprid surpassing 5 µg per liter (data not shown). Mean nymphal counts fell to 4.4 (±1.4) by week 8 compared with 30.49 (±4.4) in the untreated control.

The decline on number of nymphs at the beginning of week 9 may be due natural mortality, emigration and emergence to the adult stage, as observed in untreated control. However, mean number of nymphs remained between 30 and 40 through week 10 in the untreated trees while mean counts dropped below 2 at week 10 in the imidacloprid-treated orange trees (Fig 4a). Season-long differences between treated and untreated nymphal counts were significantly (P < 0.0001) high based on a repeated measures MANOVA. The decline in nymphal counts in 2001 corresponded nicely with the rise of imidacloprid titers in oranges, but this was not apparent in lemons or during the evaluation the following year when nymphal counts were so low (Catle et al., 2005). However the variability in the application through the irrigation system and/or the rate of uptake could also accounted for the significant variation in number of nymphs or adults among trees (Catle et al., 2005).

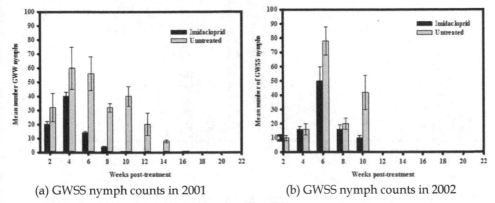

(a) GWSS nymph counts in 2001   (b) GWSS nymph counts in 2002

Fig. 5. Mean (±SE) number of (a) GWSS nymphs in 2001 and (b) in 2002 on oranges trees treated with imidacloprid compared with untreated trees. Sample size each week was n=12 trees using a bucker sampler thrusted a t five different locations per tree.

4.2 GWSS adult counts in citrus

GWSS Adults were not observed through the first eight weeks post treatment (Fig 4b). With maturation of the first nymphs and emergence to adults, numbers of adults rapidly increased in both treated and untreated oranges trees between weeks 9 and 12 (15 June–6 July). Contributing to the influx of young adults into the imidacloprid-treated oranges was the absence of any buffer zones between the treated and untreated trees (two rows of treated trees only). By week 14, however, a divergence in the mean number of adults caught in each treatment had begun, reaching its greatest difference in week 18 (17 August). Difference in adult densities was greater throughout week 25 (5 October) after which treated and untreated GWSS adult counts began to converge (Fig 4b). The mean number of GWSS adults between 15 June and 30 November 2001 was significant ($P < 0.0001$) between treated and untreated orange trees.

In contrast to the mean number of nymphs observed in 2001 in orange trees, the difference in number of GWSS nymphs in lemon trees in 2002 was inconsistent treated and untreated lemon trees (Fig 5a). Numbers of GWSS nymphs in untreated lemons were especially erratic, thus making it difficult to observe any clear treatment effect ($P = 0.40$). However, a very similar pattern to the orange trees was observed for GWSS adult counts in treated and untreated lemon trees (Fig 5b).

It is clear that protection by imidacloprid was not sporadic or spatially uneven, but rather was confronted by a phenomenon where mass emergence of adults coupled with heightened flight activity simply overwhelmed both treated and untreated trees in the orchard (Catle et al., 2005). After a few weeks, GWSS adult numbers began to decline and population remained consistently and significantly lower than the untreated orange and lemon trees. Similarly, a rapid increase in nymphal densities occurred in both treated and untreated orange trees in 2001, much as they were observed in Temecula in 2000 (Catle et al., 2005). The antifeedant effects of imidacloprid on other herbivores belonging to Hemiptera (Sternorrhyncha) are well established (Nauen et al., 1999).

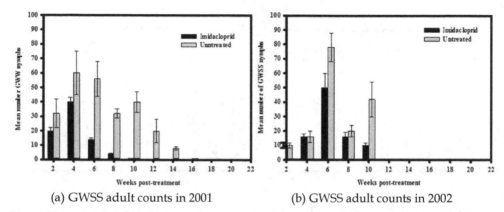

(a) GWSS adult counts in 2001   (b) GWSS adult counts in 2002

Fig. 6. Mean (±SE) number of (a) GWSS adults in 2001 and (b) in 2002 on oranges trees treated with imidacloprid compared with untreated trees. Sample size each week was n=12 trees using a bucker sampler thrusted a t five different locations per tree.

Results from these experiments showed that a recruitment of adults from surrounding orchards during week increased densities in imidacloprid and untreated controls but at levels that were significantly lower for imidacloprid treated–trees. Moreover, Variability in the application through the irrigation system and/or the rate of uptake could have accounted for the significant variation observed among trees. Also, substantial reductions in GWSS nymphs and adults in imidacloprid treated-trees observed during 2001 production year were sustained for 4-5 months post treatment. Overall, GWSS infestations were reduced in imidacloprid-treated trees than in untreated-trees. Data on the number of GWSS adult and nymph counts from grapevines and citrus studies confirmed that citrus is the primary and preferred host for GWSS as observed in Temecula valley and when given other choices such as grape, almond, cherry, stone fruit, the insect feeds on these hosts as Kern County provides a perfect variety of hosts (Tubajika et al., 2007; Blua et al., 1999; Purcell & Saunder, 1999; Raju et al., 1980).

5. Pierce's Disease incidence

Our previous study on the analysis of the spatial patterns of PD incidence in the lower San Joaquin Valley, CA indicated that GWSS may not be infective but their movements within the vineyard after arrival are important in the spread of the infection, which resulted in the symptoms we recorded (Tubajika et al., 2004).

Experimental plots consisting of Thompson Seedless, Flame Seedless and Chenin Blanc cultivars which were approximately 15-years-old were assessed for Pierce's disease (PD) incidence using a visual evaluation of disease symptoms such as stunted shoot growth, leaf scorch, and persistent petioles, a condition that occurs when the leaf blades scald and abscise, leaving only petioles attached to the shoot (Anon, 1992; Purcell, 1974). Plants were assessed at 30-d intervals. Ten leaf samples from each plant were bulked and assayed for the presence of *X. fastidiosa* using ELISA and IC-PCR and PD incidence was assessed as previously described.

When results were averaged across years, the incidence of PD was significantly (P < 0.05) lower (6%) in plots treated with kaolin than in plots treated with conventional insecticides (14%) (Tables 3, 4). There was no significant difference in the incidence of PD among grape cultivars in both years (P = 0.67). Also, chemical by cultivar interaction did not affect PD incidence in either year (Tables 3, 4).

Mean (± SE) percent PD incidence[xy]		
Cultivars	Kaolin	Conventional insecticides[z]
Thompson Seedless	8.4 ±1.1 aB	18.7 ± 2.4 aA
Flames Seedless	8.1 ± 1.1 aB	19.2 ± 2.1 aA
Chenin Blanc	7.8 ± 0.5 aB	17.5 ± 2.9 aA
Mean	8.1 ± 0.4	18. 5 ± 2.2

[x]Values are means of eight observations (Four replications x two years). Within rows (small letters) and across rows (capital letters), means followed by the same letter did not differ significantly (P< 0.05) according to Fisher's least significant difference test.
[y]Pathogen identity was confirmed by ELISA and IC-PCR assays. *Xylella fastidiosa* strain Temecula (ATCC 700964) collected from grape in Temecula, California was used as a reference control.
[z]dimethoate (Dimethoate 400) applied at 4.67 liter per ha, methomyl (Lannate LV) applied at 2.33 liter per ha and naled (Dibrom 8E) applied at 0.74 liter per ha.

Table 3. Effects of kaolin and conventional insecticides on Pierce's Disease incidence on Thompson Seedless, Flame Seedless, and Chenin Blanc cultivars during the 2001 production year at Bakersfield, CA.

Overall incidence of PD was 43% higher in 2001 than in 2002. All of the infected grape plants were removed from the plots. The inoculum sources for the 2002 season were subsequently reduced (Tubajika et al., 2007). In 2001, PD incidence was 18% on plants treated with insecticides and 4% in kaolin-treated plants (Table3). In 2002; the incidence of PD was 8% on plants treated with insecticides and 4% on kaolin-treated plants (Table 4).

Our previous study on the spatial patterns of incidence of PD in the lower San Joaquin Valley, CA indicates that GWSS may not be infective but their movements within the vineyard after arriving into the vineyard are important in the spread of the infection, and resulted in the symptoms we recorded (Tubajika et al., 2004). Field studies showed that plants treated with kaolin were less likely to become infected with *X. fastidiosa* and had a lower incidence of PD symptoms than untreated control plants (Tubajika et al., 2007). There are limited reports on application of particle film to control plant diseases caused by the vector population in the fields (Glenn et al., 1999; Blua et al., 1999; Puterka et al., 2000; Puterka et al., 2003). Also, data showed that the GWSS had a lower rate of survival following exposure to kaolin-treated plants for 48 h. This finding is similar to previous reports where kaolin completely protected plants from insect feeding (Blua et al., 1999; Puterka et al., 2000; Puterka et al., 2003). They suggested that the kaolin protects hosts against GWSS by camouflaging the plant with a white coating making them visually unperceivable, or by reflecting sunlight, which repels leafhoppers as well as aphids.

Mean (± SEM) percent PD incidence[xy]		
Cultivars	Kaolin	Conventional insecticides
Thompson Seedless	4.3 ±0.5 aB	8.8 ± 1.4 aA
Flames Seedless	4.2 ± 0.5 aB	8.1 ± 1.1 aA
Chenin Blanc	3.9 ± 0.2 aB	7.9 ± 0.9 aA
Mean	4.1 ± 0.3	8.3 ± 1.3

[x]Values are means of eight observations (Four replications x two years). Within rows (small letters) and across rows (capital letters), means followed by the same letter did not differ significantly ($P < 0.05$) according to Fisher's least significant difference test.
[y]Pathogen identity was confirmed by ELISA and IC-PCR assays. *Xylella fastidiosa* strain Temecula (ATCC 700964) collected from grape in Temecula, CA was used as a reference control culture.
[z]dimethoate (Dimethoate 400) applied at 4.67 liter per ha, methomyl (Lannate LV) applied at 2.33 liter per ha and naled (Dibrom 8E) applied at 0.74 liter per ha.

Table 4. Effects of kaolin and conventional insecticides on Pierce's Disease incidence on Thompson Seedless, Flame Seedless, and Chenin Blanc cultivars during the 2002 production year at Bakersfield, CA.

6. Conclusion

The GWSS was recently introduced into California and poses a serious threat to the grape industry because it is a very effective vector of the bacterium that causes Pierce's disease. This introduction of the GWSS has changed the epidemiology of *Xf* diseases affecting important agronomic and horticultural crops as well as landscape ornamental and native trees in CA by infesting these crops with the bacteria. The epidemic of PD in the vineyards brought into focus the urgent need to control GWSS populations, especially around vineyards. A new technology, called particle film and a systemic insecticide, imidacloprid were assessed both in greenhouse and field.

Results showed that kaolin protects plants from insect feeding, oviposition, and infestation by coating the plant surfaces with a protective mineral barrier. In caged field studies, we found that GWSS nymphs and adults were highly repelled by lemon trees treated with kaolin. In field studies that compared three biweekly kaolin treatments to six weekly contact insecticide treatments, kaolin performed as well as insecticides in reducing GWSS adult numbers and oviposition. A good coverage of plant leaves with kaolin is important when using this material for insect control. In greenhouse studies, GWSS adult counts were reduced greatly in kaolin-treated plants versus untreated control trees.

Based on data on GWSS nymph and adults counts in citrus, the persistence of imidacloprid in citrus varied as near-peak levels of imidacloprid were sustained for 6–10weeks before gradually declining as substantial reductions in GWSS nymphs and adults were observed in imidacloprid-treated trees during the 2001 trial and were sustained for 4–5months post-treatment. Imidacloprid effect on GWSS nymphs was not as well pronounced in the 2002 trial, when overall GWSS infestations were much reduced from the previous year. However, consistently lower adult infestations were observed in 2002 for imidacloprid compared with untreated trees.

The application of kaolin and imidaclorprid impacted GWSS mortality and PD incidence. Higher GWSS mortality rates were observed on kaolin-treated plants than on untreated-plants. The reduction in PD incidence observed in field studies suggest that either kaolin treatments as barriers to GWSS infestation or imidacloprid can be effective in reducing GWSS population and PD symptoms, and can be valuable tools for PD management. Both treatments strategies could also be used where insect resistance to imidacloprid would be a concern since insect resistance to particle films would not be a concern. Additionally, these treatments could be combined with other PD management approaches in integrated pest management.

The costs of chemicals and applications by either commercial applicators or growers were not determined in these studies. However, the choice of chemical to apply (alone or in combination) is important and the level of net returns will depend on this as well as other factors such as frequency of applications, disease intensity, and growth stage at which the chemical sprays are initiated. Additional work assessing these and other potential benefits are needed to fully determine the economic value of these treatments for grape production.

7. References

Almeida, R. P. P., & Purcell, A. H. (2003). Transmission of *Xylella fastidiosa* to grapevines by Homalodisca coagulata (Hemiptera cicadellidae). J. Econ. Entomol. 96, 264–271.

Anon. (1992). *Grape pest management, 2nd ed. Univ. of California, Div. Nat. Res., Okland, CA. Publ. No. 3343.*

Blua, M. J., Phillips, P. A., & Redak, R. A. (1999). A new sharpshooter threatens both crops and ornamentals. *Calif. Agric., 53*, 22-25.

Castle, S. J., F.J, B., Jian, B. L., & Toscan, N. C. (2005). Spatial and temporal distribution of imidacloprid and thiamethoxam in citrus and impact on *Homalodisca coagulata* populations. . *Pest Manag. Sci, 61*, 75-84.

Costa, H. S., Blua, M. S., Bethke, A., & Redak, R. A. (2000). Transmission of *Xylella fastidiosa* to oleander by the glassy-winged sharpshooter, *Homalodisca coagulata*. HortScience 35, 1265–1267.

Glenn, D. M., Puterka, F. G., van der Zwet, T., Byers, R. E., & Feldhake, C. (1999). Hydrophobic particle films: a new paradigm for suppression of arthropod pests and plant diseases. J. Econ. Entomol. 92, 759–771.

Glenn, D., and G. J. Puterka. (2005) Particle Films: A new technology for Agriculture. Pp. 1-44. In J. Janick [ed.] HortReviews. Vol.31. John Wiley & Sons, Inc., Hoboken, NJ.

Hemingway, J., & Ranson, H. (2005). Chemical Control of Vectors and Mechanisms of Resistance. In W. C. Marquardt (Ed.), *Biology of Disease Vectors* (second ed., pp. 627-637). San Francisco: Elsevier Academic Press.

Hill, B. L., & Purcell, A. H. (1995). Multiplication and movement of *Xylella fastidiosa* within grape and four other plants. Phytopathology 85, 1368–1372. .

Krewer, G., Dutcher, J. D., & Chang, C. J. (2002). Imidacloprid 2F insecticide slows development of Pierce's disease bunch grapes. J. Entomol. Sci. 37, 101–112. .

Minsavage, G. V., Thompson, C. M., Hopkins, D. L., Leite, B. C., & Stall, R. E. (1994). Development of a polymerase chain reaction protocol for detection of *Xylella fastidiosa* in plant tissue. Phytopathology 84, 456–461. .

Nauen, R., Koob, B., & Elbert, A. (1999). Antifeedant effects of sublethal dosages of imidacloprid on *Bemissia tabaci. Entomol. Exp. Appl., 88*, 287-293.

Perring, T. M., Farrar, C. A., & Blua, M. J. (2001). Proximity to citrus influences Pierce's disease in Temecula Valley vineyards. Calif. Agric. 55, 13–18.

Pooler, M. R., Myung, I. S., Bentz, J., Sherald, J., & Hartung, J. S. (1997). Detection of *Xylella fastidiosa* in potential insects by immunomagnetic separation and nested polymerase chain reaction. Lett. Appl. Micro¬biol. 25, 123–126.

Purcell, A. H. (1974). Spatial patterns of Pierce's disease in the Napa Valley. Am. J. Enol. Viticutt. 25, 162–167. .

Purcell, A. H., & Finley, A. H. (1979). Evidence for noncirculative transmis¬sion of Pierce's disease bacterium by sharpshooter vectors. Science 206, 839–841.

Purcell, A. H., Saunders, R. S., Hendson, M., Grebus, M. E., & Henry, M. J. (1999). Causal role of *Xylella fastidiosa* in Oleander leaf scorch disease. Phytopathology 89, 53–58.

Purcell, A. H., & Saunders, S. R. (1999). Glassy-winged sharpshooters expected to increase plant disease. Calif. Agric. 53, 26–27.

Puterka, G. J., Glenn, D. M., Sekutowski, D. G., Unruh, T. R., & Jones, S. K. (2000). Progress toward liquid formulations of particle films for insect and disease control in pear. Environ. Entomol. 29, 329–339.

Puterka, G. J., Reinke, M., Luvisi, D., Ciomperlik, M. A., Bartels, D., Wendel, L., et al., (2003). Particle film, Surround WP, effects on glassy-winged sharpshooter behavior and its utility as a barrier to sharpshooter infestations in grapes. Plant Health Prog.Online, doi: 10.1094/PHP-2003-0321-RS. .

Puterka, G. J., and D. M. Glenn. (2008) Kaolin-Based Particle Films for Arthropod Control. pp. 2075-2080, In (J. L. Capinera, Ed.) Encyclopedia of Entomology, Kluwer Academic Publishers.

Raju, B. C., Nome, S. F., Docampo, D. M., Gohen, A. C., Nyland, G., & Lowe, S. K. (1980). Alternative hosts of Pierce's disease of grapevines that occur adjacent to grape growing areas in California. Am. J. Enol. Viticult. 31, 144–148.

Romoser, W. S., & Stoffolano, G. J. (1998). *The Science of Entomology*. Boston, MA: WCB/McGraw-Hill.

Sisterson, M. S., Liu, Y. B., Kerns, D. L., & Tabashinik, B. A. (2003). Effects of kaolin particle film on oviposition, larval mining, and infestation of cotton by pink bollworm (Lepidoptyera: Gelechiidae). *J. Econ. Entomol., 96*(3), 805-810.

Sorensen, S. J., & Gill, R. J. (1996). A range extension of *Homalodosca coagulata* (Say) (Hemiptera: Clypeorrhyncha: Cicadellidae) to southern California *Pan-Pac Entomol., 72*, 160-161.

Tubajika, K. M., Civerolo, E. L., Ciomperlik, M. A., Hashim, J. M., & Luvisi, D. A. (2004). Analysis of the spatial patterns of Pierce's disease incidence in the lower San Joaquin Valley in California. Phytopathology 94, 1136–1144

Tubajika, K. M., Civerolo, E. L., Puterka, G. J., Hashim, J. M., & Luvisi, D. A. (2007). The effects of kaolin, harpin, and imidacloprid on development of Pierce's disease in grape *Crop Protection, 26*, 92-99.

Turner, W. F., & Pollard, H. N. (1959). Life histories and behavior of five insect vectors of phony peach disease. USDA Techn. Bull.1188.

Unruh, T. R., Knight, A. L., Upton, J., Glenn, D. M., & Puturka, G. J. (2000). Particle-film for suppression of the codling moth in apple and pear orchards. *J. Econ.Entomol., 93*, 737-743.

8

The Conundrum of Chemical Boll Weevil Control in Subtropical Regions

Allan T. Showler
USDA-ARS, Kika de la Garza Subtropical
Agricultural Research Center, Weslaco, Texas
U.S.A.

1. Introduction

Originally a Mesoamerican insect, the boll weevil, *Anthonomus grandis grandis* Boheman (Coleoptera: Curculionidae), has spread from the tropics, where it evolved on cotton, *Gossypium hirsutum* L., and other malvaceous plant species (Burke et al., 1986; Brubaker & Wendel, 1994; Showler, 2009b), north to temperate cotton producing areas of the United States and south to northern provinces of Argentina (Cuadrado, 2002; Showler, 2009b). The pest was first detected in United States cotton in 1892 (Parencia, 1978) and infested the Cotton Belt such that by 1917, every cotton-producing county in Georgia, for example, was infested (Hunter, 1917). Adults oviposit inside cotton buds or "squares" (usually one egg per square), and the hatched larva causes the square to abscise before it can flower (Showler and Cantú, 2005). If an egg is deposited within a young boll (older bolls become too hard to penetrate), or if mouthparts penetrate the rind of squares to the inner reproductive portion, fiber-producing locks can be injured or completely destroyed, but not necessarily all four locks (Showler, 2006a; Showler & Cantú, 2008). Boll weevil losses have been valued at $83.34 billion and insecticide-based control costs at $18.67 billion between 1893 and 1999, and infestations became so injurious that cotton-free winter periods were instituted by law in some areas (Haney, 2001). Later, insecticide-based eradication programs were launched in the United States and in Argentina (Dickerson & Haney, 2001; Haney et al., 2001; Johnson & Martin, 2001; Texas Department of Agriculture, 2002; Carmona et al., 2003).

Natural enemies indigenous to the United States are not considered to be important as mortality factors against boll weevils (Jones & Sterling, 1979; Showler & Greenberg, 2003), although the imported fire ant, *Solenopsis invicta* Buren, native to South America (Buren et al., 1974; Lofgren, 1986), can account for up to 58% of boll weevil mortality in relatively wet regions where the predator thrives (Sturm & Sterling, 1990). In one study imported fire ant predation on immature boll weevils averaged 84% compared with 0.14% and 6.9% mortality caused by parasitism and desiccation, respectively (Fillman & Sterling, 1983). But in drier cotton growing areas, lack of sufficient predation to help govern populations in some new habitats outside Mesoamerica (Showler, 2007) permitted rapid dispersal (Burke et al., 1986; Showler, 2009a). While certain cultural practices, such as early planting (Showler et al. 2005) can help avoid large populations that typically accumulate in the summer (Showler, 2003, 2005), chemical intervention against building infestations has been the chief control tactic.

In subtropical south Texas, before the boll weevil eradication program was re-instated in the fall of 2005 (after a halted attempt in 1995), crop protection against boll weevils was approached using three tactics: mandatory cotton stalk destruction before 1 September, prohibition on planting until 1 February and elimination of volunteer cotton during the cotton-free winter period (Texas Department of Agriculture, 2002). Insecticides were the only in-season control approach (Showler, 2007). Some growers sprayed 2–3 "pre-emptive" treatments starting at the "pinhead" square size (1–2-mm diameter) (Heilman et al. 1979) followed by insecticide applications (often azinphosmethyl, cyfluthrin, bifenthrin, or oxamyl) whenever 10% of randomly selected medium (3–5.4-mm diameter) or large (5.5–8-mm diameter) squares (Showler, 2005; Showler et al., 2005) had oviposition punctures (Showler et al., 2005). While Heilman et al. (1979) suggested that pre-emptive spraying delays insecticide applications later into the season, other research found no beneficial effect (Showler, 2004a) and the economic value of the practice is debatable. Pre-emptive sprays might kill some adult boll weevils that have entered the field after overwintering elsewhere, but the sparse numbers of weevils at that time and the presence of less-preferred and nutritionally inferior small squares contribute relatively little to field-level population buildups, and injury to such small squares has negligible impact on lint yield (Showler, 2004b; Showler et al., 2005). Late-season spraying for immediate crop protection (not eradication) purposes is also of questionable utility because, although feeding and oviposition punctures can be abundant on bolls, older bolls ($\geq$14 days old) are less vulnerable to attack (because they harden) than younger ($\leq$10 days old) bolls and bolls do not abscise in response to boll weevil oviposition (Showler, 2006a). When injury to a boll does occur, usually because of prior adult feeding during the square stage or larval infestation of the boll, damage is often limited to individual lint-bearing locks, of which there are four (Showler 2006a). Insecticides applied in the context of crop protection after cut-out (Guinn, 1986; Cothren, 1999), when bolls predominate, generally fail to measurably suppress boll weevil infestations (Showler & Robinson, 2005; Showler, 2008a).

When cotton is forming medium and large squares, which are most vulnerable to and useful for boll weevil reproduction (Showler, 2004b), the 10% spray intervention threshold is compromised by variability in total numbers of squares over time (Showler, 2007). Declining abundance of squares coupled with surges in boll weevil populations contribute toward the likelihood of triggering interventions based on randomly sampled squares; hence, spraying later protects fewer and fewer squares (Showler, 2007). A better estimate of infestation would involve comparing numbers of oviposition-punctured squares to total squares within, for example, three-meter (or some other length) sections of rows (Showler, 2007). In a study in south Texas, the standard approach, including three pre-emptive sprays, involved nine applications that failed to increase yield and economic return (Showler & Robinson, 2005). Once 10% of the squares harbor a boll weevil egg, protecting it from contact insecticides (Showler & Scott, 2004), it is too late to expect good control.

In temperate areas of the United States, the boll weevil eradication program has had remarkable successes since its beginning in North Carolina and Virginia in 1978 to Georgia to California (Dickerson et al., 2001) and the pest has been eradicated from northern and central regions of Texas as well (USDA-APHIS, 2007; Texas Boll Weevil Eradication Foundation, 2011). The boll weevil is "functionally" eradicated in other areas of Texas (USDA-APHIS, 2007), whereby <0.001 weevils/trap/week were found during the most recently completed growing season, indicating that boll weevils are not reproducing or causing economic damage in an area [*e.g.*, >1.5 million ha of cotton in 2010 (Texas Boll

Weevil Eradication Foundation, 2011)]. Eradication involves three techniques over a 4–5-year period: pheromone trap detection, reduction of boll weevil food supply, and malathion treatments (USDA-APHIS, 2007). The process starts with a series (often seven) of fall malathion applications that were once [and probably erroneously (Showler, 2010)] termed "diapause sprays" (McKibben et al., 2001). The aim is to reduce late season populations that will be further, and dramatically, reduced during cold temperate winters when food is unavailable and temperatures can be lethal (Showler, 2009b,c, 2010). Pheromone traps are then deployed around all cotton fields during the following spring planting and spraying of each field begins 5–6 weeks later based on trap captures; the process is repeated annually until the boll weevil is no longer found (USDA-APHIS, 2007). Still, there are ≈262,000 ha of cotton in east and subtropical south Texas where eradication has not been achieved (Texas Boll Weevil Eradication Foundation, 2011). While the chance of boll weevil dispersal from infested areas on wind and in hurricane systems exists (Texas Boll Weevil Eradication Foundation, 2011), particularly from nearby Mexico where eradication efforts are not underway, there are other overriding factors contributing to the pest's persistence in the subtropics (Showler, 2007, 2009b). Misconceptions relating to boll weevil ecology and biology (Showler, 2009c), while immaterial in cold temperate areas, appear to underlie the challenges to eradication efforts under subtropical conditions (Showler, 2009b, 2010).

2. Misunderstandings

There have been misconceptions pertaining to fundamental aspects of boll weevil survival outside its native Mesoamerican region that involve dietary habits, overwintering, and diapause, all of which interrelate (Showler, 2007, 2009b,c, 2010), presenting obstacles to temperate eradication approaches when used in the subtropics. Ultimately, the problem resides in numbers of boll weevils (including offspring from overwintering weevils) that can survive cotton-free winters to feed and reproduce in large cotton plantings of the following season. In south Texas, large end-of-season populations can be observed by trapping at the edges of cotton fields that are disrupted by defoliant application, consequent host plant desiccation, harvest, and stalk shredding (Showler, 2003). Those populations move into surrounding habitats where, under temperate winter conditions, the boll weevils that survived the first-year series of late-season eradication program sprays must survive frequently severe and extended cold conditions for which the tropical insect had not evolved, as well as starvation due to lack of viable winter plant hosts (Showler, 2009b,c). Boll weevils have long been assumed to feed solely on pollen of certain malvaceous plants (Burke & Earle, 1965; Cate & Skinner, 1978), and later, pollens of other plants were recognized (Jones et al., 1992, 1993; Hardee et al., 1999), but recent research has revealed that adult boll weevils can consume cotton leaves and bracts, citrus and cactus fruit, and likely nectar (Showler & Abrigo, 2007; Showler, 2009b). In the subtropics, adult boll weevils can survive and reproduce during the winter on small patches of volunteer cotton that, despite surveillance, are overlooked, and adults can be trapped in substantial numbers around grapefruit, *Citrus paradisi* Macfad., and orange, *C. sinensis* (L.) Osbeck., orchards (Showler, 2006b). The edible endocarps of grapefruits and oranges of those citrus species can sustain up to 25% of adult boll weevils in nonreproductive condition for longer than five months (completing the cotton-free period); the maximum longevity (246 days) was only seven days less than boll weevils fed large cotton squares (Showler & Abrigo, 2007). The fruit of prickly pear cacti, *Opuntia* spp. [114 species in Mexico alone (Vigueras & Portillo, 2001)], which is widespread and abundant in south Texas, can support 10% of adult boll weevils over the winter period, and there are likely other as yet unreported food sources (Showler, 2009b). Hence, in subtropical areas of North and South America where cotton is grown in proximity with citrus, persistent boll weevil populations have been reported even after cotton growing was eliminated or where eradication programs have begun (Cuadrado, 2002; Carmona et al., 2003; Mas et al., 2007; Texas Boll Weevil Eradication Foundation, 2011). Despite the availability of *Opuntia* spp. and other host plants in Mexico and south Texas (Gaines, 1935; Lukefahr, 1956; Lukefahr & Martin, 1962; Stoner, 1968; Cross et al., 1975; Vigueras & Portillo, 2001), cotton in the Lower Rio Grande Valley remained free of boll weevils for 30 years of commercial production beginning ≈1860 (Garza & Long, 2001) even though cotton around Monclova, Coahuila, Mexico, ≈45 minutes latitude north and 220 km west of the Lower Rio Grande Valley, was so heavily infested that the crop was abandoned in 1862 (Howard, 1897). Boll weevil food sources under orchard conditions are concentrated and support substantial active populations through winter (Showler, 2006b) because endocarps are accessible through cracks, holes, or lesions while the fruit is attached to the plant or fallen (Showler, 2007; Showler & Abrigo, 2007). Establishment of boll weevils in Lower Rio Grande Valley cotton during the early 1890s (Parencia, 1978; Haney, 2001) may have been connected to a simultaneous citrus industry boom (Waibel, 1953). The author has witnessed, in mid January, large flying populations of boll weevils in and around nonsanitized (fallen fruit on the orchard floor not removed) orange and grapefruit orchards in south Texas that were so abundant that they were a nuisance. Boll weevils are also known to reproduce in volunteer cotton during Lower Rio Grande Valley winters (Summy et al., 1988), which also contradicts widely accepted, but apparently erroneous, dogma regarding the existence of winter diapause (Showler, 2009c, 2010).

For more than 50 years, boll weevils have been assumed to enter a state of winter diapause (Brazzel & Newsom, 1959), but diapause-induction studies involved weak experimental methods and dubious interpretations of results, and recent research in the subtropics indicates that boll weevils, being of tropical origin, did not evolve a diapause mechanism for surviving temperate winters (Showler, 2007, 2009c, 2010). Sterling and Adkisson's (1966) finding that boll weevils in the Texas High Plains "diapause" earlier and in greater percentages than in Central Texas (at a lower latitude) implies that boll weevil dormancy is not seasonal (a criterion for diapause), but it is instead responsive to dormancy-triggering conditions whenever they occur (Showler, 2010). Brazzel and Newsom (1959), however, claimed that, in the instance of boll weevils, diapause could be a "facultative" response to harsh, unfamiliar, conditions such as cold temperate winters. It is more likely, however, that the response is merely a metabolic and locomotory slowing caused by declining temperature (Fye et al., 1969; Jones & Sterling, 1979; Watson et al., 1986), giving the appearance of being facultative. As winter temperatures cool, a threshold for quiescence (Koštál, 2006; Guerra et al., 1984) or some other nondiapause expression of dormancy is reached first, followed later, if temperatures become sufficiently cold, by mortality (Showler, 2010). Whatever words are employed to describe the insect's response to temperate winters, eradication strategy involving "diapause spraying" has been effective where temperate winter attrition is substantial even if "diapause" might not be the technically correct term (Showler, 2010).

The inescapable point is that under subtropical conditions, particularly in the presence of relatively large plantings of citrus throughout the agricultural landscape, boll weevil mortality is not as great as in cold-winter temperate areas because winters are generally warm and can support populations with food until the spring (Showler, 2009b). In February, when cotton can be planted, boll weevil numbers near south Texas citrus orchards were found to be substantial (Showler, 2006b). Loss of major winter attrition as an eradication tool will likely require adjustments to the customary approach. Chance movement of boll weevils on wind or farm vehicles into active eradication program areas might cause setbacks to eradication, but the ecological reasons for the boll weevil's persistence in subtropical areas presents broader and more difficult challenges.

3. Chemical tactics: no easy answers

3.1 Insensitive trigger

The spray regimen for cotton crop protection against boll weevils and the reasons it was sometimes not sufficient across all growing areas have been discussed, but aspects of eradication involving insecticide application are also weakened in the subtropical context. Monitoring in-season boll weevil populations, for example, is important for determining whether to intervene and to assess efficacy. It is surprising that boll weevil surveillance fails to account for in-season changes in adult boll weevil response to grandlure largely predicated by cotton plant phenology and associated volatiles. One change occurs as cotton begins to square; then, even while boll weevils are accumulating in cotton fields, few are collected in the traps (Parajulee et al., 2001), presumably a result of competing plant volatiles from large fields of cotton versus a point pheromone source. Further, the trap's physical design presents a series of obstacles that boll weevils must negotiate before finding their way into a plastic cap on top where the weevils are counted (Showler, 2007). At low ambient populations in south Texas, differences in numbers of boll weevils captured in the conventional trap versus a sticky board trap were not detected, both traps using the same pheromone lure, but at higher populations sticky board traps collected ≥9-fold more weevils than the conventional trap, and 30% of the conventional traps collected no boll weevils when corresponding sticky boards accumulated from 82 to 511 weevils at the same locations and time; on one occasion, the conventional trap had two boll weevils compared with 2,228 on a sticky board (Showler, 2003). This is not to suggest that sticky board traps should replace the conventional trap unless their deployment can be made less labor intensive, but a more sensitive trap design would refine spray timing for greater effect as a result of more accurate population detection.

3.2 Spray timing

Because the boll weevil's life cycle includes ≈18 days in immature life stages protected within squares (Showler & Cantú, 2005), commonly-used insecticides with relatively short residual effects (≤4 days) can miss that cohort (Showler & Scott, 2004). To ensure lethal exposure to a larger proportion of the population, such insecticides would have to be sprayed *at least* once every four days. Yield increases in experimental plots were reported where some were sprayed "proactively" every 7–8 days starting when ≈2% of randomly selected squares were large (Showler & Robinson, 2005). It is unlikely, however, that the proactive spray regime would be as effective in larger commercial fields on an area-wide scale; in the study, applications were meticulous and tractor-mounted drop nozzles provided complete coverage even when the plants were high. For large boll weevil populations likes those encountered in the Lower Rio Grande Valley (Showler 2003), insecticides would have to be applied every three or four days from the time medium-sized squares (3–5.4-mm- diameter) first develop (before 2%) until cut-out when square production declines rapidly (Guinn, 1986; Cothren, 1999).

Under subtropical field conditions, feeding on pinhead- and match-head-sized squares is negligible, and large squares are preferred to medium-sized squares regardless of planting date (Showler, 2005). Boll weevil feeding punctures on large squares were 7.8- and 25-fold more abundant compared with match-head squares and bolls, respectively (Showler, 2004b). In terms of nutritional value, medium and large squares promote greater egg production and longevity of adult boll weevils than any other stage of cotton fruiting body (Showler, 2008b), and in terms of providing enough food and space for the immature stages of the boll weevil to develop, pinhead and match-head squares are generally too small (Showler, 2004b). Hence, spraying insecticides well before medium and large square sizes are available is of little value to crop protection and for impeding boll weevil reproduction, which agrees with the recommendation by Norman and Sparks (1998) for beginning boll weevil control in the Lower Rio Grande Valley when one-third-grown squares appear. Once large squares blossom and form post-bloom, young, and hardened older bolls, the nutritional value for longevity and egg production declines to nil when the rind can no longer be penetrated (Showler, 2004b). This explains why adult boll weevil populations plateau following cut-out through harvest (Showler et al., 2005). While spraying during the late season, particularly the series of late season eradication program sprays that occur in the first year (USDA-APHIS, 2007), can likely reduce boll weevil numbers, warm winters with plentiful food can ensure the survival of many until after spring cotton planting.

Scott et al. (1998) reported that, in the Lower Rio Grande Valley, early- and medium-maturing cotton varieties produce the best yields. In a similar vein, square production in early-planted cotton is lower than in later plantings and avoids the high numbers of weevils occurring in later-planted cotton (Showler et al., 2005). Although late-planted cotton produces more squares than early-planted cotton, this advantage is off-set by losses from heavy boll weevil infestations (Showler et al., 2005). The best time for planting was found to be intermediate between early and late for an optimal balance between increasing square production while avoiding the greatest accumulations of boll weevils, thereby reducing insecticide applications as well (Showler et al., 2005).

Harvest timing can also influence insecticide use. From a crop protection perspective, although harvesting late (at 75% boll splitting) rather than earlier (at 40% boll split) can require an extra insecticide application where using the proactive approach, particularly when boll weevil populations were relatively large, but harvesting late captures greater quantities of lint when more bolls have matured, resulting in better economic return, even if the late season insecticide treatment is superfluous (Showler & Robinson, 2008). Mixing the defoliant with an insecticide was found to be relatively ineffective and unreliable (Showler, 2008a).

3.3 Resistance

Boll weevil tolerance to organophosphorus, carbamate, and pyrethroid insecticides was reported by Kanga et al. (1995), but analyses of field populations have not detected resistance to malathion. It is conceivable, however, that under continual insecticide pressure from malathion only, resistance might develop (Bottrell et al., 1973), and because the boll weevil eradication program relies exclusively upon malathion, exposed boll weevil populations should be assessed intermittently for signs of resistance. For the time being, malathion remains toxic to boll weevils, even at reduced rates (Showler et al., 2002).

4. Possibilities

There are a number of ways in which chemical boll weevil control might be improved. First, a more sensitive trap would permit increasingly timely responses to the early in-season presence of adult boll weevils (but not while squares are still match-head size). At that point, spraying should provide continuous protection of vulnerable and nutritious medium- and large-sized squares. Even if sprays occur weekly, achieving acceptable control on an area-wide scale is improbable, which suggests that using a more sensitive trap design could result in more appropriately-timed, and likely increased, spray applications for the subtropics (unless spraying is conducted at ≤4-day intervals between late match-head to cut-out stages) where overwintering populations are relatively large (Showler, 2007). Both crop protection strategies and the eradication approach should evolve to incorporate emerging information on boll weevil ecology to find tactics that can help mitigate population buildups, such as avoiding late planting, use of earlier-maturing varieties, and development of longer-residual insecticides to reduce numbers of applications and to enhance protection of squares. Because subtropical boll weevil populations are active during winter and can sustain themselves on citrus, removal of such plentiful food through post-harvest orchard sanitation would augment the ban on cotton. Another overlooked tactic is plant resistance. While cotton has been bred for a variety of traits, no cultivars have been developed to resist boll weevil attack. Efforts in this direction might include altering square rind thickness or consistency to make the inner portion, where the immature weevil stages develop, less accessible, or changing the availability of certain nutrients that can affect egg production (Showler, 2009a). Eradicating a tropical pest like the boll weevil in temperate areas was achievable, but the subtropics are more akin to the insect's native habitat in terms of temperature and host plants. For this reason, adjustments to the temperate eradication strategy might have to involve tailoring insecticides, application timing, and the circumstances under which they are applied (e.g., as influenced by planting dates and phenological stages of the crop) for extending prophylactic crop protection and decimating boll weevil populations as selectively as possible to avoid the possibility of for secondary pest outbreaks.

However, even were all of the issues surrounding subtropical boll weevil eradication to be resolved, the feasibility of remaining boll weevil-free in areas along international borders is compromised by boll weevil populations breeding on the other side of the border where attention to eradication, for a complex of reasons, may not be in synchrony. Hence, the success of eradication in subtropical border areas depends to a great extent on the coordinated efforts of both countries. In the instance of a somewhat analogous pest, the desert locust, *Schistocerca gregaria* (Forskål), which can move long distances as massive swarms in Africa and the Middle East, breeding in one country can put crops in neighboring countries at risk, resulting in perpetually reactive and increasingly insecticide-based, rather than preventive maintenance strategies (Steedman, 1988; Showler & Potter 1991; Showler, 1995).

5. References

Bottrell, D.G.; Wade, L.J. & Bruce, D.L. (1973). Boll weevils fail to develop resistance to malathion after several years of heavy exposure in Texas High Plains. *Journal of Economic Entomology* 66: 791-792, ISSN 0022-0493.

Brazzel, J.R. & Newsom, L.D. (1959). Diapause in *Anthonomus grandis* Boh. *Journal of Economic Entomology* 52: 603-611, ISSN 0022-0493.

Brubaker, C.L. & Wendel, J.F. (1994). Reevaluating the origin of domesticated cotton (*Gossypium hirsutum*; Malvaceae) using nuclear restriction fragment length polymorphisms (RFLPs). *American Journal of Botany* 81: 1309-1326, ISSN 00029122.

Buren, W.F.; Allen, G.E.; Whitcomb, W.H.; Lennartz, F.E. & Williams, R.N. (1974). Zoogeography of the imported fire ant. *Journal of the New York Entomological Society* 82: 113-124, ISSN 0028-7199.

Burke, H. R.; Clark, W.E.; Cate, J.R. & Fryxell, P.A. (1986). Origin and dispersal of the boll weevil. *Bulletin of the Entomological Society of America* 32: 228-238, ISSN 1046-2821.

Burks, M.L. & Earle, N.W. (1965). Amino acid composition of upland cotton squares and Arizona wild cotton bolls. *Journal of Agricultural and Food Chemistry* 13: 40–43, ISSN 0021-8561.

Carmona, D.; Huarte, M.; Arias, G.; López, A.; Vincini, A.M.; Castillo, H.A.; Manetti, P.; Capezio, S.; Chávez, E.; Torres, M.; Eyherabide, J.; Mantecón, J.; Cichón, L. & Fernández, D. (2003). Integrated pest management in Argentina. In: *Integrated Pest Management in the Global Arena*, K. M. Maredia, D. Dakouo & D. Mota-Sanchez, (eds.), pp. 303-326, CABI Publishing, ISBN 0-85199-652-3, Wallingford, England.

Cate, J.R. & Skinner, J.L. (1978). The fate and identification of pollen in the alimentary canal of the boll weevil. *Southwestern Entomologist* 3: 263-265, ISSN 0147-1724.

Cothren, J.T. (1999). Physiology of the cotton plant. In: *Cotton: Origin, History, Technology, and Production*, C.W. Smith (ed.), pp. 207-268. John Wiley and Sons, ISBN 9780471180456, New York, New York, USA.

Cross, W.H.; Lukefahr, M.J.; Fryxell, P.A. & Burke, H.R. (1975). Host plants of the boll weevil. *Environmental Entomology* 4: 19-26, ISSN 0046-225X.

Cuadrado, G.A. (2002) *Anthonomus grandis* Boheman (Coleoptera: Curculionidae) in central and southwest area of Misiones, Argentina: pollen as feeding source and their relationship with the physiological state in adult insects. *Neotropical Entomology* 31: 121-132, ISSN 1519-556X.

Dickerson, W.A. & Haney, P.B. (2001). A review and discussion of regulatory issues. In: *Boll Weevil Eradication in the United States Through 1999*, W.A. Dickerson, A L. Brashear, J T. Brumley, F.L. Carter, W.J. Grefenstette, & F.A. Harris (eds), pp. 137-155. The Cotton Foundation, ISBN 0939809060, Memphis, Tennessee, USA.

Dickerson, W.A.; Brashear, A.L.; Brumley, J.T.; Carter, F.L.; Grefenstette, W.J. & Harris F.A. (2001). *Boll Weevil Eradication in the United States Through 1999*. The Cotton Foundation, ISBN 0939809060, Memphis, Tennessee, USA.

Fillman, D.A. & Sterling, W.L. (1983) Killing power of the red imported fire and [Hym.: Formicidae]: a key predator of the boll weevil [Col.: Curculionidae]. *Entomophaga* 28: 339-344, ISSN 0013-8959.

Fye, R.E.; McMillian, W.W.; Hopkins, A.R. & Walker, R.L. (1959). Longevity of overwintered and first generation boll weevils at Florence, S.C. *Journal of Economic Entomology* 52: 453-454, ISSN 0022-0493.

Gaines, R.C. (1934). The development of the boll weevil on plants other than cotton. *Journal of Economic Entomology* 27: 745-748, ISSN 0022-0493.

Garza, A.A. & Long, C. (2001). *The Handbook of Texas Online*. ww.tsha.utexas.ede/handbook/online/ articles/CC/hcc4.html.

Guerra, A.A.; Garcia, R.F.; Bodegas, P.R. & De Coss, M.E. (1984). The quiescent physiological status of boll weevils (Coleoptera: Curculionidae) during the noncotton season in the tropical zone of Soconusco in Chiapas, Mexico. *Journal of Economic Entomology* 77: 595-598, ISSN 0022-0493.

Guinn, G. (1986). Hormonal relations during reproduction. In: *Cotton Physiology*, J.R. Mauney & J.M. Stewart (eds.), pp. 113-136. The Cotton Foundation, ISBN 0939809-01-X, Memphis, Tennessee, USA.

Haney, P.B. (2001). The cotton boll weevil in the United States: impact on cotton production and the people of the Cotton Belt. In: *Boll Weevil Eradication in the United States Through 1999*, W.A. Dickerson, A.L. Brashear, J.T. Brumley, F.L. Carter, W.J. Grefenstette, and F.A. Harris (eds). The Cotton Foundation, ISBN 0939809060, Memphis, Tennessee, USA.

Haney, P.B.; Herzog, G. & Roberts, P.M. (2001). Boll weevil eradication in Georgia. In: *Boll Weevil Eradication in the United States Through 1999*, W.A. Dickerson, A.L. Brashear, J.T. Brumley, F.L. Carter, W.J. Grefenstette, and F.A. Harris (eds), pp. 259-289. The Cotton Foundation, ISBN 0939809060, Memphis, Tennessee, USA.

Hardee, D.D.; Jones, G.D. & Adams, L.C. (1999). Emergence, movement, and host plants of boll weevils (Coleoptera: Curculionidae) in the Delta of Mississippi. *Journal of Economic Entomology* 92: 130-139, ISSN 0022-0493.

Heilman, M.D.; Namken, L.N.; Norman, J.W. & Lukefahr, M.J. (1979). Evaluation of an integrated production system for cotton. *Journal of Economic Entomology* 72: 896-900, ISSN 0022-0493.

Howard, L.O. (1897). Insects affecting the cotton plant. *USDA Farmers' Bulletin* No. 47: 1-31.

Hunter, W.D. (1917). The boll weevil problem with special reference to means of reducing damage. *USDA Farmers' Bulletin* No. 848: 1-40.

Johnson, D.R. & Martin, G. (2001). Boll weevil eradication in Arkansas. In: *Boll Weevil Eradication in the United States Through 1999*, W.A. Dickerson, A.L. Brashear, J.T. Brumley, F.L. Carter, W.J. Grefenstette, and F.A. Harris (eds), pp. 225-233. The Cotton Foundation, ISBN 0939809060, Memphis, Tennessee, USA.

Jones, D. & Sterling, W.L. (1979). Manipulation of red imported fire ants in a trap crop for boll weevil suppression. *Environmental Entomology* 8: 1073-1077, ISSN 0046-225X.

Jones, R.W.; Cate, J.R.; Hernandez, S.M. & Navarro, R.T. (1992). Hosts and seasonal activity of the boll weevil (Coleoptera: Curculionidae) in tropical and subtropical habitats of northeastern Mexico. *Journal of Economic Entomology* 85: 74-82, ISSN 0022-0493.

Jones, R.W.; Cate, J.R.; Hernandez, E.M. & Sosa, E.S. (1993). Pollen feeding and survival of the boll weevil (Coleoptera: Curculionidae) on selected plant species in northeastern Mexico. *Environmental Entomology* 22: 99-108, ISSN 0046-225X.

Kanga, L.H.B.; Plapp, F.W., Jr.; Wall, M.L.; Karner, M.A.; Huffman, R.L.; Fuchs, T.W.; Elzen, G.W. & Martinez-Carrillo, J.L. (1995). Monitoring tolerance to insecticides in boll weevil populations (Coleoptera: Curculionidae) from Texas, Arkansas, Oklahoma, Mississippi, and Mexico. *Journal of Economic Entomology* 88: 198-204, ISSN 0022-0493.

Koštál, V. (2006). Eco-physiological phases of insect diapause. *Journal of Insect Physiology* 52: 113-127.

Lofgren, C. S. (1986). History of imported fire ants in the United States. In: *Fire Ants and Leaf-cutting Ants: Biology and Management*, C.S. Lofgren & R.K. Vander Meer (eds.), pp. 36-47. Westview Press, ISBN 9780813370712, Boulder, Colorado, USA.

Lukefahr, M.J. (1956). A new host of the boll weevil. *Journal of Economic Entomology* 49: 877-878, ISSN 0022-0493.

Lukefahr, M.J. & Martin, D.F. (1962). A native host plant of the boll weevil and other cotton insects. *Journal of Economic Entomology* 55: 150-151, ISSN 0022-0493.

Mas, G.E.; Grilli, M.P. & Ravelo, A. (2007). Esudio de la dinámica publacional del picudo del algodonero (Anthonomus grandis Boheman) mediante un sistema de información. http://www.geogra. uah.es/inicio/web_11_confibsig/PONENCIAS/2-035-Mas-Ravelo-Grilli.pdf

McKibben, G.H.; Villavaso, E.J.; McGovern, W.L. & Grefenstette, B. (2001). United States Department of Agriculture – research support, methods development and program implementation. In: *Boll Weevil Eradication in the United States Through 1999*, W.A. Dickerson, A L. Brashear, J T. Brumley, F.L. Carter, W.J. Grefenstette, & F.A. Harris (eds), pp. 101-135. The Cotton Foundation, ISBN 0939809060, Memphis, Tennessee, USA.

Norman, J.W., Jr. & Sparks, A.N., Jr. (1998). Managing cotton insects in the Lower Rio Grande Valley. *Texas Agricultural Extension Service Bull. 1210*, Texas A&M University, College Station, TX.

Parajulee, M.N.; Slosser, J.E.; Carroll, S.C. & Rummel, D.R. (2001). A model for predicting boll weevil (Coleoptera: Curculionidae) overwintering survivorship. *Environmental Entomology* 30: 550-555, ISSN 0046-225X.

Parencia, C.R., Jr. (1978). One hundred twenty years of research on cotton insect in the United States. *USDA Agricultural Handbook* No. 515: 1-17: 62-68.

Scott, A.W. Jr.; Lukefahr, M.J.; Cook, C.G.; Spurgeon, D.W.; Raulston, J.R. & Kilgore, T.P. (1998). The strategy of delayed planting with rapid fruiting cotton for the Lower Rio Grande Valley of Texas. In: *Proceedings of the Beltwide Cotton Conferences*, National Cotton Council, pp. 578-579. Memphis, Tennessee, USA, ISSN 0032-0889.

Showler, A.T. & Potter, C.S. (1991). Synopsis of the 1986-1989 desert locust (Orthoptera: Acrididae) plague and the concept of strategic control. *American Entomologist* 37: 106-110, ISSN 1046-2821.

Showler, A.T. (1995). Locust (Orthoptera: Acrididae) outbreak in Africa and Asia, 1992-1994: an overview. *American Entomologist* 41: 179-185, ISSN 1046-2821.

Showler, A.T.; Sappington, T.W.; Foster, R.N.; Reuter, K.C.; Roland, T.J. & El-Lissy, O.A. (2003). Aerially applied standard rate malathion against reduced rates of malathion + cottonseed oil for boll weevil control. *Southwestern Entomologist* 27:45-58, ISSN 0147-1724.

Showler, A.T. (2003). Effects of routine late-season field operations on numbers of boll weevils (Coleoptera: Curculionidae) captured in large-capacity pheromone traps. *Journal of Economic Entomology* 96: 680-689, ISSN 0022-0493.

Showler, A.T. & Greenberg, S.M. (2003). Effects of weeds on selected arthropod herbivore and natural enemy populations, and on cotton growth and yield. *Environmental Entomology* 32: 39-50, ISSN 0046-225X.

Showler, A.T. (2004a) Assessment of pre-emptive insecticide applications at pinhead square size for boll weevil control. In: *Proceedings of the Beltwide Cotton Conferences*, pp. 1689-1692. National Cotton Council, Memphis, Tennessee, USA, ISSN 0032-0889.

Showler, A.T. (2004b). Influence of cotton fruit stages as food sources on boll weevil (Coleoptera: Curculionidae) fecundity and oviposition. *Journal of Economic Entomology* 97: 1330-1334, ISSN 0022-0493.

Showler, A.T. & Scott, A.W., Jr. (2004). Effects of insecticide residues on adult boll weevils and immatures developing inside fallen cotton fruit. *Subtropical Plant Science* 56: 33-38, ISSN 1009-7791.

Showler, A.T. (2005). Relationships of different cotton square sizes to boll weevil (Coleoptera: Curculionidae) feeding and oviposition in field conditions *Journal of Economic Entomology* 98: 1572-1579, ISSN 0022-0493.

Showler, A.T. & Cantú, R.V. (2005). Intervals between boll weevil (Coleoptera: Curculionidae) ovipostion and square abscission, and development to adulthood in Lower Rio Grande Valley, Texas, field conditions. *Southwestern Entomologist* 30: 161-164, ISSN 0147-1724.

Showler, A.T. & Robinson, J.R.C. (2005). Proactive spraying against boll weevils (Coleoptera: Curculionidae) reduces insecticide applications and increases cotton yield and economic return. *Journal of Economic Entomology* 98: 1977-1983, ISSN 0022-0493.

Showler, A.T.; Greenberg, S.M.; Scott, A.W. Jr. & Robinson, J.R.C. (2005). Effects of planting dates on boll weevils (Coleoptera: Curculionidae) and cotton fruit in the subtropics. *Journal of Economic Entomology* 98: 796-804, ISSN 0022-0493.

Showler, A.T. (2006a). Boll weevil (Coleoptera: Curculionidae) damage to cotton bolls under standard and proactive spraying. *Journal of Economic Entomology* 99: 1251-1257, ISSN 0022-0493.

Showler, A.T. (2006b). Short-range dispersal and overwintering habitats of boll weevils (Coleoptera: Curculionidae) during and after harvest in the subtropics. *Journal of Economic Entomology* 99: 1152-1160, ISSN 0022-0493.

Showler, A.T. (2007). Subtropical boll weevil ecology. *American Entomologist* 53: 240-249, ISSN 1046-2821.

Showler, A.T. & Abrigo, V. (2007). Common subtropical and tropical nonpollen food sources of the boll weevil (Coleoptera: Curculionidae). *Environmental Entomology* 36: 99-104, ISSN 0046-225X.

Showler, A.T. (2008a). Efficiency of tank-mixing insecticide with defoliant against adult boll weevil (Coleoptera: Curculionidae) populations as determined by late-season field disturbance trapping. *Subtropical Plant Science* 60: 58-65, ISSN 1009-7791.

Showler, A.T. (2008b). Longevity and egg development of adult female boll weevils fed exclusively on different parts and stages of cotton fruiting bodies. *Entomologia Experimentalis et Applicata* 127: 125-132, ISSN 0013-8703.

Showler, A.T. & Cantú, R.V. (2008). Effect of adult boll weevil feeding on cotton squares. In: *Proceedings of the Beltwide Cotton Conferences*, pp. 1419-1421. National Cotton Council, Memphis, Tennessee, USA, ISSN 0032-0889.

Showler, A.T. & Robinson, J.R.C. (2008). Cotton harvest at 40% versus 75% boll-splitting on yield and economic return under standard and proactive boll weevil (Coleoptera: Curculionidae) spray regimes. *Journal of Economic Entomology* 101: 1600-1605, ISSN 0022-0493.

Showler, A.T. (2009a). Free amino acid profiles in reproductive and rind portions of cotton fruiting bodies. *Subtropical Plant Science* 61: 37-48, ISSN 1009-7791.

Showler, A.T. (2009b). Roles of host plants in boll weevil range expansion beyond tropical Mesoamerica. *American Entomologist* 55: 228-236, ISSN 1046-2821.

Showler, A.T. (2009c). Three boll weevil diapause myths in perspective. *American Entomologist* 55: 42-50, ISSN 1046-2821.

Showler, A.T. (2010). Do boll weevils really diapause? *American Entomologist* 56: 100-105, ISSN 1046-2821.

Steedman, A. (1988). *The Locust Handbook*. Overseas Development Natural Resources Institute, ISBN 0859542815, London, England.

Sterling, W.L. & Adkisson, P.L. (1966). Differences in the diapause response of boll weevils from the High Plains and Central Texas and the significance of this phenomenon in revising present fall insecticidal control programs. *Texas Agricultural Experiment Station Bulletin 1047*, College Station, Texas, USA.

Stoner, A. (1968). *Sphaeralcea* spp. as hosts of the boll weevil in Arizona. *Journal of Economic Entomology* 61: 1100-1102, ISSN 0022-0493.

Sturm, M.M. & Sterling, W.L. (1990). Geographical patterns of boll weevil mortality: observations and hypotheses. *Environmental Entomology* 19: 59-65, ISSN 0046-225X.

Summy, K.R.; Cate, J.R. & Hart, W.G. (1988). Overwintering strategies of boll weevil in southern Texas: reproduction on cultivated cotton. *Southwestern Entomologist* 13: 159-164, ISSN 0147-1724.

Texas Boll Weevil Eradication Foundation (2011). *Texas Boll Weevil Eradication Foundation, Inc.* http://www.txbollweevil.org/.

Texas Department of Agriculture (2002). Administrative Code, Title 4, Part 1, Chapter 20, Subchapter A, Rule 20.1, Texas Department of Agriculture, Austin, Texas, USA.

USDA-APHIS (United States Department of Agriculture-Animal and Plant Health Inspection Service). (2007). *Boll Weevil Eradication*. USDA-APHIS, Washington, D.C., USA.

Vigueras, A.L. & Portillo, L. (2001). Uses of *Opuntia* species and the potential impact of *Cactoblastis cactorum* (Lepidoptera: Pyralidae) in Mexico. *Florida Entomologist* 84: 493-498.

Waibel, C.W. (1953). Varieties and strains of citrus originating in the Lower Rio Grande Valley of Texas, pp. 18-24. In: *Proceedings of the Seventh Annual Rio Grande Valley Horticultural Institute*, G.P. Wene, pp. 18-24. Weslaco, Texas, Jan. 21, 1953, USA (no ISSN).

Watson, T.F.; Bergman, D. & Palumbo, J. (1986). Effect of temperature and food on developmental time of the boll weevil in Arizona. *Southwestern Entomologist* 11: 243-248, ISSN 0147-1724.

9

Management of Tsetse Fly Using Insecticides in Northern Botswana

C. N. Kurugundla[1], P. M. Kgori[2] and N. Moleele[3]
[1]Water Affairs, Private Bag 002, Maun
[2]Animal Health, Box 14, Maun,
[3]Biokavango Project, Okavango Research Institute, Maun,
Botswana

1. Introduction

The tsetse fly (*Glossina* spp.) is only found in Africa and carries trypanosomes (the disease agents causing human sleeping sickness and animal trypanosomosis- *Nagana*), reaching their southern limits (particularly the *morsitans* group) in Botswana and Kwazulu Natal in South Africa. The tsetse fly transmitted disease is a serious problem in Sub-Saharan Africa and it is estimated that the removal of this disease could double livestock production and markedly increase cultivation levels. It is estimated that potential distribution of tsetse fly in Africa is 300,000 km^2 (Mathiessen & douthwaite, 1985).

Generally, the economic and social impacts of *nagana* and sleeping sickness on animal production and human health are severe, estimates put annual cattle production losses at US$2.7 billion and >55, 000 people dying from sleeping sickness annually (Budd, 1999). The disease is usually of chronic and debilitating form. It is essentially a problem in rural areas, where the threat and burden of such is a significant contributor to rural poverty and malnutrition.

In Botswana, the tsetse infested area covered ≤ 5% but has had significant impact on livestock and human populations, particularly in Ngamiland and Chobe regions, largely because of the wet areas in an otherwise dry country. Prior to the rinderpest pandemic of 1896 which significantly reduced tsetse populations in much of east and southern Africa as a result of the critical loss of food source, the distribution of tsetse fly in Botswana had reached its historical limits of approximately >20,000 km^2 (Ford, 1971; Jordon, 1986). As the tsetse populations recovered from the rinderpest epizootic the incidence of trypanosomosis increased. For instance, between 1949 and 1960, the cattle populations in Chobe District declined by ≤ 95% (Lambrecht, 1972). Ploughing along the flood plains is a common local practice in these areas, and whilst beneficial in the sense of optimal exploitation of soil moisture, this practice carries the risk of potential exposure of farm workers to tsetse fly bites. An average of 50 (geometric mean range = 13-272) trypanosomosis cases was recorded between 1957 and 1977 in Maun Hospital each year. Today the Okavango Delta, Kwando-Linyanti-Chobe areas are the most important destinations for international tourists and one of the major sources of revenue for the country. The risk of sleeping sickness within these areas is perceived as potential threat to the tourism industry, both locally and nationally (RTTCP, 1995).

However, at the end of the rinderpest epidemic, a rebound of the tsetse population in northern Botswana and the subsequent disease challenges led to the introduction of intensified control efforts in the mid-twentieth century using conventional and improved insecticide based methods. This chapter therefore reviews the history of management of tsetse fly control in Botswana, with specific focus on the factors that influence tsetse fly distribution in Botswana, methods of insecticide applications to control tsetse fly, effectiveness of the control methods, and monitored environmental impacts.

2. History, distribution, control and management of tsetse fly in Botswana

Comprehensive historical review of tsetse distribution in Botswana up to the 1980s is provided by Davies (1980). Early records largely from European explorers suggest that tsetse occurred in a continuous belt from Angola, through Caprivi Strip (Namibia) and then the Kwando-Linyanti and Chobe River systems, then via the Selinda Spillway to cover most of northern Botswana's Okavango Delta and its immediate surroundings including Nxai Pan N. P (Figure 1). The history of control and management of tsetse in Botswana is detailed below.

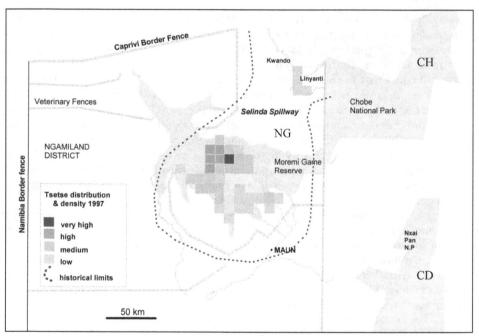

Fig. 1. Distribution of tsetse flies in Botswana (NG = Ngamiland District – Okavango Delta, Maun, Kwando Linyanti; CH = Chobe District – Chobe and Kasane; and CD = Central District – Nxai Pan N. P)

2.1 1940-1960 early methods of tsetse control
The disease Trypanosomosis (sleeping sickness and *Nagana*) is the primary reason for controlling tsetse fly in Botswana. In the early 1940s, cattle *Nagana* in parts of Ngamiland was so bad that a government tsetse control department was formed to combat the disease.

The earliest methods used for reducing tsetse numbers in Botswana (and throughout Africa), were large-scale clearing of bush and vegetation where tsetse fly rested, thus denying them shelter, and elimination of wild game animals which provided the natural food source of the fly. However, both these methods had huge environmental implications and were gradually phased out when alternative methods became available. Nonetheless, bush clearing continued as a complementary method until the mid 1960's (Davis, 1980). Game fencing also emerged as alternative, especially to limit the incursion of game animals into the settled areas of the Okavango Delta fringes.

2.2 1960-1972 residual ground spraying

The method of controlling tsetse using chemicals became common in the 1940s following the discovery of chlorinated hydrocarbon insecticides. In Botswana, residual insecticides such as DDT were introduced in 1967 by applying it to selected tsetse resting sites using knapsack spraying machines. This technique, known as *selective ground spraying*, targeted only about 20% of all potential tsetse resting sites in the woodland, including tree trunks, lower large branches, rot holes and holes on the ground such as ant-bear holes, springhare and hyena dens (Davies, 1980). The residual insecticides would remain available (even after spraying) to tsetse which emerged from underground pupal sites for 2 to 3 months after spray treatment. Given the accessibility challenges within the Okavango Delta, the ground spraying method of control got restricted only to the peripheral areas around the western Delta, the 'Maun Front' or along the Savuti Channel, and it was found to be unsuited to the swamps and island mosaics of the delta interior.

2.3 1970-1990 non-residual aerial spraying

Following preliminary trials carried out in 1971, Botswana's tsetse control strategy switched almost entirely to aerial spraying of insecticide using the Sequential Aerosol Technique (SAT). Very low dosages of endosulfan, and later a cocktail of endosulfan and synthetic pyrethroid insecticide were applied several times over the tsetse habitat to cover all the emerging tsetse pupal period. At best, the early SAT campaign by aerial spraying (Davies, 1979) reduced the tsetse distribution limits from 20,000km^2 to 5,000km^2 but still tsetse could not be eliminated completely despite that being the primary objective from the onset.

2.4 1990-2000 traps and targets

In 1992 Botswana (following other African countries) adopted the use of chemically-impregnated tsetse screens or targets and traps known as the *odour-bait technique* (Vale and Torr, 2004). The system was pioneered in Zimbabwe in the 1980s and became wildly used due to its perceived environmental sensitivity. Targets were treated with synthetic pyrethroid insecticides, particularly deltamethrin suspension concentrate formulation of 0.6% (w/v) following the treatment procedure (Kgori et al., 2006). In the technique the insecticide is applied only to the target screens and not the surrounding vegetation. All traces of the insecticide would therefore disappear once the targets are removed. The concept was well received and particularly suited for the Okavango's pristine wilderness.

Overall, the effectiveness of targets depended on the management and institutional capacity of the government's Tsetse Control Department. Effective distribution of targets and their regular maintenance in the Okavango Delta became a problem as the access was severely impeded by vegetation and terrain. However, a long period of drought ending in 1999 allowed some 25,000 targets to be deployed throughout much of the usually inaccessible parts of the Okavango Delta. These gains were to be reversed when good rains returned in 1999/2000, thus again putting the effectiveness of this tsetse control measure at question. With the above average rainfall, the tsetse fly was able to recover and disperse beyond the confines of the Okavango Delta, taking the threat of trypanosomosis once again back to the people and livestock. Cattle deaths resulting from *Nagana* increased during this period (Sharma et al., 2001).

2.5 2000 onwards - aerial spraying; a reversal in strategy

In the year 2000, the Botswana government initiated a new programme to control tsetse fly and trypanosomosis in and around the Okavango Delta. However, the primary objective remained unchanged; to eliminate tsetse and trypanosomosis from the Okavango Delta and the adjacent Kwando and Linyanti. Tsetse surveys in 2001 showed that the tsetse distribution limits had extended from 5,000 km^2 to about 12,000 km^2. Some 30,000 cattle within the villages surrounding the Okavango delta were clearly at risk of *Nagana* and they were subsequently treated with prophylactic trypanocides.

Reintroduction of aerial spraying became the cornerstone of the new campaign, but this time around targets were used as protective barriers between successive operations to stop tsetse fly from reinvading treated areas. Two aerial spraying operations were planned and executed in succession (in 2001 and 2002) to cover all the infestation in the Okavango Delta (Kgori et al., 2006; Allsopp & Phillemon-Motsu, 2002). In 2001, about 7,000 km^2 (upper box Figure 2) of the northern part of the Delta was aerial sprayed, followed in 2002 by 8,650km^2 (Lower Box Figure 2) of all the remaining infestation in the south (Figure 2). The insecticide of choice was deltamethrin (0.35% (w/v) ulv formulation applied at night using four turbo thrush fixed-wing aircraft. The aircraft were all guided by previously unavailable advanced navigation guidance system (Kgori et al., 2009).

3. Effectiveness of control measures

Apart from limited ground spraying in accessible peripheral areas such as Savuti Channel in the Chobe District, aerial spraying using ulv applications of endosulfan was the only method used to control tsetse fly in Botswana in the 1970s and 80s. Unfortunately for Botswana, since tsetse fly control was conducted largely in aquatic and pristine environment, endosulfan concentrations had to be kept lower than usual; it was applied at maximum of 12 g ha^{-1} compared to operations in other countries where higher levels of 16 to 20 g ha^{-1} were used to achieve comparatively good results (Douthwaite et al., 1981).

Typically, a series of aerial spray treatments were applied at night-time in winter with stable air and temperature inversion conditions in place, and using Piper Aztec or Cessna 310 aircraft from a height of about 15 m above the tree canopy. Under such inversion conditions, the tiny insecticide droplets would drift through the tsetse habitat in order to effectively kill adult tsetse fly. The insecticide was dispersed through a micronair atomizer set to give droplets in the range of 30–40 microns. This ensured that only small quantities of the insecticide were applied to the tsetse habitat. During these early years of aerial spraying, SAT operations were flown without the benefit of advanced aircraft guiding systems which are available today. Therefore spray distribution relied upon far less sophisticated methods of guidance such as ground marker parties with hand held miniflares positioned at each end of the spray run in order to guide the pilots. Also the formulation was non-residual in

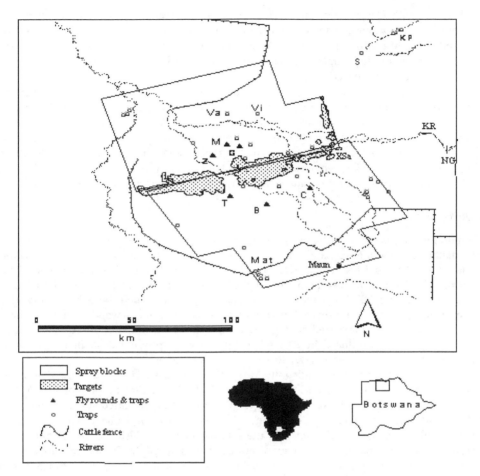

Fig. 2. Operations to control tsetse in the Okavango Delta region of Botswana conducted between 2001 and 2002, and the locations of tsetse and other non-target agents' monitoring sites. Sites indicated by solid triangles indicate the permanent sites at Zimbiri (Z), Mombo (M), Guai (G), Thapagadi (T), Bobo (B) and Chief's Island (C). Open circles indicate occasional monitoring sites used at various times during the spraying operations, including Kings Pool (KP), Selinda (S), Vumbura (Va), Vundumxiki (Vi), and Matsibe (Mat). Source: Kgori et al (2006). Xakanaxa Salvinia monitoring sites (XSa), Khwai River (KR) and North Gate (NR) – Source: Kurugundla and Serumola (2007).

nature; the insecticide deposition remained in the environment only for a short period of time and therefore was non-persistent.

Large-scale trials (1,000 – 4,000 km^2) of aerial spraying with endosulfan in Botswana began in 1973 as a pilot programme to assess the feasibility of even bigger operations to eliminate tsetse fly from the whole of the Okavango Delta (Davies, 1979). By the mid 80s, the scale of the annual operations had increased to about 6,000 – 9,000 km^2 and very good control was achieved. The distribution limit was reduced significantly from 20,000 km^2 to 5,000 km^2 and the threat of both *Nagana* and human sleeping sickness was effectively controlled due to sequential aerial spraying of insecticide, reduced water levels with less flooded areas and increased drought conditions in Botswana. Sequential spraying was necessary because the tsetse (i.e., young adults) would continue to emerge from the underground pupal stage for several days after each preceding cycle.

However, despite repeated applications complete eradication was not possible. Several possible reasons may have contributed to the failure by early SAT operations to achieve eradication in Botswana - including; low dosages used, lack of boundary protection since odour baited targets were unavailable then, poor navigation and random rather than systematic treatments. The navigation challenges observed often resulted with localized over-spraying which raised environmental concerns, notably fish kill (Merron & Bruton, 1991), but the low dosages used ensured that only limited off-target effects were possible.

4. Deltamethrin

When endosulfan-based tsetse aerial spraying campaign was ended in 1992, targets almost immediately became the next and only preferred option for tsetse fly control in Botswana. At the time, pyrethroid insecticides were also becoming firmly established in the insecticide market and Deltamethrin was the preferred choice for use in the treatment of odour-baited targets (Vale & Torr, 2004). For the next ten years, targets remained the only method used for controlling tsetse in northern Botswana's Okavango Delta, Kwando-Linyanti-Chobe River systems. A new integrated strategy involving the reintroduction of sequential aerial spraying was later introduced in 2001 when targets were proving difficult to effectively implement in Northern Botswana due to accessibility problems.

Prior to commencement of spraying in May 2001, tsetse surveys indicated that tsetse distribution had increased significantly to about 12,000 km^2 from the previous 5,000 km^2 since the end of the last aerial spraying campaign in the 1980s. Increased spread of tsetse fly could be attributed to a long period of no effective control and above average rainfall coupled with increased flood inflows in the Okavango Delta. Also the use of motorized boats as well as airplanes used to transport tourists in and out of the Delta contributed to the spreading of tsetse flies. This was evidenced by the occasional sighting of the tsetse flies in Maun which became a real concern to authorities, and hence the need for concerted effort to reverse the situation.

A two year aerial spraying operation (2001 and 2002, Figure 2) for the north and south of the Okavango Delta using Turbo Thrush Aircraft was planned and implemented in succession to cover the entire Okavango Delta. At the interface of the two spray blocks a target barrier was deployed to prevent tsetse crossing from one unsprayed area to another sprayed area between the successive spray operations. The revised strategy also included a component of the sterile insect technique (SIT) as technical backstop, should SAT fail to eliminate tsetse as was previously the case in the 1980s (Feldmann, 2004).

Improvement in the reintroduction of aerial spraying in 2001, 2002 and 2006 was the availability of latest navigation systems which could ensure precision placement of spraying material and eliminate overdosing or even under-dosing through erratic track guidance. For instance all aircraft were fitted with GPS-guided SATLOC navigation and spray management equipment accurate to about 1m. The system therefore had control on the distribution of the spray application and automatically cut off the spray if the aircraft wondered out of the spray block and indeed the prescribed flight path. The aircraft insecticide dispersal units involved two boom-mounted micronair AU 4000 rotary atomizers operated at cage speeds of 10,800 rpm and flowrate of 7.6-8.6 l/km². Such technology development used in recent SAT operations had positive implications on the distribution and deposition of formulated insecticide droplets and, ultimately on the efficiency and effectiveness of the spray application. During 2001 and 2002 operations, and later in 2006 at Kwando and Linyanti, deltamethrin insecticide (0.35% (w/w) ulv formulation was applied at 0.26 -0.30 g active ingredient (a.i.) ha⁻¹. The higher dose rate of 0.3 g a.i. ha⁻¹ was used for the first two cycles when tsetse fly population would normally be at its highest density (Saunders, 1962).

Initially, two successive spray treatments were conducted in 2001 and 2002 to cover approximately 16,000 km² in the Okavango Delta (Figure 2). A similar and follow up operation took place in 2006 covering an additional 10,000 km² of the Kwando and Linyanti border area, north of the Okavango Delta which also extended across Namibia's eastern Caprivi border into Southern Angola (Figure 3) in order to guarantee complete removal of all the remaining tsetse fly infestation in northern Botswana. This approach ensured that the northern tsetse infestation along the Kwando and Linyanti Rivers did not re-infest the Okavango Delta.

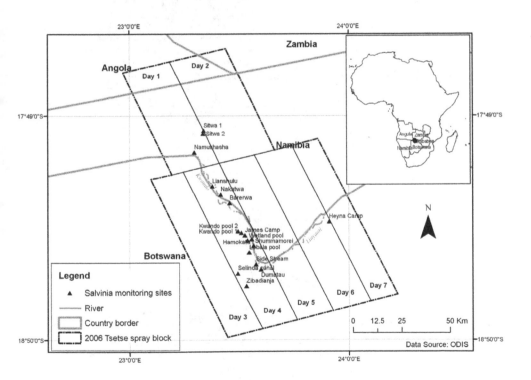

Fig. 3. Area of deltamethrin spraying in 2006. Sites of bioassay experiments and field monitoring areas to assess the impacts of deltamethrin spray on salvinia weevils and other aquatic invertebrates in Kwando-Linyanti River System.

Spraying of the block (7000 km²) commenced on 3 June 2001. Figure 4 shows the mean daily catch from the five fly-rounds (Figure 4A) and 18 traps within the spray block (4C), and the fly-rounds outside it (Figure 4B). The results from the fly-rounds within the block show that the mean tsetse catch up to the first day of spraying was 44.6 tsetse day⁻¹. Immediately after each cycle the daily catch declined to zero, but then recovered to peaks of 12.8 tsetse day⁻¹ after the 1st cycle and 20.6 tsetse day⁻¹ after the 2nd cycle. Thereafter recovery was less marked, peaking at 6.3 tsetse day⁻¹ after the 3rd cycle, 1.2 tsetse day⁻¹ after the 4th cycle and finally no tsetse was found after the 5th cycle. By contrast, the daily catches from fly-rounds outside the spray block (Figure 4B) increased significantly over the period of spray operation.

A. 2001 - Fly-rounds within spray block

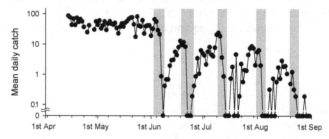

B. 2001 - Fly-rounds outside spray block

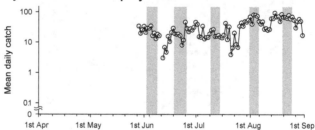

C. 2001 - Traps within spray block

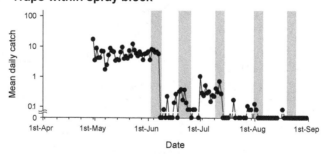

Fig. 4. Detransformed mean daily catch of tsetse from fly rounds and traps within (Mombo, Guai and Zimbiri) or outside (Chief's Island, fly rounds only) the area sprayed during the 2001 control operation. Vertical grey bars indicate the timing of the five spray cycles. Source: Kgori et al., (2006).

Spraying commenced in the south block (8650 km^2) on 16 May 2002. Following the start of spraying, the catches declined dramatically (Figure 5). For instance, the mean daily catch from fly-rounds was 101.7 (±0.026, n=148) for the period 10 April - 16 May compared to 0.23 (±0.019, n=66) for the period 24 May - 3 June, this being the period between the end of 1st spraying cycle and the beginning of the second cycle (Figure 5A) and after the 4th cycle no tsetse were found. The catches from traps showed a similar decline (Figure 5B). During 2006 campaign, the tsetse surveys started on 13th May 2006 (Day 1 = 13 May 2006. Figure 6 shows that the abundance of tsetse was higher than 100 with fluctuations in Kwando-Linyanti prior to spraying and it reduced to zero after 2nd, 3rd, 4th and zero catches after the 5th spraying cycle (Kgori et al., 2006, Kgori et al., 2009, VEEU-TCD/DAHP 1998).

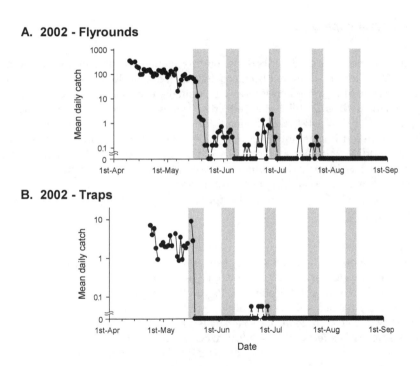

A. 2002 - Flyrounds

B. 2002 - Traps

Fig. 5. Detransformed mean daily catch of tsetse from fly rounds and traps within (Bobo Island, Thapaghadi and Chief's Island) the area sprayed during the 2002 control operation. Vertical grey bars indicate the timing of the five spray cycles. Source: Kgori et al., (2006).

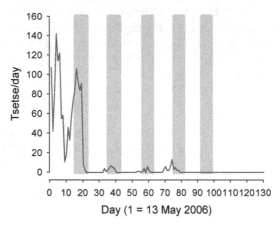

Fig. 6. Tsetse survey results at Kwando/Linyanti 2006 in five spray cycles.

5. Environmental implications

5.1 DDT
'Silent Spring' (Carson, 1962) described the environmental impacts of the indiscriminate spraying of DDT (Dichloro-diphenyl-trichloroethane) in the world and its effects on ecology or human health. It has high potential to bioaccumualte especially in predatory birds and magnify through food; and still highly effective against disease vectors such as malaria parasite (Thomas, 1981). DDT was banned for agricultural use worldwide under the Stockholm Convention but its limited use in disease vector control still continues in some parts of the world to this day and remains controversial (Connel, 1999). Its persistence ranges from 22 days to 30 years depending on the conditions of the ecosystem. It is insoluble in water, less biodegradable and degrades by means of aerobic, anaerobic and photolysis (Thomas, 1981). No specific monitoring of impacts of DDT spraying in Botswana was ever carried out.

5.2 Dieldrin
Deldrin was used world-wide to control locusts and tropical disease vectors, such as tsetse fly and mosquitoes prior to the 1970s, and has persistence of 5-25 years. It has similar characteristics to DDT (Hunter & Robinson, 1968). Botswana was one of the first countries in Africa to monitor the side-effects of chemical control of tsetse fly. In 1964 riverine forests were sprayed with Dieldrin against *G. morsitans* in an area near Maun in the Okavango Delta, Botswana. The effects of the motorized spraying (mist blower) on non-target organisms were observed for a period of 10 days following spraying (Allsopp, 1978). Among the animals found dead were many birds, mammals, reptiles and fish (Graham, 1964). Casual observations of the aerial spray showed no difference between the wildlife populations between the sprayed and unsprayed areas (Davies & Bowles, 1979).

5.3 Endosulfan
Endosulfan degrades more rapidly than DDT and dieldrin; 9% degradation in water between 5-25 days, 15-30 days from vegetation surface, but remains more than 100 days to

one year in soil as residue (Maier-Bode, 1968). It is being phased out globally and a global ban on the manufacture and use of endosulfan was initiated under the Stockholm Convention in April 2011. The ban will take effect in mid 2012 and more than 80 countries became part of the ban (http://en.wikipedia.org/wike/endosulfan). Chemical control methods have been used in anti-tsetse campaigns in at least 20 countries in Africa. Side-effects on non-target organisms have been studied to a greater or lesser extent in only about half of the 20 countries (Table 1).

Country	Year (s)	Type of insecticide + method of application	Reference
Botswana	1964	Dieldrin - Ground spray (Mist blower)	Graham, 1964
	1975-78	Endosulfan ULV Fixed-wing	Ali, 1978; Fox et al., 1979
	2001-02	Deltamethrin – Aerial spray	Perkins and Ramberg, 2004; 2004a
	2006	Deltamethrin - Aerial spray	Bonyongo & Mazvimavi, 2007; 2008
Cameroon	1979	Dieldrin – Helicopter	Muller et al., 1980
Chad	1972-1974	DDT- Ground spray	Tibayrence and Gruvel, 1977
Ivory Coat	1979	Dieldrin – Ground spray	Koeman and Pennings, 1970
	1968-69	Endosulfan and Decamethrin – Helicopter	Everts, 1979
Kenya	1970-72	Dieldrin – Fixed wing	Allsopp, 1978
Namibia	2006	Deltamethrin – Aerial spray	Bonyongo & Mazvimavi, 2007; 2008
Niger	1968	DDT – Ground spray	Koeman & Pennings, 1970
	1969-70	Dieldrin –Ground spray	Koeman et al., 1971
	1974-76	Dieldrin – ULV Helicopter	Koeman et al., 1978
	1977	Endosulfan – Helicopter	Dortland et al., 1977
Nigeria	1975-76	Dieldrin/Endosulfan ULV helicopter	Dortland et al., 1977
	1976	DDT/Endosulfan – Ground spray	Koeman et al., 1978
	1976	Endosulfan – ULV fixed-wing	Takken et al., 1976
	1977-78	Pyrethroids –Ground and Helicopter spray	Smies et al., 1980
Tanzania	1961-71	Dieldrin – Ground spray	Sserunjoji & Tjell 1971
	1979	DNOC (2 methyle-4, 6-dinitro-ortho-oresol), Bush clearing	Tarimo & Palloti, 1979
Uganda	1963-73	Endosulfan-pyrethroids – Helicopter	Wilson, 1972
Zambia	1968	Endosulfan – ULV fixed-wing	Magadza, 1979
Zimbabwe	1978	Endosulfan – ULV aerial	Cockbill. 1979

Table 1. Anti-tesetse fly campaigns in Africa and impacts studied on the non-target agents.

Deposition

The insecticide deposition in the various habitats in the Okavango Delta is presented in Table 2. However, the range and frequency of individual values are insignificant between them (Douthwaite et al., 1981). Endosulfan concentration in different types of aquatic habitats nine hours after spraying at 9.5 g ha^{-1} varied with depths and the values were insignificant between the various habitats: In marshes $\leq$ 0.5 m = 0.81 µg l^{-1}; in main river between 1-2 m depth = 1.16 µg l^{-1} and in open pool at $\leq$ 0.5 m depth = 1.54 µg l^{-1} where as persistence of the insecticide in pools of water declined from 1.42 µg l^{-1} nine hours after spraying and undetectable levels within five days (Douthwaite et al., 1981). Its half life in aquatic environment is between one and five days (Moulton, 1973).

	Open water	Water in grass swamp	Open grassland	Grassland under tree canopy
Mean (Range)	20.7 (1.4-42.9)	17.7 (3.6-45.9)	21.1 (4.5-43.8)	14.3 (2.4-24.5)
Variance	116.1	193.8	188.3	54.2

Table 2. Quantities of endosulfan (µg) drift found on aluminum sheets placed in different habitats and analyzed one hour after spraying 9.5 g ha^{-1} (4 µg per sheet = 1 g ha^{-1}). Source: Douthwaite et al., (1981).

5.3.1 Fish

Bioaccumulation

The major route of uptake in aquatic invertebrates is probably via the digestive system (Roberts, 1975), whereas fish absorb most endosulfan directly from the surrounding water injuring gills, thus reducing the oxygen consumption and disrupting the osmoregulatory function of aquatic organisms (Saravana & Geraldine, 2000). Dortland et al., (1977) have shown that endosulfan applied at 900 g ha^{-1} for tsetse control near West African Rivers produced residue levels of 1.4 and 37.3 µg g^{-1} fish muscle and liver respectively. Experimental applications of endosulafan to fish in paddy fields (Moulton, 1973) showed that gouramis, *Tricho gaster pectoralis* could accumulate the pesticide in abdominal organs to concentrations over 1000 times those in surrounding water. It was found that the absorbed endosulfan in fish was rapidly metabolized to the endosulfan sulphate (Ali, 1978).

The edible fish species tested for bioaccumulation during the endosulfan spraying in Botswana include species of *Clarias, Serranochromis, Schilbe, Haplochromis, Tilapia, Marcusenius*. Endosulfan mean residue concentration with respect to percentage lipid 5 days after the spraying was 0.19 µg g^{-1} wet tissue in muscle, while the maximum found in whole dead small fish was 1.5 µg g^{-1} wet tissue. These values refer to the combined concentration of alpha+beta + endosulfan sulfate - all three compounds are equally toxic to the fish (Anon, 1973). The USA has set a tolerance limit for endosulfan in meat of 0.2 µg g^{-1} fresh weight; this therefore means muscle tissue from living fish in the Okavango would be considered safe (Douthwaite et al., 1981).

Fatty species tend to accumulate the most endosulfan (Douthwaite et al., 1981), and one might perhaps expect these groups would show greater mortality in the field. Such relationship was however, not apparent in the study because the very fatty insectivore *Marcusenius macrolepidotus* was never found dead after spraying as they accumulated higher residue concentration of 1.002 µg g^{-1} wet weight in its viscera lipid. So the bioaccumulated insecticide can do little damage to vital metabolic processes. Indeed, the laboratory experiments with *Hepsetus* suggest that the fatty lean fish is susceptible and succumb most readily to poisoning to endosulfan and thus fish weight is probably the dominant factor determining the survival, i.e. fry and small individuals are at greater risk. Tests with the predators of fish such as crocodile *Crocodilus niloticus*, fish eagle *Heliaeetus vocifer*, Kingfisher *Ceryle rudis* accumulated endosulfan residues, although these never exceeded 0.2 µg g^{-1} wet weight, except in the visceral fat of the crocodile with 0.783 µg g^{-1} wet weight suggesting again that fatty organs tended to accumulate the highest residue concentrations (Douthwaite et al., 1981).

Mortality

During ultra-low-volume aerial applications of endosulfan at 14g ha^{-1} for control of tsetse fly in savanna woodland, Zimbabwe, no deleterious effect of the insecticide was demonstrated, other than fish (*Tilapia* spp.) in shallow water (Cockbill, 1979). In general, endosulfan killed small fish first, (*Alestes lateralis*) although almost all species were ultimately affected by the 10 g ha^{-1} spray. An endosulfan spray done in the Okavango Delta in 1978 showed that the fish mortality extended over a period of 3 days after spraying in four cycles (Table 3). Fish species affected included *Tilapia, Hemichromis, Haplochromis, Pseudocrenilabrus, Serranochromis, Clarias, Barbus, Schilbe and Hepsetus*. Among them, *Tilapia* was the dominant species affected with 65% poisoning followed by *Pseudocrenilabrus* with 13% (Douthwaite et al., 1981).

Fish mortality per hectare (n = 9 sites)			
Cycle 1	Cycle 2	Cycle 3	Cycle 4
12 g ha^{-1}	12 g ha^{-1}	9 g ha^{-2}	6 g ha^{-2}
542.7 ±339.6 (0-3250)	544.9 ±266.5 (0-2344)	672.1 ±336.5 (0-2531)	2.0 ±1.6 (0-16)

Table 3. Fish mortality resulting from endosulfan spray in Khwai River in the Okavango Delta. (SE ± = sd/√n). Figures in the parantheses indicate minimum and maximum mortality (Source: Douthwaite et al., 1981).

The general symptoms of endosulfan or other pesticides' poisoning in fish are epithelial lifting (Bucke et al., 1996, Choudary et al., 2003), Hyperplasia (Munshi et al., 1996), muscle hypertrophy and aedema, liver necrosis (Bucher & Hofer, 1993). Samples of liver and brain were tested from *Hepsetus odoe, Tilapia sparrmanii, T. rendalli and Sarotherodon andersoni* throughout the 1978 spraying and during the ensuing 7 months. Whereas liver samples alone were collected from *Clarias ngamensis and C. gariepinus*. In general endosulfan spray induced liver focal (peripheral) necrosis, brain aedema leading to encephalitis, lining in cephalic tissue during the spray periods as well as 15 days after cycle 6 in the species tested. However, the fish showed healing areas of focal necrosis in liver, absence of aedema in brain in the post spray periods after 7 months. Among the species, *Hepsetus odoe* was tolerant to endosulfan showing only slight change in the fatty liver few days after the spray (Douthwaite et al., 1981).

5.3.2 Aquatic invertebrates

Among the Oligochaeta, Chironomidae, Trichoptera and Ephemeroptera groups the numbers of Chironomid larvae reduced significantly ($P \geq$ 0.001), Oligochaets and Ephemeroptera reduced in abundance while Trichoptera showed fluctuations during the 1978 spray. The decrease in Chironomid larvae in the sprayed lagoon was unexpected as the reported LC50 values for the larvae were considerably higher than the residue levels recorded in the study indicating a normal seasonal change. The Ephemeroptera, Oligochaeta and Chironomidae remained almost constant in pre spray, mid spray while in post spray they increased significantly reflecting positive response to the falling water levels rather than related to the endosulfan effects. With the exception of *Hexarthra* sp. the species namely *Filina, Brachionus, Keratella* and *Polyarthra* perecentage representation was considerably lower in the lagoons following the spraying season (Douthwaite et al., 1981).

5.3.3 Terrestrial invertebrates

The abundance of terrestrial invertebrates in floodplains, grassland and in dry land under tree canopies, *Colophospermamum mopane* canopies in the sprayed and unsprayed areas was determined in the Okavango Delta. The major groups in the studies included Chironomidae, Cicadellidae, Diptera, Hymenoptera, Orthoptera, Araneae, Formicidae, Coleoptera. Only in the case of adult *Chironomids* was there a large reduction in numbers in the sprayed site. However their abundance was almost identical in the post spray periods with the 'control sample' attributing doubts over the spray effects. Three of the major groups namely Formicidae, Tenebrionidae and Diptera showed reduction in activity and the most affected genus was the *Pheidole* in Formicidae following the spray. There was a significant decrease in numbers ($P \geq$ 0.05) of spiders in the sprayed site which correspond to the significant increase ($P \geq$ 0.001) in the sprayed site in the 3rd cycle (Douthwaite et al., 1981).

5.3.4 Flying insects

The behaviour of nocturnal insects during the spraying in the Okavango Delta was caught in the water trays in the grassland along the margins of the permanent swamps before and after each spray application in both sprayed and unsprayed areas. In the first spray cycle a sharp peak in mortality was observed in the adult *Chironomid* insects trapped in the sprayed site. However similar results were not obtained during subsequent spray cycles. This could be due to the emergence of adults from the larvae as the adults have short life-span. Among other groups, only Culicidae, some *Nematocera* and *Diptera spp.* were abundant in water traps and fluctuated in numbers from night to night in both sprayed and control sites, thus showing that there was no consistent evidence for either decreases or increases following each spray cycle (Douthwaite et al., 1981).

5.3.5 Birds

Endosulfan is neither toxic nor cumulative in birds and therefore unlikely to be lethal to them. However, there could be indirect effects on the insectivorous and piscivorous birds through disruption of their food supplies. The occurrence of birds in *Acacia* woodland was monitored before and after spraying in 1978 using transects. Besides similarities in the species in four transects, each transect lost and gained some species between sampling periods, but no species disappeared from both sprayed and unsprayed transects. A comparison of 39 species was grouped by diet, and by site and method of feeding. To determine the source of the heterogeneity comparisons were made within groups of insectivores, frugivores and granivores and birds with mixed diets. Significant heterogeneity occurred only in the granivores such as *Lamprotornis* starlings that increased in the unsprayed transects but declined in the sprayed transects, and the Grey Hornbill *Tockus nasutus* and Sengegal Coucal *Centropus senegalensis*, which did the opposite. The other insectivore species showed divergent occurrences include Meyer's Parrot *Poicephalus meyeri*, a granivore, Brubru *Nilaus afer* and Red-billed Wood-hoopoe *Phoeniculus purpureus*. These changes in occurrence could, in view of the varied habitats of the species concerned, be as well explained by rainfall, vegetation re-growth between sprayed and unsprayed transects as by the effect of spraying.

The major diet for kingfisher (*Ceryle rudis*) is fish ranging from 28-112 mm and 0.2 to19.1 g in weight. The diet consists mostly of *Tilapia, Haplochromis, Barbus and Pseudocrenilabrus*. These species accumulated the endosulfan in liver and brain and the feeding of these fish by kingfisher may have some concentrations of endosulfan. The study showed the total concentration of alpha and beta endosulfan and endosulfan sulfate in the pooled samples from three kingfisher birds as 0.012 µg g^{-1} wet weight in the liver and 0.205 µg g^{-1} weight in the brain. These concentrations are no higher than the levels found in fish and agree with earlier observations (Maier-Bode, 1968).

5.4 Deltamethrin

Deltamethrin is a broad spectrum insecticide, relatively stable but less persistent in the environment than the organochlorine pesticides (Grant & Crick, 1987). It has been in wide use in various crops, in gardens, indoors, outdoors for controlling pests such as mites, ants, weevils, beetles, leafhoppers the world over (Tomlin, 2006). Its persistence ranges from 8 to 48 hours (Erstfield, 1999) in water and 5.7 to 209 days (EFASP, 1999) in terrestrial habitats. Reported concentrations in water are rarely greater than 20 ng l^{-1} (Amweg et al., 2006) and were only found to be toxic to honey bees (*Apis* sp.) (Tomlin, 2006). The 2001 deltamethrin spraying over the upper Okavango Delta was done with very little systematic monitoring of the impacts/effects on the non-target organisms (Perkins & Ramberg, 2004). However, the subsequent spraying in 2002 period was done with adequate regular monitoring from the 1st to the 5th cycle. The analysis of data has been assessed by two methods.

i.   Sampling was carried out just before each cycle spraying event and in the following day after the spraying. This assumed that significant changes in abundance of individuals in any taxa are likely to be caused by the spraying.

ii. Analysis of trends in numbers of any taxa is compared from cycle 1 to 5 to show the tolerance and susceptibility.

In 2002, the spray deposition as determined by rotating slides was between 23 and 867 drops cm^{-2} and varied considerably within and between sites (Wolski & Huntsman-Mapila, 2002), which could be due to habitat variation, wind, temperature and distance from the flight lines. The wool strands that were exposed to the spraying absorbed the insecticide and revealed that for tsetse fly lethal concentrations were likely to have lasted for four to five days after the spray event (Perkins & Ramberg, 2004). However deltamethrin deposition determined in water, sediments and soil in 2002 and 2006 yielded almost insignificant results except one sediment sample that had low levels of the insecticide (Bonyongo &

Mazvimavi, 2007). Nevertheless, the insecticide deposition as captured by spreading aluminium foil sheets during the monitoring of salvinia weevils gave between 0.2 µg m^{-2} and 6.9 µg m^{-2} in 2002 and 2006 spraying occasions (Kurugundla & Serumola. 2007; Kurugundla et al., 2010).

5.4.1 Aquatic Invertebrates

Impacts - 2002

It is well known that flowing and still water usually differ in biotype composition and hence aquatic macro invertebrate studies were undertaken in still water of pools and flowing channels in Xakanaxa. The results were compared with those of the control site of Khwai River at North Gate. A total of 695 macro invertebrate samples and 200 zooplankton samples were collected, and 64 macro invertebrate families were identified. In channels, abundance declined by 46% after five spray cycles (927 individuals to 520) and in lagoons it declined by 25% (1230 individuals to 917). However, in the reference control site in Khwai River at North Gate, the abundance increased slightly over five spray cycles (789 individuals to 839) (Palmer, 2002).

Samples fixed in 75% ethanol were identified to family in the field using Gerber & Gabrial, (2002) and Davies & Day, (1998). A total of 47 taxa were recorded in channels and 49 taxa in lagoons and in total 65 taxa were identified, of which 23 taxa commonly occurred consistently in samples before and after the spraying. Out of these common taxa 6 showed distinct rapid declines after the first spray cycle and had disappeared completely after the fifth cycle. This corresponds to a loss of 26% of common taxa and was likely caused by the deltamethrin spray deposition. Whereas at North Gate on the Khwai River control site, the number of taxa increased from 29 to 30 after five spray cycles.

It has been shown that the mortality differs between the channels and lagoons and also between the taxonomic groups. Among 64 taxa, more than 24 have been found to be susceptible while few taxa (11) were found to be resistant to the insecticide spraying. Mollusks are probably more physiologically resistant to spraying than insects for obvious shell protection. The other survivor insect families include Chironomidae (non-biting midges), Ceratopogonidae (biting midges), Libellulidae (hairy dragon flies) and Caenidae (crawling mayflies). All these entire insect families live in the sediment, which may function as a protection against the insecticide spray, which is less bioavailable as the pyrethroids could be partitioned and adsorbed to various organic sediments (Muir et al. 1985). On the other hand, the elimination rate recorded for the Hemiptera (water bugs) Carduliidae (dragonflies), for most Ephemeroptera (mayflies) families and some other families could be understood by their active behavior in free water surfaces, sediment and vegetation surfaces. In particular the air breathing behaviour exhibited by most Hemiptera and Coleoptera (beetles) force them to come into contact with deltamethrin and oil-based carrier, paraffin that accumulate on water surface as a thin film (Perkins & Ramberg, 2004; Bonyongo & Mazvimavi, 2007).

Recovery - 2003

The recovery monitoring was measured in 2003 at the same periods and similar sites to those used in 2002 in order to assess recovery at community, family and morphospecies levels. The total abundance after a year was still significantly lower: 39% in channels and

60% in lagoons. The spraying caused significant reductions in abundance of many sensitive taxa often specific for respective habitats like lagoons and channels. What remained in both habitats was a common group of resistant species. The distinct feature is that the reappearance of some of the sensitive taxa that was affected in 2002 spraying was found again during the 2003 recovery studies. At the family level, the most negatively affected families were Atyidae (shrimps) which are characteristic of areas of permanent flow, and Pleidae (pygmy backswimmers), that are found in more seasonal areas. The abundance of Naucoridae (creeping water bugs) was significantly lower than the benchmark as well. Out of 39 identified morphospecies four (10%) were classified as sensitive as they did not reappear during the 2003 recovery study. Three species belonged to the family Notonectidae (Backswimmers) and one to Naucoridae (creeping water bugs), which might reflect a loss from the system. It is however difficult to assign this to persistent post-spray impacts of deltamethrin as the natural variation and abundance in aquatic invertebrates from year to year in the Delta is unknown (Palmer & Davies-Coleman, 2003)

Impact – 2006

Samples were collected at 21 sites in the main Kwando- Linyanti Rivers, Zibadianja and Dumatau lagoons, floodplains and on the vegetation of *Phragmites*, papyrus (*Cyperus papyrus*), hippo grass (*Vossia cuspidata*) and water lilies, (*Nymphaea* spp.) using kick nets (Picker et al 2002). The abundance of macro-invertebrates reduced between 10 and 90% in all the sample sites and affected marginally in the rivers, lagoons and the floodplains. About 70% macro-invertebrates were reduced in numbers from the vegetation types. Abundance of the invertebrate taxa reduced between 40 and 90% except *Chironomids* which appeared to increase during the spraying period. Twenty-four families of macro invertebrates were knocked down almost completely when compared between pre-spray and at the end of spray. About 11 of the total families survived the five spraying cycles (Masundire & Mosepele, 2006).

Recovery – 2007

An average of about 25% of the macroinvertebrates in 24 taxa (Table 4) that were present before spraying did not reappear one year after spraying. Perkins & Ramberg (2004a) reported 26% loss in taxa following spraying in the Okavango Delta in 2002. The mean abundance of all macro-invertebrates was only 11% in Kwando River while more than 100% in Linyanti River, Floodplains, Zibadianja and Dumatau lagoon areas. The *Phragmites* – and hippo grass dominated habitats had still below 50% of pre-spray levels of abundance and in other sites the invertebrates recovered well. Families such as Atydae and Dytiscidae were still well below pre-spray levels while Corixidae, Hirudinea and Chironomidae exceeded their pre spray levels of abundance. Planorbinae appear to have been little affected by the spraying (Masundire, 2007).

5.4.2 Terrestrial invertebrates

Impact – 2002 and 2006

In 2002, sampling was done by collecting the terrestrial invertebrates that had been knocked down by the aerial spray, under tree crowns of *Kigelia Africana*, *Lonchocarpus capassa* and *Combretum imberbe* on plastic sheets with an area of 3 m² in the riparian zone and under *Colophospermum mopane* in the drier land (Dangerfield, 2002). In 2006, similar studies were conducted in Kwando during the spraying in the woodland dominated by *Croton megalobotrys, Colophospermum mopane, Combretum imberbe, Lonchocarpus nelsii* and in open vegetation dominated by *Pechuel-loeshea leubnitziae* and *Cynodon dactylon* (riparian grass land).The results were compared to the control sites in unsprayed Khwai River at North Gate. Percent declines in terrestrial invertebrate abundance in combined woodland types in 2002 (Dangerfield, 2002) and 2006 were 64% and 63% respectively. All other common taxa showed significant declines as well with the exception of Flies (Diptera Families, Chironomidae and Ceratopogonidae) that have their larval stages in water and therefore produced mass swarming during spring and early summer.

During 2006, 21 taxa groups belonging to 14 families were identified to species level compared to 31 families of flies identified to morphospecies during 2002 spraying. There was marked decrease (50%) in arthropod species richness in the spray block among the woodlands as compared to pre-spray and after spray cycle 5. Significant difference in the composition of families of flies was recorded between the 2002 and 2006 monitoring. The most common families of flies during the 2002 and 2006 monitoring were Tabanidae, Tipulidae, Muscidae, Calliphoridae, Anthomyiidae, Syrphidae, Tephritidae. Spider families sampled in 2002 and 2006 were the same with the exception of family Theraphosidae which was first recorded during 2006. Family Heteropodidae was recorded in 2006 but not in 2002. Family Oxyopidae disappeared after 5th cycle of spraying in both 2002 and 2006. The crickets, particularly *Gryllus bimaculatus* was lost after the 1st cycle and was not captured throughout the remaining spraying cycles. Grasshoppers, *Lithdiopsis carinatus, Truxaloides* sp., *Ailopus thalassinus* and *Eucoptacara exguae* disappeared from the riparian grasslands (Chikwenhere & Shereni, 2006)

Common name	Taxa order	% Reduction – 2002	% Reduction – 2006
Ants	Hymenoptera	11	74
Flies	Diptera	41	60
Mosquitoes	Diptera	71	80
Beetles	Coleoptera	84	43
Wasps	Hymenoptera	23	65
Leaf hoppers & bugs	Hemiptera	46	45
Spiders	Arachnida	30	24

Table 4. Effects of deltamethrin on terrestrial arthropods dwelling in woodland types after spray cycle 2 in 2002 and 2006 (Source: Bonyongo and Mazvimavi, 2007).

In 2006, mosquitoes suffered the highest reduction of 80% followed by ants with 74%, the wasps had 65% reduction, while flies reduced by 60%. The rest of the groups listed declined below 50% and the least reduced was spiders at 24%. Conversely, ants were the least abundant while maximum reductions were found in beetles at 84% followed by mosquitoes with 71% reduction in 2002 (Table 4).

Recovery - 2003 and 2007

In May and August 2003 and 2007 sampling of terrestrial invertebrates in the selected woodland tree types yielded interesting results in the Okavango and Kwando respectively. In the key insect groups, the spider abundance recovered in all cases to pre-spray levels; beetle abundance had not recovered on *K. africana*, but recovered on other tree species; there was no change in fly or ant abundance between 2002 and 2003 while Hemeptera abundance was significantly greater in 2003 than before the spray events. Fifty-seven morphospecies were found for the first time in 2003 indicating that the species that were lost from 2002 had started to recover in 2003. This shows that there is great variation between years in the composition of invertebrates (Dangerfield & McCulloch, 2003). There was sharp increase in the flies and mosquitoes in Kwando/Linyanti areas in the recovery periods of 2007. Beetle abundance was highest at above 60% and wasps and ants increased more than 100% a year after the spraying. However crickets, *Gryllus bimaculatus* were not detected in 2007 where as Grasshoppers *Lithdiopsis carinatus, Truxaloides sp., Ailopus thalassinus* and *Eucoptacara exguae* that had disappeared from the riparian grasslands started to recover in 2007 (Chikwenhere, 2007)

5.4.3 Flying insects
About 7,500 individuals, mostly flies, were sampled in 2002. Within each cycle catches were lower in the days after the spray event but increased between the spray cycles and much greater in the subsequent pre-cycle catches. Increase in the abundance of flies over the cycles was most likely due to increasing temperatures and the arrival of the annual flood in the sampled areas.

5.4.4 Fish
Using gill nets, the abundance of fish in the Kwando and Linyanti Rivers were studied. In general, a comparison of relative abundance before and after the five spray cycles remained stable. Relative abundance in fish in a given water body reflects the seasonal events, water levels, feeding and breeding behaviour (Mosepele, 2006). However there was decrease in species diversity from the pre spray period until cycle 2 and diversity remained stable after cycle 5. Being an opportunistic predator (Teferra et al., 2003, Mosepele et al., 2005) slight decrease in *Schilbe intermedius* is caused by feeding of knocked down terrestrial and aquatic invertebrates. Increased competition for food between the species does occur. *Hepsetus odoe* and *S. intermedius* were pscivorous predators (Merron & Bruton, 1991) and deltamethrin would have increased inter-specific competition for food between these two species as the feeding rate of *S. intermedius* progressively decreased from the pre spray periods until cycle 5. *Brycinus lateralis* and *B. poechii* feed on small aquatic invertebrates and therefore it was possible that their abundance decreased in the river systems due to decreased food supply caused by deltamethrin spray. The lack of significant change in the feeding behavior of *Marcusenius macrolepidotus* during the spraying suggests that its dominant prey item (Chironomids) were not significantly affected by deltamethrin.

5.4.5 Birds
Using circular point counts for monitoring forest birds, transects for *Acacia* thorn-veld species, and boat surveys for water dependent species, sampling was carried out on three occasions in the Okavango Delta. In summary, there were some local changes in bird populations during the spraying, but these changes could not be attributed to effects of deltamethrin (Pendleton, 2002; Perkins & Ramberg, 2004). The spraying of deltamethrin over the Kwando-Linyanti areas in 2006 did not have discernible negative effects on any of the bird species monitored (Slaty Egret, Arnot's Chat, raptors and

vultures). The only significant effect recorded was increased feeding success of Slaty Egrets during spraying due to the negative effect of the deltamethrin on its prey. This was short-lived and the temporarily depressed fish stocks did not cause the Slaty Egrets to vacate the area.

6. Aquatic plants (*Salvinia molesta*) and biological control weevil *Cyrtobagous salviniae*

The distribution of salvinia and weevil populations varies widely under field conditions. Therefore, using field and static short-term toxicity bioassay methods (Reish & Cshida, 1987) in the Okavango (Figure 2) and iron cages representing open conditions in Kwando (Figure 3), the impacts of the insecticide on the weevil were determined during the five sequential aerial spray cycles. The controls were maintained in unsprayed area of Khwai River at North Gate. Assessments of the impacts were carried out at 12, 36 and 60 hours after the spray in each cycle from basins and cages as well as in field conditions before and after each spraying cycle (Kurugundla & Serumola, 2007, Kurugundla et al., 2010).

Mean survival of the 50 adult salvinia weevils extracted using Berlese funnels (Boland & Room, 1983) in the controls was generally in the range of 45.5 (mortality = 09.0%) to 48.4 (mortality = 03.2%) in number. The range of weevils' survival in closed basins exposed to deltamethrin in 2002 was between 25.0 (mortality = 47.0%) and 44.7 (mortality = 04.9%) while it was 33.7 (mortality = 27.2%) and 39.3 (mortality = 16.4%) in cages in 2006 with respect to controls (Table 5). However, in the 4th cycle of 2006 spray, the weevil mortality was 52.5% (live weevils = 22.0 ±1.8) (Table 5), which was obviously due to the formation of ice crystals and the cold conditions in the weevil extraction cups ($\leq$ 50C) increasing the weevil mortality significantly (P $\leq$ 0.05), but not necessarily as a result of deltamethrin spray drift. Cesida (1980) found that pyrethroid toxicity increased at low temperatures.

The percentage deltamethrin collected on the aluminium foil sheets was in the range of 0.2% to 6.9% of the applied rate and the insecticide drift is varied depending on the inversion, wind direction and temperature (Perkins & Ramberg, 2002; Bonyongo & Mazvimavi, 2007). Significant difference in weevil mortality was observed between 2.3% and 6.9% (Table 5) deltamethrin deposition and did not show significant mortality at less than 2% deposition as determined from the aluminium foils (Kurugundla & Serumola, 2007; Kurugundla et al., 2010). The fluctuations in the abundance of weevils in the field Salvinia infestations at the sites of Xakanaxa (Figure 2) and Kwando (Figure 3) after praying was due to the spatial distribution of weevils and to less breeding during winter (Forno et al., 1983; Naidu et al., 2000) despite some spray effects. In all five cycles the abundance of weevils in 20 standard plants was higher than one (Forno, 1987) showing that weevils maintained their equilibrium at Shummamori, Hamokata, Lebala and Selinda (Figure 3) during the spray programme.

Although deltamethrin deposition caused significant weevil mortalities, the insecticide did not affect the weevil ability to control salvinia. It might be difficult to relate the declines of weevils in the field to the deltamethrin toxicity, as toxicity is influenced by factors such as temperature, season of spraying, habitat conditions and a protective mechanism possessed by the life stages of weevils (Schlettwein & Giliomee, 1990). In aquatic habitat conditions deltamethrin aerosols could be diluted, partitioned and adsorbed onto various organic sediments (Muir et al., 1985), which would reduce the toxicity not only to the weevils but several aquatic invertebrates. It is also suggested that the vegetation might act as a limiting factor for insecticide deposition on target surfaces, unlike in the open areas. On cold nights the weevils normally hide in buds, roots and beneath the leaves. They deposit their eggs in buds and underneath leaves, and emerging larvae normally feed inside the rhizome (Forno et al 1983). Therefore adults and larvae would often be protected from contact with the insecticide (Schlettwein & Giliomee, 1990).

Cycle	Okavango Delta - Basins – 2002				Kwando-Linyanti - Cages- 2006			
	Control	12 hours	36 hours	60 hours	Control	12 hours	36 hours	60 hours
1	47.2 ±1.2 (5.6%)	*32.0 ±1.3 (32.3%)	*30.5 ±1.4 (35.4%)	*25.0 ±1.0 (47.0%	47.0 ±1.6 (06.0%)	39.3 ±2.4 (16.4%)	37.4 ±4.5 (20.4)	35.7 ±1.8 (24.0%)
2	48.4 ±0.5 (03.2%)	31.7 ±0.9 (34.5%)	38.2 ±1.0 (21.1%)	38.8 ±1.3 (19.9%)	46.5 ±0.6 (07.0%)	39.1 ±2.5 (16.0%)	*35.2 ±3.3 (24.3%)	*35.8 ±2.0 (23.0%)
3	46.0 ±0.6 (08.0%)	42.3 ±0.6 (08.1%)	33.5 ±0.9 (27.2%)	31.0 ±1.4 (32.6%	45.5 ±0.6 (09.0%)	37.7 ±3.3 (17.2%)	*34.4 1.4 (24.6%)	*35.0 ±1.9 (23.1%)
4	47.0 ±0.6 (06.0%)	37.0 ±1.1 (21.3%)	44.7 ±1.3 (04.9%)	37.0 ±1.4 (21.3%	46.3 ±0.9 (07.1%)	*35.9 ±3.1 (22.5%)	*33.7 ±1.7 (27.2%)	*22.0 ±1.8 (52.5%)
5	45.4 ±0.5 (09.2%)	*30.3 ±0.7 (33.3%)	*30.5 ±0.9 (32.8%)	*32.2 ±1.5 (29.1%)	46.3 ±1.3 (07.4%)	*36.1 ±1.7 (22.0%)	36.7 ±1.8 (20.7%)	*34.3 ±3.1 (26.0%)
	Okavango Delta- mean of 5 cycles				Kwando-Linyanti - Mean of 5 cycles			
	Deltamethrin deposition (% m⁻²)		Weevil mortality (%)		Deltamethrin deposition (% m⁻²)		Weevil mortality (%)	
Average	2.7 ±0.5		26.5 ±1.0		4.1 ±0.8		29.7 ±5.1	

Table 5. Mean survival of 50 weevils (SE = sd√n) in the Okavango Delta and Kwando-Linyanti in response to deltamethrin spray deposition (% m⁻²). Figures in parentheses indicate corrected percent mortality with respect to controls in each cycle. * Probability ≤0.05 with reference to controls.

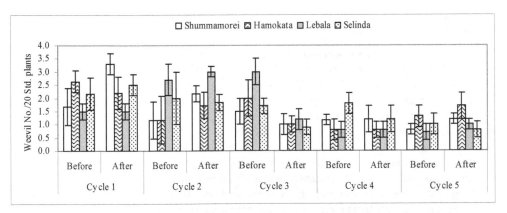

Fig. 7. Number of weevils in 20 plants by standard plant method before and after the spray in five cycles.

Average weevil mortalities are typically 26.5% in 2002 and 29.7% in 2006 at 2.7% and 4.1% deltamethrin respectively (Kurugundla& Serumola 2007; Kurugundla et al., 2010). It was also observed that Paradise pool and Lebala pool were completely covered with the salvinia two months after the end of spraying, yet the weevils controlled the infestation after 7 to 8 months. Aerial spraying of deltamethrin for controlling tsetse fly in any given area is not a continuous process and it is applied only in winter, when the breeding rate of the weevils generally low. The two important monitoring studies conducted in 2002 and 2006 confirm that, although C. *salviniae* was affected negatively by the aerial spraying of deltamethrin, it recovered thereafter as shown by the subsequent effective control of salvinia in Paradise and Lebala pools in the Okavango delta and Kwando River respectively.

7. Socio-economic implications

No side effects on human health was reported and people expressed their appreciation about the programme. However, there were sporadic reports of irritation to eyes during the spraying as reported by the humans. During the spraying campaign, people continued utilizing crops, fish and wild veld products. No short-term land use changes were observed and no disturbances to domestic live stock (Bendsen et al., 2006). People who moved from the core of tsetse infested areas during 1960s and 1970s have now settled permanently in Caprivi region. The changes in land use did not become apparent during the spraying and in the post spray periods. Botswana is one of the prime wilderness tourism destinations and there were no direct or indirect impacts on the tourism inflow as the result of spraying. The successful eradication of the flies has created an enabling environment for livestock development. No stock losses due to *nagana* have occurred after the spray of 2001, 2002 and 2006 and the carrying capacity of the rangelands has increased. Eighty-two commercial livestock farms, at a size of 2000 ha each, have already been sanctioned by the Namibian Government in the western section of the Spray Block area in Caprivi. Only 8% of the lodges were against the spraying and 92% of the tour operators appreciated the tsetse eradication. The successful eradication of tsetse fly would save the Botswana Government the recurrent costs that were invested annually for the control of flies through the maintenance of 10,000 odour baited targets in the tsetse dominated areas (Bonyongo & Mazvimavi, 2008).

8. Conclusions

The incidence of *nagana* in northern Botswana as a result of tsetse fly spread increased between 1950 and 1960. Besides large scale clearing of bush and vegetation, ground sprays using insecticides such as DDT, Dieldrin, endosulfan and deltamethrin have been used to control the spread of tsetse fly. Application of non-residual spraying of endosulfan in the Okavango Delta, coupled with odour bait technique in the northern wetlands in 1992, reduced the tsetse fly distribution from 20,000 km^2 to 5000 km^2. By exploiting the improved aerial spraying techniques (fitted with GPS-guided spray equipment fixed to the aircraft), Botswana Government sprayed deltamthrin in 2001 and 2002 in the Okavango Delta and in 2006 in the Kwando-Linyanti systems. Almost 10 years following the end of spraying in the Okavango Delta, tsetse fly have still not been found and the threat of cattle trypanosomosis has been quelled. Furthermore, the tsetse frontiers involving the northern tsetse fly distribution along the Kwando and Linyanti Rivers bordering Caprivi region in Namibia – which is part of the continental common tsetse fly-belt has been effectively pushed back into Southern Angola. As such, the threat of reintroduction of tsetse fly back into northern Botswana has been greatly reduced.

Endosulfan aerial spraying did not produce serious harm to terrestrial invertebrates and no significant difference between seasonal and spraying effects was found in aquatic invertebrates at 12 g ha^{-1} of endosulfan applications. Possible exceptions included adult Chironomidae and Hymenoptera other than ants, both of which showed some declines in the spraying season. Endosulfan had possible influence on migration of fish and *Tilapia rendalli* abundance declines in shallow vegetated areas. Residue of endosulfan was highest in *Schilbe mystus* at 0.04 ppm in muscle and 0.28 ppm in viscera in gram wet weight. The spraying influenced the feeding in king fisher due to behavioural changes. However, physiological studies in fish showed that surviving fish became significantly debilitated although recovery followed cessation of spraying. However, several groups of invertebrates, especially arthropods, are susceptible to the deltamethrin and deltamethrin spraying caused significant reductions in abundance of sensitive aquatic and terrestrial taxa. The results indicate that the surface dwelling arthorpods were affected in great deal rather than the groups such as leaches, snails, pond damsels and others that live in sediments. High elimination rate recorded for the order Hemiptera (water fly), Ephemeroptera (may flies) and Coleoptera (beetles) as they are active in free water and on vegetation surfaces. However deltamethrin did not affect the fish and birds as the result of deltamethrin spraying.

9. References

Ali, S. (1978). Degradation and environmental fate of endosulfan isomers and endosulfan Sulphate in mouse, insect and laboratory model ecosystems. *Diss. Abstr. Int. B* 39: 2117

Allsopp, R. (1978). The effect of Dieldrin, sprayed by aerial application for tsetse control on game animals. *Journal of Applied Ecology* 15: 117–127.

Allsopp R & Phillemon-Motsu, T.K. (2002). Tsetse Control in Botswana – a reversal in strategy. *Pesticide Outlook* 13: (2) 73-76

Amweg, E. L. Weston, D.P., You, J. & Lydy, M. J. (2006). Pyrethroid Insecticides and Sediment Toxicity in Urban Creeks fromCalifornia and Tennessee. *Environ. Sci. Technol.* 40 (5), 1700-1706.

Anon (1973) FAO/Who Evaluation of Endosulfan Residues in Food. FAO/AGP: CP/103, Rome

Bendsen, H., Mmopelwa, G. & Motsholapheko, M. (2006). Socio-economic issues. *In*: Environmental Monitoring of May-August Aerial Spraying of Deltamethrin for Tsetse Fly Eradication in the Kwando-Linyanti and Caprivi Region – 2007. M. C. Bonyongo & D. Mazvimavi (Eds.). HOORC, University of Botswana, pp 224 – 283

Boland, N. P & Room, P.M. (1983). Estimating population density of *Cyrtobagous singularis* Hustache (Coleoptera: Curculionidae) on the floating weed *Salvinia molesta* using Berlese Funnels. *Journal of Australian Entomological Society* 22: 353-354.

Bonyongo, M. C. & Mazvimavi.D (2007). Environmental Monitoring of May-August 2006 Aerial Spraying of Deltamethrin for Tsetse fly Eradication in the Kwando-Linyanti and Caprivi Region. Department of Animal Health and Production. M. C. Bonyongo and D. Mazvimavi (Eds), HOORC, University of Botswana,. 283pp.

Bonyongo, M. C. & Mazvimavi, D. (2008). Recovery Monitoring After 2006 Aerial Spraying of Deltamethrin for Tsetse fly Eradication in the Kwando-Linyanti and Caprivi Region. Department of Animal Health and Production. M. C. Bonyongo and D. Mazvimavi (Eds.)-2008, HOORC, University of Botswana, 105pp

Bucher, F. &Hofer, R. (1993). The effects of treated domestic sewage on three organs (gills, kidney, liver) of brown trout *(Salmo trutta)*, *Water Research* 27: 255-261

Bucke, D, Vethaak, D. Lang, T. & Mellargaad, S. (1996). International council for the Exploration of the Sea Techniques in Marine Environmental Science (Copenhagen), Common Diseases and Parasites of Fish in the North Atlantic, Training Guide for Identification

Budd, L. (1999). DFID-Funded Tsetse and Trypanosome Research and Development since 1980. Economic Analysis (vol. 2) Department for International Development, UK.

Carsen, R. (1962). *Silent Spring.* Houghton Tifflin (Eds.), New York, USA.

Cesida, J.E. (1980). Pyrethrum flowers and pyrethroid insecticides. *Environmental Health Perspective* 4: 189-202.

Chikwenhere, C. & Shereni, W. (2006). Terrestrial invertebrates. *In*: Environmental Monitoring of May-August Aerial Spraying of Deltamethrin for Tsetse Fly Eradication in the Kwando-Linyanti and Caprivi Region – 2007. M. C. Bonyongo & D. Mazvimavi (Eds.). HOORC, University of Botswana, pp 63-126

Chikwenhere, C. (2007). Terrestrial invertebrates. *In*: Recovery Monitoring After 2006 Spraying Deltamethrin for Tsetse Fly Eradication in the Kwando-Linyanti and Caprivi Region – 2008. M. C. Bonyongo & D. Mazvimavi (Eds.). HOORC, University of Botswana, pp 15-45

Choudary, N., Sharma, M., Verma, M.P. & Joshi, S.C. (2003). Hepato and neurotoxicity in rat exposed to endosulfan *Journal of Environmental Biology* 24: 305-8

Cockbill, G.F. (1979). Do ultra low volume applications of endosulfan present a serious hazard to the environment? Zimbabwe Rhod. *News Science* 13: 136–137, 139 (TTIQ abstract No. 628)

Connell, D. (1999). *Introduction to Toxicology*, Blackwell Science. p 68.

Dangerfield, J. M. (2002) Monitoring of terrestrial invertebrates. *In*: Environmental Monitoring of Tsetse Fly Spraying Impacts in the Okavango Delta. Perkins JS and Ramberg L (Eds.) -2004. Okavango Report Series No. 2. HOORC, University of Botswana, Botswana. pp 52-110.

Dangerfield, J. M. & McCulloch, K. (2003). Monitoring of terrestrial invertebrate recovery. *In*: Environmental Recovery Monitoring of Tsetse Fly Spraying Impacts in the Okavango Delta Perkins J.S and Ramberg L (Eds.) - 2004. Okavango Report Series No. 3. HOORC, University of Botswana, Botswana. pp 101- 157

Davies J.E. (1979). Evaluation of aerial spraying against tsetse (Glossina morsitans centralis MACHADO) in Botswana. Report No. 209. XLVIIth General Session of the O.I.E Committee. 21-26 May 1979. 40pp. Paris

Davies J.E (1980). A History of Tsetse Control in Botswana – pre 1980. Tsetse Control Division, Maun Botswana. pp 165

Davies, J. E. & Bowles, J. (1979). Effects of large-scale aerial applications of endosulfan on tsetse fly, *Glossina morsitans centralis* Machado, in Botswana. Miscellaneous Report No. 45. Center for Overseas Pest Research, London

Davies B., & J. Day. (1988) *Vanishing Waters.* UCT Press, Rondenbosch Cape Town

Dortland, R. J., Van Elsen, A.C., Koeman, J.H. & Quirijns, J.K. (1977). Observations on side-effects of a helicopter-application of endosulfan against tsetse flies in Niger. Unpublished Report, Department of Toxicology, Wageningen University, Holland

Douthwaite, R.J., Fox, P.J., Matthiessen, P & Russel-Smith, A. (1981). Environmental impact of aerosols of endosulfan, applied for tsetse fly control in the Okavango Delta,

Botswana. Final Report of the Endosulfan Monitoring Project, Overseas Development Administration, London. 141pp

EFASP. (1999). *Environmental Fate Assessment for the Synthetic Pyrethroids*; U.S. Environmental Protection Agency, Office of PesticidePrograms, Environmental Fate and Effects Division, U.S. Government Printing Office: Washington, DC

Erstfeld, K. M. (1999). Environmental fate of synthetic pyrethroids during spray drift and field runoff treatments in aquaticmicrocosms. *Chemosphere* 39: (10), 1737-1769.

Everts, J.W. (ed.). (1979). Side-effects of aerial insectivide applications against tsetse flies near Bouaflé, Ivory Coast. Dept. Toxicology, Agric. Univ., Wageningen, The Netherlands. 115 pp.

Feldmann, U. (2004). The sterile insect technique a component of area-wide integrated pest management of tsetse. In: Maudlin, I, Holems, P and Miles, M (Eds.) The Trypanosomiases. CABI Publising,Wallingford, UK, pp.565-582

Ford, J.U. (1971). The Role of the Trypanosomiasis in African Ecology. A Study of the Tsetse Fly Problem. Clarendon Press, Oxford, UK, 568 pp

Forno I. W. (1987). Biological control of the floating fern *Salvinia molesta* in north-eastern Australia: plant-herbivore interactions. *Bulletin Entomological research* 77: 9-17

Forno I. W., Sands D. P. A., & Sextone, W. (1983). Distribution, biology and host specificity of *Cyrtobagous singularis* Hustache (Coleoptera: Curculionidae) for the biological control of *Salvinia molesta*. *Bulletin of Entomological Research* 73: 85-95

Fox, P.J., Matthiessen, P., & Douthwaite, R.J. (1979). ODM Endosulfan Monitoring Project. Interim Report. Maun, Botswana, 25 pp.

Gerber, A & Gabriel, M. J. M. (2002). Aquatic Invertebrates of South African Rivers. Field Guide WQS, DWAF, Private Bag X313, Pretoria.

Graham, P. (1964). Destruction of birds and other wild life, by dieldrex spraying against tsetse fly in Bechuanaland. *Arnoldia* 10 (1): 1–4.

Grant I. F & Crick , H. O. P. (1987). Environmental impact of sequential applications of deltamethrin aerosols applied for tsetse control in Zimbabwe. London. Tropical Development and Research Institute

Hunter C.G. & Robinson, J. (1968). Aldrin, dieldrin and man. *Food Cosmetics and Toxicology* 6:253-260.

http://en.wikipedia.org/wike/endosulfan. Endosulfan 17pp. Retrieved 20 June 2011

Jordon, A. M. (1986). Trypanosomiasis Control and African rural Development, Longman Group Limited, Harlow, UK, 357 pp

Kgori P. M, Modo, S. & Torr, S. J. (2006). The use of aerial spraying to eliminate tsetse from the Okavango Delta of Botswana. *Acta Tropica* 99: 184-199

Kgori P.M, Orsmond G. and Phillemon-Motsu T.K (2009) Integrating GIS and GPS-assisted navigation systems to enhance the execution of SAT-based tsetse elimination project in the Okavango Delta (Botswana) In Cecchi and Mattioli Edn; *Geospatial datasets and analyses, PAAT T&S Series 9*; 61-67

Koeman, J.H. & Pennings, J. H. (1970) An orientational survey on the side-effects and environmental distribution of insecticides used in Tsetse Control in Africa. *Bull. Environm. Cont. Toxicology* 5: 164–170.

Koeman, J.H., Rijksen, H.D., Smies, M., Na'isa, B.K., & MacLennan, K.J.R. (1971). Faunal changes in a swamp habitat in Nigeria sprayed with insecticide to exterminate *Glossina*. Netherlands. *Journal of Zoology* 21 (4): 434–463.

Koeman, J.H., Boer, W.M.J., Den, A. F., Feith, H. H. Longh, P. C., de Spliethoff, B., Na'isa, K. & Spielberger, U. (1978). Three years observation on side-effects of helicopter applications of insecticides to exterminate *Glossina* species in Nigeria. *Environmental Pollution* 15: 31–59.

Kurugundla C. N & Serumola. O. (2007). Impacts of deltamethrin spray on adults of the gaint salvinia bio-control agent, *Cyrtobagous salviniae*. *Journal of Aquatic Plant Management* 45: 124-129

Kurugundla, C. N., Bonyongo, M.C. & and Serumola, O. (2010). Impact of deltamethrin aerial spray on adult Cyrtobagous salviniae in Botswana. *South Africa Journal of Aquatic Science* 35(3): 259-265

Lambrecht, F. L. (1972). The tsetse fly: a blessing or a cause? In: Farver, M. T. Milton, J. P. (Eds.), The Carless Technology-Ecology and International Development. The Natural History Press, Garden City, New York, USA, pp 726-741.

Magadza, C.H.D. (1969). Aerial application of Thiodan for tsetse eradication-an ecological assessment. A.R.C. of Zambia, unpublished report.

Maier-Bode, H. (1968). Properties, effect, residues and analytics of the insecticide endosulfan, *Residue Reviews* 22: pp 1-44.

Masundire, H. M. & Mosepele, B. (2006). Aquatic invertebrates. *In*: Environmental Monitoring of May-August Aerial Spraying of Deltamethrin for Tsetse Fly Eradication in the Kwando-Linyanti and Caprivi Region – 2007. M. C. Bonyongo & D. Mazvimavi (Eds.). HOORC, University of Botswana, pp 127 – 154

Masundire, H. M. (2007). Aquatic invertebrates. *In*: Recovery Monitoring After 2006 Aerial Spraying of Deltamethrin for Tsetse fly Eradication in the Kwando-Linyanti and Caprivi Region - 2008. Department of Animal Health and Production. M. C. Bonyongo and D. Mazvimavi (Eds.), HOORC, University of Botswana, pp 46-73

Matthiessen, P. & Douthwaite, R. J. (1985). The impact of tsetse fly control campaigns on African wildlife. *Oryx*, 19: 202-209

Merron G.S & Bruton, M. N. (1991). The physiological and toxicological effects of Arial spraying with insecticides on the fish stocks of the Okavango Delta, Botswana. Interim Rep. No4, WWF, 12pp, Gland, Switzerland

Mosepele, K., Williams, L. & Mosepele, B. (2005). assessment of the feeding ecology of the Silver catfish (Schilbe intermedius Ruppal 1832) in a seasonal floodplain of the Okavango Delta, Botswana. *Botswana Notes and Records* 37: 208-217

Mosepele, K. (2006). Fish. *In*: Environmental Monitoring of May-August Aerial Spraying of Deltamethrin for Tsetse Fly Eradication in the Kwando-Linyanti and Caprivi Region – 2007. M. C. Bonyongo & D. Mazvimavi (Eds.). HOORC, University of Botswana, pp 172-195

Moulton, T.P. 1973. The effects of various insecticides (especially thiodan and BHC) on fish in the paddy fields of West Malaysia. *Journal of Malaysian Agriculture* 49: 224-253

Muir D.C.G, Rawn, G. P., Townsend, B. E. & Lockhart, W. L. (1985). Bioconcentration of cypermethrin, deltamethrin, fenvalerate and permethrin by *Chironomus tentans* larvae in sediment and water. *Environmental Toxicological Chemistry* 4: 51-61

Müller, P., Nagel, P. & Flacke, W. (1980). Ecological side effects of Dieldrine application against tsetse flies in Adamaoua, Cameroon. Report presented at Expert Consultation on Envir. Impact of Tsetse Control 9–12 June 1980, Rome.

Munshi, J.S.D, Dutta, H.M., Brusle,J., Gonzalez, I. & Anandon, G. (1996).The structure and function of fish liver in: Fish Biology Munshi, J.S.D., H. M. Dutta (eds.) Science Publishers Inc. New York

Naidu K.C., Muzila, I., Tyolo, I. & Katorah, G. (2000). Biological control of *Salvinia molesta* in some areas of Moremi Game Reserve, Botswana. *African Journal of Aquatic Science* 25:152-155.

Palmer, C. G. (2002). Monitoring of Aquatic invertebrates In: Environmental Monitoring of Tsetse Fly Spraying Impacts in the Okavango Delta-2002, Perkins JS and Ramberg L (Eds.) 2004. Okavango Report Series No. 2. April 2004, HOORC, University of Botswana, Botswana. pp31 – 51

Palmer, C. G & H. D. Davies-Coleman (2003) Recovery monitoring of aquatic invertebrates. In: Environmental Recovery Monitoring of Tsetse Fly Spraying Impacts in the Okavango Delta, Perkins J.S and Ramberg L (Eds.) - 2004. Okavango Report Series No. 3, University of Botswana, Botswana. pp 52-100

Pendleton, F. (2002). Monitoring of birds. In: Environmental Monitoring of Tsetse fly Spraying Impacts in the Okavango Delta, Final Report. Perkins JS and Ramberg L (Eds.) -2004. Okavango Report Series No. 2. HOORC University of Botswana, Botswana. pp 111-130.

Perkins J. S and L. Ramberg L (2004) Environmental Monitoring of Tsetse fly Spraying Impacts in the Okavango Delta-2002. Okavango Report Series No. 2., HOORC,, University of Botswana, Botswana. 150pp

Perkins J. S and L. Ramberg L (2004a) Environmental Recovery Monitoring of Tsetse fly Spraying Impacts in the Okavango Delta-2003. Okavango Report Series No. 3., HOORC,, University of Botswana, Botswana. 157pp

Picker, M., Griffiths, C. & Weaving, A. (2002). Field Guide to insects of South Africa. Struil Publishers. 444pp

Reish D. L. & Cshida, P.S. (1987). Manual methods in aquatic environment research 247, part – 10, Short term static bioassay, FAO Food and Agriculture Organization of the United Nations, Fisheries Technical Paper, Rome 62 pp

Roberts, D. (1975). Differential uptake of endosulfan by the tissue of Mytilus edulis. Bull. Environ. Contam. Toxicol. 13: 170-176

RTTCP (1995). Study of the tsetse fly problem common to Angola, Botswana, Namibia and Zambia. Accounting No. 6 ACP PRP 468. 174-pp. RTTCP, Harare, Zimbabwe

Saravana, B. P. & Geraldine, P. (2000). Histopathology of the hepatopancreas and gills of the prawn Macrobrachium malcolmsonii exposed to endosulfan. *Aquatic Toxicology* 50: 331-339

Saunders, D.S. (1962). Age determination for female tsetse flies and the age composition of samples of G. pallidipes, G.p. fuscipes and G. brevipalpis. *Bulluten of Entomological Research 53: 579-595*

Sharma, S.P, Losho, T., Malau, C., Mangate, K.G., Linchwe, K.B., Amanfu, W. & Motsu, T.K. (2001). The resurgence of trypanosomosis in Botswana. *J.S. Afr.Vet. Assoc.* 72, 232 – 234.

Schlettwein C. H. G. & Giliomee, J.H. (1990). The effects of different dosages of the insecticide mixtures endosulfan/alphamethrin on adults of the biological control agent Cyrtobagous salviniae (Coleoptera: Curculionidae) against Salvinia molesta. *Madoqua* 17(1): 37-39

Smies, M., Evers, R.H.J., Piejnenburg, F.H.M. & Koeman, J. H. (1980) Environmental aspects of field trials with pyrethroids to eradicate tsetse fly in Nigeria, *Ecotoxicology, Environment and Safe* 4: 114-128

Sserunjoji, S.J.M. & Tjell, T.C. (1971). Insecticide Residues following Tsetse Control in Uganda. Paper No. 13 of a joint FAO/IAEA panel, Vienna, October 1971.

Takken, W., Bruijckere, F. L. G. & Koeman J. H. (1976). Environmental impact studies concerning the fixed-wing aircraft endosulfan applications in the Southern Guinea Savanna zone in Nigeria. Report Dept. Toxicology Agriculture University. Wageningen, the Netherlands.

Tarimo, C.S. & Pallotti, E. (1979). Residues of aerially sprayed DNOC in biological materials. OAU/STRC, 15th meeting ISCTR 1977: 572-578 (TTIQ abstract No. 776).

Teffera, G., Feledi, B. & Motlhabane , D. (2003). The effects of rainfall on the composition and quality of food ingested by two species of fish *Schilbe intermedius* and *Oreochromis massambicus* in Gaborone Dam. *Botswana Notes and Records* 53: 179-186

Tibayrenc, R. & J. Gruvel (1977) La campagne de lutte contre les glossines dans le bassin du Lac Tchad II. Contrôle de l'assainissement glossinaire. Critique technique et financière de l'ensemble de la campagne. Conclusions générales. *Rev. Elev. Méd. Vét. Pays trop.* 30: 31-39.

Thomas D. R. (1981). *DDT: Scientists, Citizens, and Public Policy.* New Jersey: Princeton University Press

Vale, G.A. and Torr, S.J. (2004). Development of bait technology to control tsetse. In The *Trypanosomiases* (Maudlin, I. *et al.*, (Eds.) pp. 533-546, CABI Publishing.

VEEU-TCD/DAHP (1998) Socio-economics of Tsetse Fly Control in Northern Botswana. In collaboration with DFID UK. 92- pp. Gaborone, Botswana

Wilson, Vivian, J. 1972 Observations on the effect of Dieldrin on wildlife during tsetse fly *Glossina morsitans* control operations in Eastern Zambia. *Arnoldia* (Rhodesia) 5: 1-12

Wolski, P and Huntsman-Mapila, P. (2002). Mateorology and spray deposition. *In*: Environmental Monitoring of Tsetse Fly Spraying Impacts in the Okavango Delta-. Perkins J.S and Ramberg L (Eds.) - 2004. Okavango Report Series No. 2. HOORC, University of Botswana, Botswana. pp 17-27

Part 2

Toxicological Profile of Insecticides

10

Trends in Insecticide Resistance in Natural Populations of Malaria Vectors in Burkina Faso, West Africa: 10 Years' Surveys

K. R. Dabiré[1], A. Diabaté[1], M. Namountougou[1], L. Djogbenou[2],
C. Wondji[3], F. Chandre[4], F. Simard[5], J-B. Ouédraogo[1],
T. Martin[6], M. Weill[7] and T. Baldet[8]

[1]IRSS/Centre Muraz, BP 390 Bobo-Dioulasso,
[2]IRSP/ Ouidah,
[3]Liverpool School of Tropical Medicine,
[4]LIN/Montpellier UMR MIGEVEC
[5]IRD/IRSS, Bobo-Dioulasso UMR MIGEVEC,
[6]CIRAD, UR HORTSYS, Montpellier,
[7]ISEM, CNRS/Université Montpellier 2, Montpellier,
[8]IRD/CIRAD/CREC, Cotonou,
[1,5]Burkina Faso
[2,8]Bénin
[3]UK
[4,6,7]France

1. Introduction

Malaria is a major threat in endemic regions causing at least 1 million deaths each year affecting poor and underserved populations living in tropical and sub-tropical regions. Diseases control ideally entails prevention and treatment of human infections. However, few vaccines are currently available and many pathogens are now resistant to anti-parasitic drugs. Additionally populations from endemic countries have less access to treatments due to economical impediments. Thus, the control of vectors in many instances is the only affordable measure (Beier *et al.*, 2008). Mosquito control is mainly achieved by using insecticides and secondarily bio-larvicides (*Bt*-H14, *B. sphaericus*), predators (fish or copepod predators) or parasitic load (fungi), and/or by modifying the physical environment (WHO, 2006). Insecticides target a vital physiological function, leading to mosquito death. Unfortunately, due to their extremely large numbers and short generation span, mosquito populations evolve very rapidly and become resistant to insecticides, leading to repeated field control failures. Resistance results from the selection of mutant individuals able to survive and reproduce in presence of insecticide, the insecticide failing to disrupt the function of its target. In 2007, more than 100 mosquito species were resistant to at least one insecticide, some species being resistant to several compounds (Whalon *et al.*, 2008). Very few classes of synthetic insecticides are available today for vector control, the most recent has been introduced 20 years ago and none are expected in the near future (Nauen, 2007). The low availability of insecticides due to resistance is further reduced in many countries by the removal from the market of compounds for public health because for their toxicity for humans or the lack of specificity in non-target species (Rogan & Chen, 2005).

Resistance is a genetic adaptation to the modification of the environment induced by insecticides. It usually appears locally, sometimes independently in different places, but may spread rapidly through migration (Brogdon & McAllister, 1998; Weill *et al.*, 2003). However, mosquito resistance is not only due to the insecticides used for mosquito control, but also to the many pesticide pollutions present in their environment which are generated by a large variety of human activities including insect control for agriculture and other house-hold protections. These pollutions may dramatically affect resistance genes dynamics and threaten vector control strategies. The overall pesticides pressure that select resistance in mosquitoes need to be clarified, both in terms of insecticides usage and quantity.

An. gambiae is a complex, with seven sibling species that are closely related and morphologically indistinguishable from each other by routine taxonomic methods (Gillies & Coetzee, 1987). These sibling species are however different with respect to ecological and behavioral characteristics and to vectorial competence. In West Africa, *An. gambiae s.s.* and *An. arabiensis* are the two main species of the complex that transmit malaria, with the former being the most efficient vector due to its high anthropophily (White, 1974, Lemasson *et al.*, 1997). Previous study carried out on the species composition in Burkina Faso indicated that *An. gambiae s.l.* was found to be a mixture of *An. gambiae s.s.* and *An. arabiensis* across the Sudan (98.3% *vs.* 1.7%), Sudan-sahelian (78.6% *vs.* 21.4%) and the Sahel (91.5% *vs.* 8.5%) ecotypes (Dabiré *et al.*, 2009a). *An. gambiae s.s.* contains two molecular forms, M and S, which co-exist in West Africa (della Torre *et al.*, 2005). The M form was predominant in permanent breeding sites such as rice fields, whereas the S form was predominant in temporary habitats notably rain-filled puddles which are productive during the wet season. In Burkina Faso, genes conferring resistance to insecticides display large frequency differences in M and S forms of *An. gambiae s.s.* and *An. arabiensis*. Resistance of *An. gambiae s.l.* to DDT and pyrethroids (PYR) is especially conferred in West Africa by mutation of the sodium channel target site, the L1014F *kdr* (Chandre *et al.*, 1999; Diabaté *et al.*, 2002; Awolola *et al.*, 2005; Nguessan *et al.*, 2007). Burkina Faso is composed of three agro-climatic zones and the use of insecticides to control agricultural and human health pests varies considerably in the different zones particularly as the main cotton cropping areas are found in the south west of the country. In this last region, the intensive use of insecticides most notably for fighting the cotton *Gossypium hirsutum* L. pest is thought to have selected insecticide resistance genes in mosquitoes whose breeding sites are exposed to pesticide runoff (Diabaté *et al.*, 2002; Dabiré *et al.*, 2009a & b). The goal of this chapter is to summarise the resistance to insecticides status mainly in *An. gambiae s.l.* populations throughout these different agro-climatic areas and to discuss how it could limit the efficacy of malaria vector control strategies in short and long terms at the country scale. Such information is vital to determine the suitability of pyrethroids used for bednet impregnation and CX or OP based-combinations for indoor residual spraying (IRS).

2. Materials and methods

In Burkina Faso country-wide surveys associating bioassays and molecular investigations were carried out from 2000 to 2010 through 26 localities and they allow updating the resistance status to DDT and pyrethroids and the distribution of L1014F *kdr* among *An. gambiae s.l.* into different agro climatic zones (table 1). We were also interested more recently from 2007 to perform bioassays with some insecticides among OP and CX and also to detect the *ace-1R* mutation 2-3 days aged females of *An. gambiae s.l.* issued from wild larvae were exposed to several molecules such as DDT 4%, permethrin 0.75%, deltamethrin 0.05%, bendiocarb 0.1%, chlorpyriphos methyl (CM) 0.04%, carbosulfan 0.04% and fenithrotion 0.04% according to the WHO tube protocol (WHO, 1998). These active molecules were chosen as they represented each family of classic insecticides commonly used in public health. Some of these molecules such as permethrin, deltamethrin and bendiocarb are now in use in Burkina Faso through impregnated bednets and IRS application.

Study sites: Burkina Faso covers three ecological zones, the Sudan savannah zone in the south and west, the arid savannah zone (Sudan-sahelian) which extends throughout much of the central part of the country and the arid land (Sahel) in the north. The northern part of the country experiences a dry season of 6-8 months with less than 500 mm of rainfall per year. Rainfall is heaviest in the south-west (5-6 months) with a relatively short dry season. The varied ecological conditions are reflected in the different agricultural systems practiced throughout the country, from arable to pastoral lands. The western region constitutes the main cotton belt extending to the south where some new cotton areas have been cultivated. All ecological zones support the existence of *Anopheles* species that vector malaria and the disease is widespread throughout the country (Figure 1).

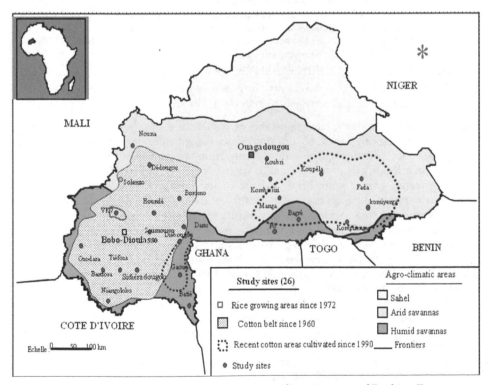

Fig. 1. Localities of the study through the three agro-climatic zones of Burkina Faso.

Study sites	Geographic references	Climatic areas	Environment	Agricultural practices	Recent date of collection
Batié	9°80'N; 2°90'W	Sudanian	rural	cereals, cotton	20/08/09
Gaoua	10°40'N; 3°15'W	Sudanian	sub-urban	cereals, cotton	20/08/09
Diébougou	10°95'N; 3°24'W	Sudanian	sub-urban	cereals, cotton	20/08/09
Dano	11°10'N; 3°05'W	Sudanian	rural	cereals, cotton	20/08/09
Banfora	10°60'N; 4°70'W	Sudanian	sub-urban	cereals, cotton	15/08/09
Sidéradougou	10°60'N; 4°25'W	Sudanian	rural	cereals, cotton	15/08/09
Tiéfora	10°50'N; 4°50'W	Sudanian	rural	cereals, cotton	15/08/09
Orodara	11°00'N; 4°91'W	Sudanian	rural	fruits, cotton	15/08/09
Dioulassoba	11°22'N; 4°30'W	Sudanian	traditionnal-urban	swamp	15/08/09
Soumousso	11°01'N; 4°02'W	Sudanian	rural	cotton	15/08/09
VK7	11°41'N; 4°44'W	Sudanian	rural	rice, cotton	08/08/09
VK5	11°24'N ; 4°23'W	Sudanian	rural	rice	08/08/09
Pô	11°20'N; 1°10'W	Sudanian	sub-urban	cereals, cotton	28/08/09
Houndé	11°50'N; 3°55'W	Sudanian	sub-urban	cotton	10/08/09
Boromo	11°75'N; 2°92'W	Sudan-sahelian	sub-urban	cotton	16/08/09
Solenzo	12°37'N; 3°55'W	Sudan-sahelian	rural	cotton	16/08/09
Dedougou	12°50'N; 3°45'W	Sudan-sahelian	sub-urban	cotton	16/08/09
Nouna	12°70'N; 3°90'W	Sudan-sahelian	sub-urban	cotton	16/08/09
Koubri	12°35'N; 1°50'W	Sudan-sahelian	rural	vegetables	28/08/09
Kombissiri	12°05'N; 1°35'W	Sudan-sahelian	rural	vegetables, cotton	28/08/09
Manga	11°66'N; 1°05'W	Sudan-sahelian	sub-urban	cereals, cotton	28/08/09
Koupela	12°20'N; 0°40'W	Sudan-sahelian	sub-urban	cotton	30/08/09
Fada	12°05'N; 3°55'E	Sudan-sahelian	sub-urban	cotton	30/08/09
Kompienga	11°30'N; 0°40E	Sudan-sahelian	rural	vegetables, cotton	30/09/09
Komiyenga	11°70'N; 0°60E	Sudan-sahelian	rural	cotton	30/09/09
Yamtenga	12°21'N ;1°31'W	Sudan-sahelian	peri-urban	swamp	28/08/09

Table 1. Main study sites across the country from where natural populations of *An. gambiae s.l.* were collected for susceptibility tests to insecticides in Burkina Faso.

Mosquitoes sampling: To evaluate the status of resistance of *An. gambiae s.l.* to insecticides in the three ecological zones of Burkina Faso, anopheline larvae were sampled in countrywide collections during the rainy season, from September to October. Larvae were collected at each locality from breeding sites such as gutters, tires, swallow wells and pools of standing water. Larvae were brought back to the insectary and reared to adulthood. When it was not possible to collect larvae because of the distance between the sampling site and the insectary or due to sampling constraints at the site such as excessive rainfall or flooding, alternative collections of adult mosquitoes were made using indoor aerosol insecticide spraying. *An. gambiae s.l.* were identified morphologically using standard identification keys of Gillies & Coetzee (1987). The results presented here summarized those of transversal studies in whole country 2000 and 2009 with particular focus on the period from September to October 2009.

Insecticide susceptibility test: Susceptibility test was performed on 2-3-day-old *An. gambiae s.l.* females provided by larva collections using the WHO standard vertical tube protocol. Three insecticide-impregnated papers were used: DDT 4%, permethrin 0.75% (cis:trans = 25:75), deltamethrin 0.05%, bendiocarb 0.1%, CM 0.04%, carbofuran 0.04% and fenithrotion 0.04%. Mosquitoes were tested against "Kisumu" a fully susceptible reference laboratory strain. Mortality controls were carried out by exposing both the "Kisumu" strain and wild populations from each site to non-insecticidal impregnated paper. After 1 h exposure,

Trends in Insecticide Resistance in Natural Populations of Malaria Vectors in Burkina Faso, West Africa: 10 Years' Surveys

199

mosquitoes were transferred into insecticide free tubes and maintained on sucrose solution. Final mortality was recorded 24 h after exposure. The threshold of susceptibility was fixed at 98% for the four active molecules according to the protocol of WHO (1998). Are considered as susceptible, suspected resistant and resistant populations with respectively 100 to 98%, 98 to 80 %and under 80% of mortality rates. Dead and survivor mosquitoes were grouped separately and stored on silicagel at -20°C for subsequent PCR analysis.

Molecular analysis: DNA extraction and PCR identification of the *An. gambiae* M and S and *An. arabienis*: Genomic DNA was extracted from individual mosquitoes according to a slightly modified version of the procedure described by Collins *et al.* (1987). After quantification of the extracted DNA, adults of *An. gambiae s.l.* tested in bioassay were processed by PCR for molecular identification of species of the *An. gambiae* complex and molecular forms respectively (Scott *et al.*, 1993; Favia *et al.*, 2001). Those survived or dead in bioassay were after processed in other PCR analysis for the detection of *kdr* and *ace-1R* mutations. For *kdr* detection, a sub-sample of 30 mosquitoes per site of the permethrin/deltamethrin-tested specimens and those collected by indoor spraying were processed by PCR for prior species identification and molecular characterisation of M and S forms of *An. gambiae s.s.* according to Scott *et al.* (1993) and Favia *et al.* (2001) respectively. The frequency of the L1014F mutation in the same samples was determined by allele-specific PCR as described by Martinez-Torres *et al.* (1998).

A*ce-1R* mutation was detected using the PCR-RFLP assay described by Weill *et al.* (2004) with minor modifications. Specific primers, *Ex3AGdir* (GATCGTGGACACCGTGTTCG) and *Ex3AGrev* (AGGATGGCCCGCTGGAACAG) were used in PCR reactions (25µl) containing 2.5µl of 10X *Taq* DNA polymerase buffer, 200µM of each desoxynucleoside triphosphate (dNTP), 0.1U of *Taq* DNA polymerase (Qiagen, France), 10pmol of each primer and approximately 1 to 10ng of DNA. PCR conditions included an initial denaturation step at 94°C for 5min followed by thirty five cycles of 94°C for 30s, 54°C for 30s and 72°C for 30s, with a final extension at 72°C for 5min. Fifteen microlitres of PCR product was digested with 5U of *Alu*I restriction enzyme (Promega, France) in a final volume of 25 µl at 37°C for 3 hours. Products were then analysed by electrophoresis on a 2% agarose gel stained with ethidium bromide and visualized under UV light.

Statistical analysis: The proportion of each species and molecular forms were compared between the study sites. The frequencies of *kdr* and *ace-1R* mutations were calculated according to the formula $p = 2AA+Aa/2n$ where AA was the number of homozygotes, Aa the number of heterozygotes and n the size of analyzed sample. It was compared between sites and between *An. gambiae* M and S molecular forms and *An. arabiensis* by chi square tests. The genotypic frequencies of ace-1R in mosquito populations were compared to Hardy-Weinberg expectations using the exact test procedures implemented in GenePOP (ver.3.4) software (Raymond & Rousset 1995)

3. Resistance to pyrethroids and organochlorine

Reports of resistance in mosquito vector populations in Burkina Faso appeared as early as the 1960 s, when *An. funestus* and *An. gambiae s.l.* populations that showed resistance to dieldrin and DDT, were described (Hamon *et al.*, 1968a; Hamon *et al.*, 1968b). More recent studies have confirmed that resistance to DDT4% is still prevailing with highest level in *An. gambiae s.l.* populations in Burkina Faso where also resistance to certain pyrethroids was increasingly reported (Diabaté *et al.*, 2002, 2004a; Dabiré *et al.*, 2009a). Indeed *An. gambiae s.l.*

populations were resistant to DDT4% in every part of the country and mortality rates below 60 % were observed at the country scale (Fig. 2). They were found also resistant to permethrin 0.75% in the Sudan climatic zone in the western region and also in several sites in the central part of Burkina Faso (Fig. 3). Surprisingly, except for the areas with a very long history of cotton cropping, the tested populations of *An. gambiae* remained susceptible to deltamethrin 0.05%, although decreased mortality values lead to suspect an emergence of resistance in the ongoing years (Dabiré *et al.*, 2009a). However, this result should be interpreted with caution as the resistance is a progressive process and recent data recorded in 2009-2010 showed 5 sites mostly located in the central region remained susceptible foci (Fig. 4).

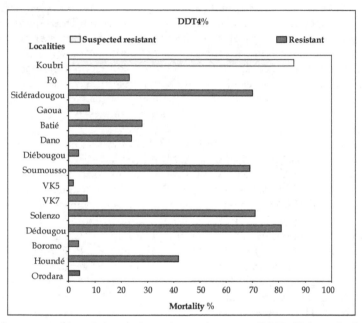

Fig. 2. Mortality rates of *Anopheles gambiae s.l.* populations to DDT 4% from Sudan, Sudan-sahelian and sahelian areas in Burkina Faso.

Resistance is the result of a limited number of physiological mechanisms; it is often monogenic and due to point mutations in a structural gene, gene amplification or changes in transcriptional regulation (Hollingworth & Dong, 2008). It results from the selection of mutant individuals able to survive and reproduce in presence of insecticide, the insecticide failing to disrupt the function of its target (Whalon *et al.*, 2008). The resistance phenotype to pyrethroids and DDT 4% observed in natural populations of *An. gambiae s.l.* was already attributed to a *kdr* mutation as it is the major mechanism involved in cross resistance to pyrethroids and DDT4% in West Africa (Chandre *et al.*, 1999; Diabaté *et al.*, 2002). Until recently, it was assumed that this mutation was the L1014F substitution in West Africa while the L1014S substitution was found in the East (Ranson *et al.*, 2000). However, we now know that both mutations coexist in some countries and are widely distributed throughout sub-Saharan continent and also in Benin and Burkina Faso (Verhaeghen *et al.*, 2006; Etang *et al.*, 2006; Djegbe *et al.*, 2011; Dabiré *et al.*, unpublished).

Trends in Insecticide Resistance in Natural Populations of Malaria Vectors in Burkina Faso, West Africa: 10 Years' Surveys

201

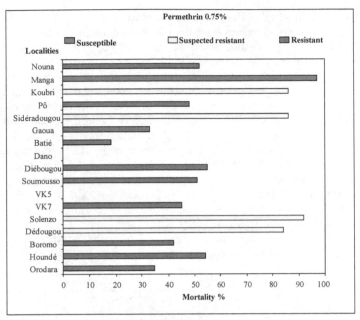

Fig. 3. Mortality rates of *An. gambiae s.l.* populations to permethrin 0.75 % from Sudan, Sudan-sahelian and sahelian areas in Burkina Faso.

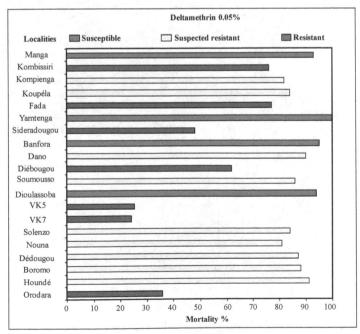

Fig. 4. Mortality rates of *An. gambiae s.l.* populations to deltamethrin 0.05 % from Sudan, Sudan-sahelian and sahelian areas in Burkina Faso.

In Burkina Faso, the frequency of the L1014F *kdr* mutation was first described in the S form of *An. gambiae s.s.* in high frequencies especially in the Western part of the country where the use of insecticides is intensive in agriculture (Chandre *et al.*, 1999, Diabaté *et al.*, 2002). But few years later it had been also found within the M form and was suspected to be the result of an introgression from the S form *An. gambiae s.s.* (Weill *et al.*, 2000; Diabaté *et al.*, 2003). Up to day the distribution of this mutation at the country scale is variable, ranging from 0.5 to 0.97 for the S form in the Sudanian region and decreasing in the Sudano-sahelian and Sahelian areas with averaged values fluctuating between 0.1 and 0.6. Compared to 2000 data (Diabaté *et al.*, 2004a), the frequency of L1014F *kdr* mutation increased notably from 2004 to 2006 before getting stable around the fixation level in some localities (Fig. 5). As mentioned above, no *kdr* was detected in 1999 in the M form (Chandre *et al.*, 1999). But early in 2000, the L1014F mutation was identified from few specimens of M form from rice growing area, peaking maximally at 0.04 (Diabaté *et al.*, 2003). Nowadays the L1014F *kdr* has increased drastically in the M form with varying frequencies between climatic areas, and reaching high frequencies (0.93) in cotton growing belts with a geographic expansion to the sudano-sahelian region where it was formerly absent (Dabiré *et al.*, 2009a). It has also increased in *An. arabiensis* (0.28) where it was formerly reported only from one specimen in 2002 (Diabaté *et al.*, 2004b) (Fig. 6).

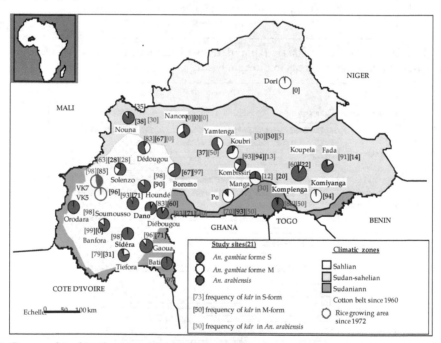

Fig. 5. Geographic distribution of L1014F *kdr* allele in *An. gambiae s.l.* populations inducing pyrethroids and DDT resistance profile in Burkina Faso in 2009 [numbers in bracket represent frequency of L1014F *kdr* allele frequencies].

Globally the distribution of DDT and pyrethroids resistance in regions of intensive cotton cultivation suggests that indirect selection pressures from the agricultural use of insecticides may be responsible for the development of resistance in *An. gambiae s.l.* populations. The

Trends in Insecticide Resistance in Natural Populations of Malaria Vectors in Burkina Faso, West Africa:
10 Years' Surveys

203

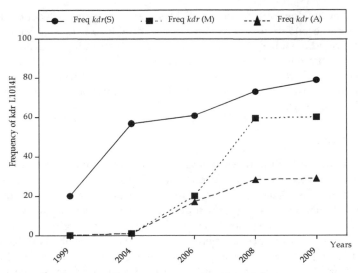

Fig. 6. Evolution of allelic frequencies (in percentages) of L1014F *kdr* in natural populations of *An. gambiae s.l.* from 1999 to 2009 in Burkina Faso[S: *An. gambiae* S form; M: *An. gambiae* M form, A: *An. arabiensis*].

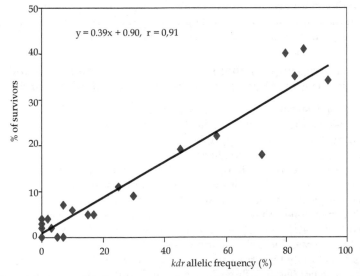

Fig. 7. Correlation between *kdr* frequency and mortality rates of *An. gambiae s.l.* tested with permethrin 0.75%.

mean frequency of the *kdr* allele was the highest in populations from the Sudanian zone where the lowest mortality rates to pyrethroids and DDT 4% were seen in bioassays. By contrast, the *kdr* allele frequency was lower in *An. gambiae s.s.* populations is central and eastern sites where cotton cultivation is recent. A majority of susceptible phenotypes were observed in wild populations of *An. gambiae* from these areas. In addition, the results of this study suggest that the domestic use of insecticides may also exert a selection on *An. gambiae* populations that is secondary to that from the agricultural insecticides. Indeed all collections made in cities located outside the cotton belt showed high mortality rates and a relative low frequency of *kdr* compared to those of cotton belt. The correlation between the high frequencies of L1014F *kdr* mutation and the proportion of surviving individuals after DDT/pyrethroids exposures (figure 7) suggests that this mutation is the main mechanism of resistance to these insecticides.

4. Metabolic resistance

Most studies conducted in Burkina Faso have focused on the modification of target sites by mutation and did investigate the occurrence and the role of metabolic resistance in the observed resistance of *An. gambiae s.l.* Recent tests performed in Dioulassoba (an old central district of Bobo-Dioulasso crossed by the Houet river) showed that *An. arabiensis* from urban polluted breeding sites was resistant to DDT 4% but fully susceptible to pyrethroids and OP/CX, suggesting an existence of metabolic resistance probably GST which is more specific to DDT acting as the main resistance mechanism (Dabiré *et al.*, unpubl.). Even more recently, preliminary results gathered only in VK7 (a sample from a rice growing area surrounded by cotton fields) showed an overexpression of detoxifying enzymes such as glutation-S-transferases, cytochrome P450 oxygenases in populations of *An. gambiae s.s.* with high *kdr* frequencies suggesting the existence of multi-resistance mechanisms to pyrethroids (Fig. 8A,B&C). But more investigations are needed to better address the role of metabolic components on the expression of resistance phenotypes observed in natural populations of *An. gambiae s.l.* especially in areas where insecticide pressure is high.

5. Resistance to organophosphates (OP) and carbamates (CX) and geographic distribution of *ace-1^R* mutations and duplicated *ace-1^D* allele

In Burkina Faso the resistance to OP/CX has been monitored since 2002 only in few sites of the Western areas of the country, and lately extended to the country scale since 2006. Although fenithrotion 0.4%, chlorpyriphos methyl (CM) 0.4%, carbosulfan 0.4% and bendiocarb 0.1 % were tested, the monitoring was well sustained only with bendiocarb 0.1% which was expected to be used in Burkina Faso indoor residual spraying to supplement the efficacy of ITNs especially in localities where *An. gambiae* is resistant to PY. Except for CM 0.4% for which *An. gambiae* populations were fully susceptible irrespectively of the locality, the other OP/CX mentioned above showed mortality rates ranging from 5% to 100% (Fig. 9). The lowest mortality rates were obtained with carbosulfan 0.4% (5%) and bendiocarb 0.1% (20%) especially in areas located in cotton belt such as Houndé, Orodara, Tiefora and Banfora. The susceptibility to bendiocarb 0.1% was also recorded in the central areas where cotton growing is recent (Fig. 10). From 2005 on, the detection of *ace-1^R* mutation involved in OP/CX resistance allowed to evaluate the distribution of this allele in field populations of *An. gambiae s.l.* The characterisation of this allele was based on a PCR-RFLP diagnostic (Weill *et al.*, 2004) that allow the identification of the amino-acid substitution, from a glycine to a serine at the position 119, in the AChE1 catalytic site (G119S). In *Culex pipiens*, there is direct and indirect evidence that the resistance allele (*ace-1^R*) entails a large fitness cost, probably due to the mutated AChE1 having a much lower level of activity. Homozygous *ace-1^R* mosquitoes survive in the presence of insecticide, but are rapidly outcompeted in the

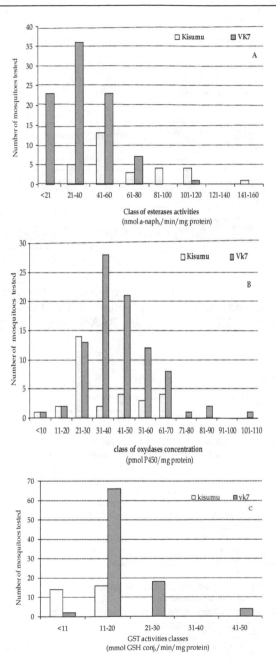

Fig. 8. Activity of detoxifying enzymes such as esterases (A), oxydases (B), and GST (C) in natural populations of *An. gambiae* from Vallée du Kou (VK7) compared to that of *An. gambiae* "Kisumu" (susceptible reference strain). Note the over-expression of oxydases and GST in the VK7 sample.

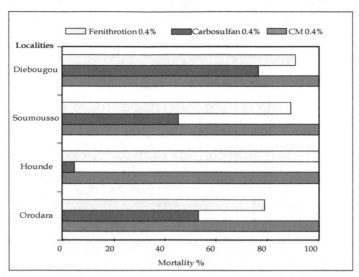

Fig. 9. Mortality rates of *An. gambiae s.l.* populations exposed to Chlorpyriphos methyl (CM) 0.04 %, carbofuran 0.04% and fenithrotion 0.04% from four sites located on the cotton belt in South west of Burkina Faso [100-98%=susceptible; 98-80%= suspected resistance; <80%=resistant].

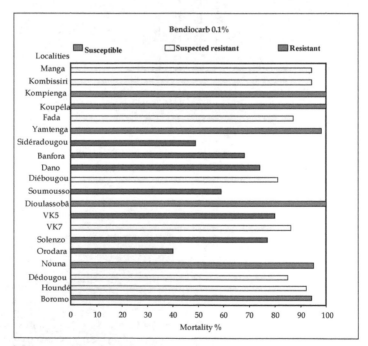

Fig. 10. Mortality rates of *An. gambiae s.l.* populations exposed to bendiocarb 0.1% from Sudan, Sudan-sahelian and sahelian areas in Burkina Faso.

Trends in Insecticide Resistance in Natural Populations of Malaria Vectors in Burkina Faso, West Africa:
10 Years' Surveys

207

absence of insecticide. There are evidences that the same phenomenon exists also in *An. gambiae s.s* (Djogbenou *et al.*, 2010). Even though the results of bioassays were more recent (2009) we presented only the *ace-1R* frequencies from 2006 to 2008 samples.

This mutation was distributed throughout the Sudan and Sudan-Sahelian localities reaching relative high frequencies (0.6) in the South-West, moderate frequencies (<50%) in the central region, and being absent in the Sahel. It was far more frequent in the S form than in the sympatric M mosquitoes (averaging in mean 0.32 for the S form *vs.* 0.036 for the M form) (Djogbenou *et al.*, 2008a). Even though the *ace-1R* mutation was spread across two climatic zones, it was recorded mostly in the cotton growing areas (Dabiré *et al.*, 2009b). Although the *ace-1R* mutation was less spread within the *An. gambiae* s.s. M form, the highest frequency (0.63) was recorded in this form at Houndé located just on the limit of the Sudan region in 2008 (Fig. 11). The observed genotypic frequencies were not significantly different from Hardy-Weinberg expectations at the 95% confidence level in populations from any site except in the *An. gambiae* s.s. S form population from Orodara, where an excess of heterozygotes was observed. S-form samples from a number of other sites also showed a higher than expected number of heterozygous genotypes including Banfora (expected 10, observed 13), Diebougou (expected 11, observed 16) and VK7 (expected 12, observed 17) although Hardy–Weinberg equilibrium was not rejected. These results suggest that a fitness cost is associated to this mutation (Labbé *et al.*, 2007), but see the next paragraph. No *An. arabiensis* was detected up today carrying the *ace-1R* mutation.

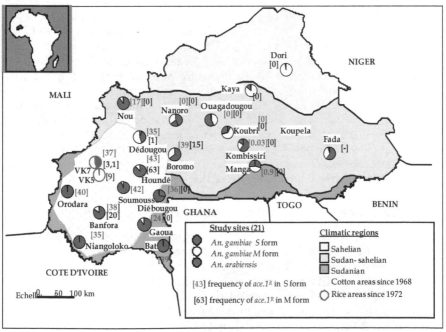

Fig. 11. Geographic distribution of *ace-1R* allele in *An. gambiae s.l.* populations inducing OP/CX resistance profile in Burkina Faso in 2008 [numbers in bracket represent percentage of *ace-1R* allele frequencies]. No *An. arabiensis* was found carrying ace-1R allele so we did not represent its frequency for this species.

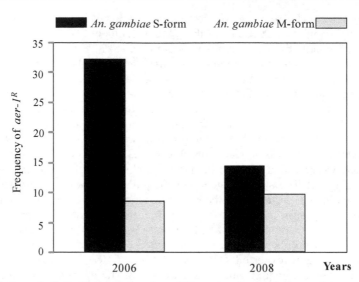

Fig. 12. Evolution of allelic frequencies of *ace-1R* in natural populations of *An. gambiae s.s.* in Sudan area (cotton belt) from 2006 to 2008 in Burkina Faso.

The frequency evolution of this allele during the two years is not regular (fig.12), but considering only the Sudan area, it seems to decrease within the S form and increase slightly in the M form from 2006 to 2008 (Fig. 12). As no data existed before 2006, we did not know the trait of evolution of this allele in the past. That should greatly contribute to explain the inverse tendency of this allele within the two forms because the resistance pattern is complex in this area where excess of heterozygous for *ace-1R* allele should probably co-exist with the duplication allele Ag-*ace-1D* (see the next paragraph). However the allele frequencies in the two forms need to be compared statistically from a solid sample sizes. Regular monitoring of the same localities with the same protocols should give a better insight of the evolution of the G119S mutation of the *ace-1* gene. Data will have to be analyzed in relation with the possible coexistence of other resistance mechanisms such as *kdr* mutation or metabolic based resistance as well as with the existence of the duplicated allele Ag-*ace-1D* which may decrease the fitness cost of this mutation (Berticat *et al.*, 2008).

6. Duplication of *ace-1^D* allele in *An. gambiae s.s.* from Burkina Faso

The G119S mutation conferring resistance to organophosphates and carbamates was distributed throughout the Sudan and Sudan-sahelian correlated with the cotton growing areas. This mutation has been identified in the *ace-1R* allele, and recently in a "duplicated allele" (*ace-1D*), putting in tandem a susceptible and a resistant copy of *ace-1* on the same chromosome. The ace-*1D* has been recorded in field populations of *An. gambiae* M and S forms and was shown to have come from the same duplication event in both forms (Djogbenou *et al.*, 2008b). A unique *ace-1D* allele has been observed in Côte d'Ivoire and Burkina Faso, with an estimated frequency >50% in some populations (Djogbenou *et al.*, 2009).

In Burkina Faso, the *ace-1D* allele frequency could reach 50% and is mainly present in the S form principally in the old cotton belt in the South West. The duplicated allele was also observed on the littoral of Ivory Coast with high frequencies in the M form, and may be present at a low frequency in Benin and Togo (Djogbenou *et al.*, 2010) (Fig. 13).

If, as suspected by Labbé *et al.* (2007), *ace-1D* allele has a lower fitness cost than *ace-1R*, it would increase dramatically the diffusion of OP and CX resistance in *An. gambiae s.s.* natural populations.

Presently, there is no simple test to characterized the duplicated allele *ace-1D* as diagnostic tests do not discriminate between heterozygotes *ace-1S/ace-1R* and genotypes with the *ace-1D* allele. Formal identification of *ace-1D* in a field female thus necessitate to cross this female with a susceptible male, and screen its offspring for CX resistance (see Labbé *et al.* 2007 for detailed procedure). This is clearly not possible on a large number of specimens and to better address the role and the impact of the *ace-1* duplication in resistance schemes, it appears urgent to build an *ace-1D* homozygous laboratory strain to investigate how the duplication modifies the fitness of its carriers.

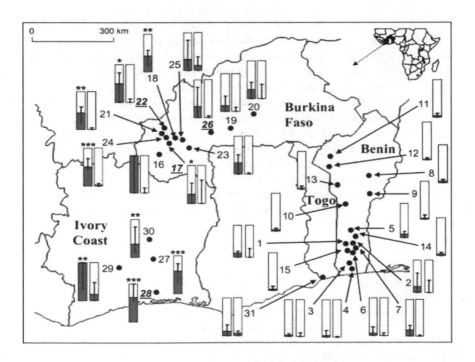

Fig. 13. Ag- *ace-1D* frequency in Western Africa. The frequency is given for each *An. gambiae* molecular form: S (red) and M blue). Samples in which *ace-1D* was detected by molecular analysis are bolded and underlined. Significant presence of the duplicated allele (before Bonferroni correction) is given with * for $P < 0.05$, ** for $P < 0.01$ and *** for $P < 0.001$. (Figure from Djogbenou *et al.*, 2009 in Malaria Journal)

7. Multiresistance status in natural populations of *An. gambiae s.l.*: the coexistence of *kdr* and *ace-1^R* mutations

It was in 2005 when the detection of *ace-1^R* (G119S) mutation was systematically performed for the first time in *An. gambiae s.s.* natural populations from Burkina Faso (Dabiré *et al.*, 2008). This mutation was found together with *kdr* mutation within the same populations. That suggests the existence of multiresistance mechanisms occurring in the same populations. The same individuals were found carrying the two genes but the *kdr* always appeared as homozygous (*kdr^R/kdr^R*). The functional links of the two genes needed to be further investigated (Dabire *et al.*, 2008). Indeed as indicated by the results of bioassays, the occurrence of such multiresistance mechanisms should explain why *An. gambiae s.s.* populations are becoming resistant to all classes of insecticides especially in the South west parts of the country. The individuals carrying the two genes appeared to be phenotypically more resistant to pyrethroids and bendiocarb than those carrying only *kdr* or *ace-1^R*, respectively. Such a synergy between *kdr* and *ace-1^R* has been observed in *Culex pipiens* (Berticat *et al.*, 2008). Assuming that the *ace-1^R* is associated to a high fitness cost (Djogbenou *et al.*, 2010), it should be interesting to investigate how the *kdr* mutation influences the fitness cost related to *ace-1^R*. Also, it was previously shown in *Culex pipiens* that mosquitoes carrying both *kdr* and *ace-1^R* mutations suffer less cost than the one carrying only *ace-1^R* (Berticat *et al.*, 2008). Both this synergy between pyrethroids and OP/CX resistance mechanisms and the spread of the *ace-1^D* allele in natural populations of *An. gambiae s.s.* could largely hamper the expected results of using OP/CX as alternative insecticides to the PYR becoming ineffective by the presence of *kdr*. It is crucial to build laboratory colonies carrying the two mutations from which the benefit of changing insecticides could be properly tested. More recently, in 2011, we also recorded metabolic based resistance in *An. gambiae* from Burkina Faso. Even though these results are preliminary, they further complicate the pattern of resistance in this country, and may represent a dramatic threat for malaria vector control in the near future.

In conclusion, all these aspects need to be properly addressed by fine fundamental research to decrypt the link between the resistance schemes and to give sense in vector control point of view.

8. Other malaria vectors

Anopheles funestus belongs to a group of no less than nine species that are difficult to distinguish based solely on morphological characters of a single life stage (Gillies and Coetzee 1987, Harbach 1994). Species identification difficulties have been recently addressed by molecular techniques based on the polymerase chain reaction (PCR) by using a cocktail of species-specific primers permitting the identification of the six most common species of the group (Koekemoer *et al.*, 2002). Recent analyses of rDNA sequences (Cohuet *et al.*, 2003) revealed the occurrence, in West and Central Africa, of a new taxon morphologically related to *An. rivulorum* Leeson, which is provisionally named *An. rivulorum*-like, thereby enlarging the number of members of the *An. funestus* group to 10 species. Among all the members of the *funestus* group, *An. funestus* s.s. is the most anthropophilic species, and it is considered as the only major malaria vector (Coetzee & Fontenille, 2004), although in a Tanzanian village the circumsporozoite protein of *Plasmodium falciparum* was detected by immunological techniques in some *An. rivulorum* specimens (Wilkes *et al.* 1996).

Trends in Insecticide Resistance in Natural Populations of Malaria Vectors in Burkina Faso, West Africa:
10 Years' Surveys

211

An. funestus like other malaria vectors is controlled by the use of insecticides such as insecticide treated materials or as indoor residual spraying (IRS). Unfortunately, *An. funestus* is increasingly developing resistance across Africa to different classes of insecticides used in public health, such as PYR, CX and DDT (Brooke *et al.*, 2001; Casimiro *et al.*, 2006; Cuamba *et al.*, 2010; Morgan *et al.*, 2010). There are alternative agrochemicals, such as fipronil that could be introduced but the potential for cross-resistance from existing mechanisms segregating in field populations needs to be more investigated.

The insecticide resistance in *An. funestus* populations was early recorded from Burkina Faso, where resistance was found to dieldrin, a cyclodiene abundantly used in Africa in the 1960s for cotton crop protection but also for malaria vector control (Hamon *et al.*, 1968b). Dieldrin resistance was also reported in *An. funestus* from Cameroon, Benin, Nigeria and Mali (Service, 1960; Toure, 1982; Brown, 1986). Recent studies have shown that *An. funestus* remains fully susceptible to all tested insecticides (DDT, PYR, OP/CX) except to dieldrin for which resistance remains high despite the fact that cyclodienes are no longer used in public health control programs (Dabire *et al.*, 2007). But the distribution of this resistance across the rest of the continent is unknown and need to be clarified. The understanding of factors explaining the persistence of high levels of resistance against cyclodienes in *An. funestus* as well as the geographical distribution of this resistance across the continent has been recently addressed by Wondji *et al.*, 2011. These studies indicated that *RdlR* mutation extensively reported in West and Central Africa should sustain dieldrin resistance in such *An. funestus* populations.

9. Pesticide pressures on disease vectors are from multiple origins

The question that remains to be clearly identified is the origin of insecticide pressures that select the resistances observed in mosquitoes from sub-Saharan Africa. In Burkina Faso, the emergence of the *ace-1R* mutation in *An. gambiae s.s.* populations is also associated with the insecticide treatment history with OP and CX of cotton. Since the mid 1990s and until recently a pest management strategy including four windows of treatment per cropping cycle using pyrethroids, OP/CX (such as chlorpyrifos, profenofos and trizophos) and organochlorines had been adopted in order to manage the pyrethroid-resistance of *Helicoverpa armigera* and *Bemisia tabaci* that emerged throughout the cotton belt. Some bioassays performed in 2003 on *An. gambiae* populations from four sites located in the cotton belt of western Burkina Faso revealed early resistance against CX and OP insecticides pre-empting the discovery of the genetic resistance mechanism revealed in further studies. In a previous study the agricultural use of insecticides was already implicated in the development of resistance to pyrethroids in *An. gambiae s.l.* populations. Then the geographical distribution of resistance decreased in *An. gambiae s.l.* populations from the Sudan savannah to Sahelian areas and the highest levels of resistance were found in sites of cotton cultivation. The areas under cotton cultivation have expanded dramatically in the last ten years (210,000 ha in 1996 to more than 520,000 ha in 2005). A corresponding increase in the level of insecticide use has also been reported reaching more than 3×10^6 litres of pesticide per cropping campaign. Furthermore, a clear knowledge concerning the practices of populations regarding the uses of insecticides in Africa is required.

10. Insecticide resistance and malaria vector control

Although several studies in Ivory Coast and Burkina Faso had shown that ITNs may still achieve good control of PYR-target resistant *An. gambiae s.s.* populations (Darriet *et al.*, 2000; Henry *et al.*, 2005; Dabiré *et al.*, 2006), recent results from experimental hut trials conducted in Southern Benin with lambdacyhalothrin (PYR) suggested that such ITNs may fail to control these field populations (Ngessan *et al.*, 2007). Failure of indoor residual house spraying (IRS) with deltamethrin (PYR) had also occurred in South Africa where the malaria vector *An. funestus* had developed PYR-metabolic resistance. A recent study in Bioko Island (Equatorial Guinea) reported a failure of indoor residual spraying with deltamethrin on *An. gambiae* populations of the M molecular form carrying the Leu-Phe *kdr* mutation at a high frequency. These *kdr*-pyrethroid resistant populations were controlled after the introduction of a carbamate insecticide in IRS (Sharp *et al.*, 2007). Thus the malaria outbreak in this country was only brought under control after reversion to DDT spraying (organochlorine insecticide).

A concern for the potential use of OP and CX as alternative for PYR is that target and metabolic resistances to these insecticides are already present in some *An. gambiae s.s.* populations in West Africa especially in Burkina Faso. Several studies have suggested that the use of agricultural pesticides, especially for cotton but also for vegetable crops, favored the emergence and facilitated the spread of insecticide resistance within mosquito populations. Other studies have given evidence for the selection of *kdr* alleles associated with the use of pyrethroids in ITNs and other domestic strategies of personal protection, especially in Kenya and Niger (Czeher *et al.*, 2008). In countries supported by PMI (President Initiative against Malaria) such as Ghana, Senegal, Mali, Benin and Liberia in West Africa the large-scale pilot interventions implemented with bendiocarb 400mg/m^2 in IRS could also contribute to select the OP/CX resistance. However, as no global health control program in Africa used OP and CX for mosquito control, it is necessary to clearly identify the origin of insecticide pressure that select these resistances.

11. Conclusions

Reports of insecticide resistance in malaria vectors in West Africa especially in Burkina Faso indicate that insecticide resistance increases year after year and highlight the threat to the effectiveness of vector control strategies. In fact *An. gambiae s.l.* populations in Burkina Faso, and more broadly in West Africa, have evolved resistance to many of the insecticides classes used for vector control. Resistance may be conferred by target-site insensitivity such as *kdr* and *ace-1R*, other metabolic mechanisms or a combination of all as *kdr* and *ace-1R* resistance mechanisms occur concomitantly in the same populations of *An. gambiae* s.s. in the South-Western region of the country.

In conclusion the geographical distribution of insecticide resistance in *An. gambiae s.l.* populations was found in sites of cotton cultivation and vegetable in urban settlement that has expanded dramatically in the last ten years. But the role of agriculture in the selection of resistance in natural mosquito's populations needs to be clarified, both in terms of insecticides usage and quantity in order to devise strategies that may help to reduce the extension of resistance. Until the discovery of new insecticides or using new formulations of existing insecticides and also the use of genetically modified mosquitoes (GMM) and sterilised males techniques (SIT), it is crucial to integrate the regional vector resistance status in the implementation of control interventions that will preserve a long term efficacy of these vector control tools.

Unfortunately reports of insecticide resistance in vector populations increase year by year and could jeopardize malaria vector control based on the use of insecticides. The use of insecticides for bednets impregnation or for IRS represents the primary means for malaria prevention worldwide. However the efficacy of such tools has been evaluated in areas where vectors are susceptible to insecticides. Moreover, mosquito resistance is not only due to the insecticides used for mosquito control, but to the many pesticide pollutions present in their environment which are generated by a large variety of human activities necessitating insect control for agriculture (large cultures, fruits and vegetables), animal and other household protections. These pollutions may dramatically affect resistance genes dynamics and threaten these strategies.

The overall pesticide pressures that select resistance in mosquitoes need to be clarified, both in terms of insecticides usage and quantity. It is also crucial to improve our knowledge on the practices of people regarding the use of insecticides and the reasons underlying their decision process based on social and cultural contexts.

Malaria vector control programs require up-to-date information on the distribution and composition of mosquito vector populations and the susceptibility of these populations to the insecticides used for control.

12. Acknowlegements

Authors are grateful to Corus 6015 and National Malaria Control Programme of Burkina Faso which supported financially this study on resistance monitoring. We thank Nicole Pasteur for critical reading of the manuscript. We thank also *Malaria Journal* to have agreed the use of figure published in *Malaria Journal* 2009, 8:70 doi:10.1186/1475-2875-8-70.

13. References

[1] Awolola TS, Oyewole IO, Amajoh CN, Idowu ET, Ajayi MB, Oduala A, *et al.*, 2005. Distribution of the molecular forms of *Anopheles gambiae* and pyrethroids knock down resistance gene in Nigeria. *Acta Tropica* 95: 204-09

[2] Beier J, Keating J, Githure J, Macdonald M, Impoinvil D, Novak R, 2008. Integrated vector management for malaria control. *Malaria Journal* 7: S4.

[3] Berticat C, Bonnet J, Duchon S, Agnew P, Weill M, Corbel V, 2008. Costs and benefits of multiple resistance to insecticides for *Culex quinquefasciatus* mosquitoes. *BMC Evolutionary Biology* 8.

[4] Brogdon WG, McAllister JC, 1998. Insecticide resistance and vector control. *Emerging infectious Diseases* 4: 605-613.

[5] Brooke BD, Kloke G, Hunt RH, Koekemoer LL, Temu EA, Taylor ME, Small G, Hemingway J, Coetzee M, 2001. Bioassay and biochemical analyses of insecticide resistance in southern African *Anopheles funestus* (Diptera: Culicidae). *Bull Entomol Res* 91: 265-272.

[6] Brown AWA, 1986. Insecticide resistance in mosquitoes: a pragmatic review. *J Am Mosq Control Assoc* 2: 123-140.

[7] Casimiro S, Coleman M, Mohloai P, Hemingway J, Sharp B, 2006. Insecticide resistance in *Anopheles funestus* (Diptera: Culicidae) from Mozambique. *J Med Entomol* 43: 267-275.

[8] Chandre F, Manguin S, Brengues C, Dossou-Yovo J, Darriet F, Diabaté A, Faye O, Mouchet J, Guillet P, 1999. Status of pyrethroid resistance in *Anopheles gambiae* sensu lato. *Bull World Health Organ* 77: 230-4.

[9] Coetzee M, Fontenille D, 2004. Advances in the study of *Anopheles funestus*, a major vector of malaria in Africa. *Insect Biochem Mol Biol* 34: 599-605.

[10] Collins FH, Mendez MA, Rasmussen MO, Mehaffey PC, Besansky NJ, Finnerty V, 1987. A ribosomal RNA gene probe differentiates member species of the *Anopheles gambiae* complex. *Am J Trop Med Hyg* 37: 37-41.

[11] Cohuet A, Simard F, Toto JC, Kengne P, Coetzee M, Fontenille D, 2003. Species identification within the *Anopheles funestus* group of malaria vectors in Cameroon and evidence for a new species. *Am J Trop Hyg* 69: 200-205.

[12] Cuamba N, Morgan JC, Irving H, Steven A, Wondji CS, 2010. High level of pyrethroid resistance in an *Anopheles funestus* population of the Chokwe District in Mozambique. *PLoS One* 5, e11010.

[13] Czeher C., Labbo R., Arzika I., Duchemin J-B., 2008. Evidence of increasing Leu-Phe knockdown resistance mutation in *Anopheles gambiae* from Niger following a nationwide long-lasting insecticide-treated nets implementation. *Malaria Journal*

[14] Dabiré KR, Diabaté A, Namountougou M, Toé KH, Ouari A, Kengne P, Bass C, Baldet T, 2009a: Distribution of pyrethroid and DDT resistance and the L1014F *kdr* mutation in *Anopheles gambiae s.l.* from Burkina Faso (West Africa). *Tran R Soc Trop Med Hyg* 103: 1113-1120

[15] Dabiré K.R, Diabaté A., Namountougou M., Djogbenou L., Kengne P., Ouédraogo J-B., Simard F., Bass C., Baldet T., 2009b. The distribution of insensitive acetylcholinesterase (*ace-1R*) in *Anopheles gambiae* s.l. populations from Burkina Faso (West Africa). *Tropical Medicine and International Health* 14 (4): 396-403

[16] Dabiré KR, Diabaté A, Baldet T, Paré L, Guiguemdé TR, Ouédraogo JB, Skovmand O, 2006. Personal protection of long lasting insecticide-treated nets in areas of *Anopheles gambiae* ss resistance to pyrethroids. *Malaria Journal* 5-12 doi: 10.1186/1475-2875-5-12.

[17] Dabiré KR, Baldet T, Diabaté A, Dia I, Costantini C, Cohuet A, Guiguemde TR, Fontenille D, 2007. *Anopheles funestus* (Diptera: Culicidae) in a humid savannah area of western Burkina Faso: bionomics, insecticide resistance status, and role in malaria transmission. *J Med Entomol* 44: 990-997.

[18] Dabiré KR, Diabaté A, Djogbenou L, Ouari A, N'Guessan R, Ouédraogo JB, Hougard JM, Chandre F, Baldet T, 2008. Dynamics of multiple insecticide resistance in the malaria vector *Anopheles gambiae* in a rice growing area in South-Western Burkina Faso. *Malaria Journal* 7: 188.

[19] Darriet F, N'Guessan R, Koffi AA, Konan L, Doannio JM, Chandre F, Carnevale P, 2000. Impact of pyrethrin resistance on the efficacy of impregnated mosquito nets in the prevention of malaria: results of tests in experimental cases with deltamethrin SC. *Bull Soc Pathol Exot* 93: 131-4

[20] Della torre A, Tu Z, Petrarca V, 2005. On the distribution and genetic differentiation of *Anopheles gambiae* ss molecular forms. *Insect Bioch Mol Biol*, 35:7055-69

[21] Diabaté A, Brengues C, Baldet T, Dabiré KR, Hougard JM, Akogbeto M, Kengne P, Simard F, Guillet P, Chandre F, 2004a. The spread of the Leu-Phe *kdr* mutation through *Anopheles gambiae* complex in Burkina Faso: genetic introgression and de *novo* phenomena. *Trop Med Int Health* 9: 1267-73

[22] Diabaté A, Baldet T, Chandre F, Akogbeto M, Darriet F, Brengues C, Guillet P, Hemingway J, Small GJ, Hougard JM, 2002. The role of agricultural use of insecticides in resistance to pyrethroids in *Anopheles gambiae* sl in Burkina Faso. *Am J Trop Med Hyg* 67: 617-22

[23] Diabaté A., Baldet T., Chandre F., Dabiré K.R., Kengne P., Guiguemdé T.R., Simard F., Guillet P., Heminway J., Hougard J.M., 2003. KDR mutation, a genetic marker to assess events of introgression between the molecular M and S forms of *Anopheles gambiae* (Diptera: Culicidae) in the tropical savannah area of West Africa. *J Med Entomol*, 40:195-198

[24] Diabaté A, Baldet T, Chandre F, Dabiré KR, Simard F, Ouédraogo JB, Guillet P, Hougard JM, 2004b. First report of kdr mutation in *Anopheles arabiensis* from Burkina Faso, West Africa. *J Am Mosq Control Ass* 20: 195-196.

[25] Djegbe I., Boussari O., Sidick A., Martin T., Ranson H., Chandre F., Akogbeto M., Corbel V., 2011. Dynamics of insecticide resistance in malaria vectors in Benin: first evidence of the presence of L1014S *kdr* mutation in *Anopheles gambiae* from West Africa. *Malaria Journal*, 10:261 http://www.malariajournal.com/content/10/1/261

[26] Djogbenou L, Labbe P, Chandre F, Pasteur N, Weill M, 2009. Ace-1 duplication in *Anopheles gambiae*: a challenge for malaria control. *Malaria Journal* 8:70.

[27] Djogbenou L, Chandre F, Berthomieu A, Dabire KR, Koffi A, Alout H, Weill M, 2008. Evidence of introgression of the *ace-1^R* mutation and of the *ace-1* duplication in West African *Anopheles gambiae* s. s. *PLoS ONE*, 3, e2172.

[28] Djogbenou L, Noel V, Agnew P, 2010. Costs of insensitive acetylcholinesterase insecticide resistance for the malaria vector *Anopheles gambiae* homozygous for the G119S mutation. *Malaria Journal*, 9:12.

[29] Etang J, Fonjo E, Chandre F, Morlais I, Brengues C, Nwane P, Chouaibou M, Ndjemai H, Simard F, 2006. Short report: First report of knockdown mutations in the malaria vector *Anopheles gambiae* from Cameroon. *Am J Trop Med Hyg* 74: 795-797.

[30] Fanello C, Akogbeto M, della Torre A, 2000. Distribution of the knockdown resistance gene (*kdr*) in *Anopheles gambiae* s.l. from Benin. *Trans R Soc Trop Med Hy* 94: 132.

[31] Favia G, Lanfrancotti A, Spanos L, Sideén-Kiamos I, Louis C, 2001. Molecular characterization of ribosomal DNA polymorphisms discriminating among chromosomal forms of *Anopheles gambiae* ss. *Insect Mol Biol* 10: 5-3

[32] Gillies MT, Coetzee M, 1987. A supplement to the Anophelinae of Africa south of the Sahara (Afrotropical region). Publications of the South African Institute for Medical Research, 1987, no. 55. SAIMR, Johannesburg.

[33] Hamon J, Subra R, Venard P, Coz J, Brengues J, 1968a. Présence dans le Sud-Ouest de la Haute Volta de populations d'*Anopheles funestus* Giles résistantes à la dieldrine. *Med Trop* 28: 221-26

[34] Hamon J, Subra R, Sales S, Coz J, 1968b. Présence dans le Sud-Ouest de la Haute Volta de populations d'*Anopheles gambiae* "A" résistantes au DDT. *Med Trop* 28: 524-28

[35] Kaminski J, 2007. Reforme de la filière cotonnière burkinabé- Retour sur dix ans de mutations: Analyse des impacts économiques et sociaux sur les producteurs et implications des organisations agricoles. Rapport Université de Toulouse 2007, 96p

[36] Harbach RE, 1994. Review of internal classiƀcation of the genus *Anopheles* (Diptera: Culicidae): the foundation for comparative systematic and phylogenetic research. *Bull Entomol Res* 84: 331-342.

[37] Henry MC, Assi SB, Rogier C, Dossou-Yovo J, Chandre F, Guillet P, Carnevale P, 2005. Protective Efficacy of Lambda-Cyhalothrin Treated Nets in *Anopheles gambiae* Pyrethroid Resistance Areas of Cote d'Ivoire. *Am J Trop Med Hyg* 73: 859-864.

[38] Koekemoer LL, Weeto MM, Kamau L, Hunt RH, Coetzee M, 2002. A cocktail polymerase chain reaction (PCR) assay to identify members of the *Anopheles funestus* (Diptera: Culicidae) group. *Am J Trop Med Hyg* 66: 804-811.

[39] Hollingworth RM, Dong K, 2008. The biochemical and molecular genetic basis of resistance in arthropods. Pp. 192 *in* M. E. Whalon, D. Mota-Sanchez, and R. M. Hollingworth, eds. Global pesticide resistance in arthropods. CAB International, Cambridge, MA.

[40] Labbé P, Berthomieu A, Berticat C, Alout H, Raymond M, Lenormand T, Weill M, 2007. Independent duplications of the acetylcholinesterase gene conferring insecticide resistance in the mosquito *Culex pipiens*. *Molecular Biology and Evolution* 24: 1056–1067.

[41] Lemasson JJ, Fontenille D, Lochouarn L, Dia I, Simard F, Ba K, Diop A, Diatta M, Molez JF, 1997. Comparison of behavior and vector efficency of *Anopheles gambiae* and *An. arabiensis* (Dipetra: Culidae) in Barkedji, a Sahelian area of Senegal. *J Med Entomol* 34: 396-403

[42] Martinez-Torres D, Chandre F, Williamson MS, Darriet F, Berge JB, Devonshire AL, Guillet P, Pasteur N, Pauron D, 1998. Molecular characterization of pyrethroid knockdown resistance (*kdr*) in the major malaria vector *Anopheles gambiae* ss. *Insect Mol Biol* 7: 179-84

[43] Morgan JC, Irving H, Okedi LM, Steven A, Wondji CS, 2010. Pyrethroid resistance in an *Anopheles funestus* population from Uganda. *PLoS One* 5, e11872.

[44] Muller P, Chouaibou M, Pignatelli P, Etang J, Walker ED, Donnelly MJ, Simard F, Ranson H, 2008. Pyrethroid tolerance is associated with elevated expression of antioxidants and agricultural practice in *Anopheles arabiensis* sampled from an area of cotton fields in Northern Cameroon. *Molecular Ecology* 17:1145-1155.

[45] Nauen R, 2007. Insecticide resistance in disease vectors of public health importance. *Pest Management Science* 63: 628-633.

[46] Nguessan R, Corbel V, Akogbeto M, Rowland M, 2007. Reduced Efficacy of Insecticide treated Nets and Indoor Residual Spraying for Malaria Control in Pyrethroid Resistance Area, Benin. *Emerging Infectious Diseases*, 2 www.cdc.gov/eid

[47] Ranson H, Jensen B, Vulule JM, Wang X, Hemingway J, Collins FH, 2000. Identification of a point mutation in the voltage-gated sodium channel gene of Kenya *Anopheles*

gambiae associated with resistance to DDT and pyrethroids. *Insect Mol Biol* 95: 491-97

[48] Raymond M, Rousset F, 1996. GENEPOP (version 1.2): Population genetics software for exact tests and ecumenicism. *J Heredity* 86: 248-49.

[49] Rogan W. J., Chen A., 2005. Health risks and benefits of bis (4-chlorophenyl)-1,1,1-trichloroethane (DDT). *The Lancet* 366:763-773.

[50] Sharp B.L., Ridl F.C., Govenderi D., Kuklinski J., Kleinschmidt I., 2007. Malaria vector control by indoor residual insecticide spraying on the tropical island of Bioko, Equatorial Guinea. *Malaria Journal*, 6:52 doi:10.1186/1475-2875-6-52

[51] Scott JA, Brogdon WG, Collins FH, 1993. Identification of single specimens of *An. gambiae* complex by polymerase chain reaction. *Am J Trop Med Hyg* 49: 520-29

[52] Service MW, 1960. A taxonomic study of *Anopheles funestus* Giles (Diptera: Culicidae) from southern and northern Nigeria, with notes on its varieties and synonyms. *Proc Entomo Soc Lond Ser B* 29: 77-84.

[53] Touré YT, 1982. Study of *Anopheles funestus* and *Anopheles gambiae* s.l. susceptibility to insecticides in a rural area of Sudan savanna in Mali. *Cahiers ORSTOM, Ser Entomologie Medicale Parasitologie* 20: 125-131.

[54] Touré YT, Petrarca V, Traoré SF, Coulibaly A, Maiga HM, Sankaré O, Sow M, Di Decco MA, Coluzzi M, 1998. The distribution and inversion polymorphism of chromosomally recognised taxa of the *Anopheles gambiae* complex in Mali, West Africa. *Parassitologia* 40: 477-511

[55] Verhaeghen K, Van Bortel W, Roelants P, Backeljau T, Coosemans M, 2006. Detection of the East and West African *kdr* mutation in *Anopheles gambiae* and *Anopheles arabiensis* from Uganda using a new assay based on FRET/Melt Curve analysis. *Malaria Journal* 5:16

[56] Weill M, Chandre F, Brengues C, Manguin C, Akogbeto M, Pasteur N, Guillet P, Raymond M, 2000. The *kdr* mutation occurs in the Mopti form of *Anopheles gambiae* s.s. through introgression. Insect Mol. Biol. 9:451-455.

[57] Weill M, Lutfalla G, Mogensen K, Chandre F, Berthomieu A, Berticat C, Pasteur N, Philips A, Fort P, Raymond M, 2003. Insecticide resistance in mosquito vectors. *Nature* 423: 136-137.

[58] Weill M, Malcolm C, Chandre F, Mogensen K, Berthomieu A, Marquine M, Raymond M, 2004. The unique mutation in *ace-1* giving high insecticide resistance is easily detectable in mosquito vectors. *Insect Molecular Biology* 13:1-7.

[59] Whalon ME, Mota-Sanchez D, Hollingworth RM, 2008. Analysis of global pesticide resistance in arthropods. Pp. 192 *in* M. E. Whalon D, Mota-Sanchez & Hollingworth R.M. eds. Global pesticide resistance in arthropods. CAB International, Cambridge, MA.

[60] White GB, 1974. *Anopheles gambiae* complex and disease transmission in Africa. *Trans R Soc Trop Med Hyg* 68:278-301.

[61] WHO, 2006. Pesticides and their application for the control of vectors and pests of public health importance.

[62] Wilkes TJ, Matola YG, Charlwood JD, 1996. *Anopheles rivulorum*, a vector of human malaria in Africa. *Med Vet Entomol* 10: 108-110.

[63] Wondji CS, Dabire KR, Tukur Z, Irving H, Djouaka D, Morgan JC, 2011. Identification and distribution of a GABA receptor mutation conferring dieldrin resistance in the malaria vector *Anopheles funestus* in Africa. *Insect Bioch Mol Biol* 41: 484-491

[64] World Health Organisation Test procedures for insecticides resistance monitoring in malaria vectors, bio-efficacy and persistence of insecticides treated surfaces. WHO/CDS/MAL, 1998, 12p

11

Genetic Toxicological Profile of Carbofuran and Pirimicarb Carbamic Insecticides

Sonia Soloneski and Marcelo L. Larramendy
Faculty of Natural Sciences and Museum, National University of La Plata
Argentina

1. Introduction

It's well known that the pesticide usages in agriculture have led increase in food production worldwide. Although the benefits of conventional agricultural practices have been immense, they utilize levels of pesticides and fertilizers that can result in a negative impact on the environment (WHO, 1988). Only for the 2006-2007, the total world pesticide amount employed was approximately 5.2 billion pounds (www.epa.gov). Their application is still the most effective and accepted method for the plant and animal protection from a large number of pests, being the environment consequently and inevitably exposed to these chemicals. Herbicides accounted for the largest portion of total use, followed by other pesticides, like insecticides and fungicides (www.epa.gov). The goal in pesticide investigation and development is identifying the specificity of action of a pesticide toward the organisms it is supposed to kill (Cantelli-Forti et al., 1993). Only the target organisms should be affected by the application of the product. However, because pesticides are designed and selected for their biological activity, toxicity on non-target organisms frequently remains a significant potential risk (Cantelli-Forti et al., 1993). The benefits in using pesticides must be weighed against their deleterious effects on human health, biological interactions with non-target organisms, pesticide resistance and/or accumulation of these chemicals in the environment (WHO, 1988). Pesticides are high volume, widely used environmental chemicals and there is continuous debate concerning their probable role in both acute and chronic human health effects (Cantelli-Forti et al., 1993; Hodgson & Levi, 1996). Among the potential risk effects of agricultural chemicals, carcinogenesis is of special concern. The genetic toxicities of pesticides have been determined by numerous factors like their biological accumulation or degradation in the environment, their metabolism in humans, and their action in cellular components such as DNA, RNA and proteins (Shirasu, 1975). It seems essential the determination of the genotoxic risks of these pesticides before they are used in agriculture. Therefore, the carcinogenic and mutagenic potential of a large amount of pesticides has been the object of an extensive and wide investigation (WHO, 1990). These results have great predictive value for the carcinogenicity of several pesticides (IARC, 1987). The International Agency for Research on Cancer (IARC) has reviewed the potential carcinogenicity of a wide range of insecticides, fungicides, herbicides and other similar compounds. Fifty-six pesticides have been classified with carcinogenic potential in different laboratory animals (IARC, 2003). Among them, and as a brief example, chemicals compounds as phenoxy acid herbicides, 2,4,5-trichlorophenoxyacetic acid (2,4,5-T), lindane, methoxychlor, toxaphene, and some organophosphates have been reported with a carcinogenic potential in human studies (IARC, 2003).

Numerous well known pesticides have been tested in a wide variety of mutagenicity as well as DNA, chromosomal, and cellular damage endpoints (IARC, 2003). Several investigations have been reported positive associations between exposure and pesticide risk (Shirasu, 1975; Bolognesi et al., 1993, 2009, 2011; Pavanello & Clonfero, 2000; Bolognesi, 2003; Clark & Snedeker, 2005; Castillo-Cadena et al., 2006).

2. Carbamic insecticides

The carbamates are chemicals mainly used in agriculture as insecticides, fungicides, herbicides, nematocides, and/or sprout inhibitors (IARC, 1976). These chemicals are part of the large group of synthetic pesticides that have been developed, produced, and used on a large scale within the last 50 years. Additionally, they are used as biocides for industrial or other applications as household products including gardens and homes (IARC, 1976).

During the last decades, considerable amounts of pesticides belonging to the class of carbamates have been released into the environment. Humans may be exposed to carbamates through food and drinking water around residences, schools, and commercial buildings, among others (IARC, 1976). Consequently, carbamates are potentially harmful to the health of different kinds of organisms (EPA, 2004). Among all classes of pesticides, carbamates are most commonly used compounds because organophosphates and organochlorines are extremely toxic and possess delayed neurotoxic effects (Hour et al., 1998). They share with organophosphates the ability to inhibit cholinesterase enzymes and therefore share similar symptomatology throughout acute and chronic exposures. Likewise, exposure can occur by several routes in the same individual due to multiple uses, and there is likely to be additive toxicity with simultaneous exposure to organophosphates (IARC, 1976).

The N-methyl carbamates are a group of closely related pesticides employed in homes, gardens and agriculture that may affect the functioning of the nervous system (EPA, 2007). Toxicological characteristics of the N-methyl carbamates involve maximal cholinesterase enzyme inhibition followed by a rapid recovery, typically from minutes to hours (EPA, 2007). Several compounds namely aldicarb, carbaryl, carbofuran, formetanate HCl, methiocarb, methomyl, oxamyl, pirimicarb, propoxur, and thiodicarb are included as members of the N-methyl carbamate class (EPA, 2007).

3. Carbofuran. Genotoxicity and cytotoxicity profiles

Carbofuran (2,3-dihydro-2,2-dimethylbenzofuran-7-yl methylcarbamate; CASRN: 1563-66-2) is one of the most widely granular employed N-methyl carbamate esters with both contact and systemic activity. Carbofuran is a derivative of carbamic acid being its chemical structure formula shown in Fig. 1.

Carbofuran is a relatively unstable compound that breaks down in the environment within weeks or months (www.inchem.org). It is registered on a variety of agricultural uses to control soil-dwelling and foliar-feeding insects, mites and nematodes on a variety of field, fruit, forage, grain, seed, and fiber crops (EPA, 2006). Carbofuran is a systemic, broad spectrum insecticide and nematocide registered N-methyl carbamate for control of soil and foliar pests. It has been reported for 2006 that nearly one million pounds of carbofuran was applied worldwide (EPA, 2006). The most sensitive and appropriate effect associated with the use of carbofuran is its toxicity following acute exposure (HSDB, 2011). On the basis of its acute toxicity, it has been classified as a highly hazardous member (class Ib) by WHO (2009) and highly toxic compound (category I) by EPA (2006) based on its potency by the oral and inhalation exposure routes. In spite of the recommendation and regulation proposed by the United States Environmental Protection Agency (EPA) concerning the use of this carbamate within the United States of America, its application has been recently cancelled all over the Northern country by the same organization since 2009 (www.epa.gov). However, the contamination of environment with this compound can by far occur, particularly taking into consideration those countries where it is still in use and the probability of long-term low dose exposure becomes increased. Due to its extensive employment in agriculture and household, contamination of food, water and air has become serious and undesirable health problem for humans, animals and wildlife. Large quantities of this carbamate are particularly applied to different environments worldwide.

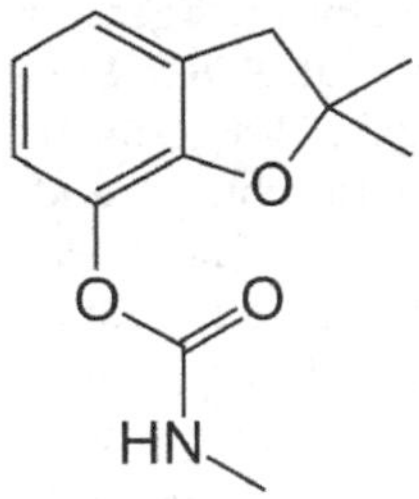

Fig. 1. Chemical structure of carborufan. Source: INCHEM (www.inchem.org).

Metabolism of carbofuran has been extensively studied in plants and animals (Dorough & Casida, 1964; Metcalf et al., 1968). In mammals, it reversibly inhibits acethylcholinesterase by carbamylation as well as others non-specific serine-containing enzymes, such as carboxylesterases and butyrylcholinesterases (Gupta, 1994). This results in accumulation of acetylcholine at nerve synapses and myoneural junctions leading to cholinergic signs and causing toxic effects (Karczmar, 1998). Epidemiological studies suggested that exposure to carbofuran may be associated with increased risk of gastrointestinal, neurological, cardiac dysfunction, and retinal degeneration (Cole et al., 1998; Kamel et al., 2000; Peter & Cherian, 2000). Carbofuran represents an acute poison when absorbed into the gastrointestinal tract by inhalation of dust and spray mist and minimally poison thought the intact skin contact (Gupta, 1994). In summary, carbofuran is reported to be teratogenic, embryotoxic and highly toxic to mammals (Gupta, 1994; WHO-FAO, 2004, 2009; WHO, 2009).

Genotoxicity and cytotoxicity studies have been conducted with this N-methyl carbamate member using several end-points on different cellular systems. A summary of the results reported so far is presented in Table 1.

The compound produced both conflicting and inconclusive results in mutagenicity tests varying according to either the end-point assessed (WHO, 1988, 2000-2002, 2009; WHO-FAO, 2004, 2009). When mutagenic activity was assessed in bacterial systems either positive or negative results have been reported. Carbofuran has been found to be non-mutagenic in *Salmonella typhimurium* since negative or weak positive response were observed in the number of mitotic recombinants regardless of the presence or absence of a rat liver metabolic activation system (Blevins et al., 1977a; Gentile et al., 1982; Waters et al., 1982; Haworth & Lawlor, 1983; Hour et al., 1998; Yoon et al., 2001). These results indicate that carbofuran cannot be considered mutagenic in bacterial systems. However, it was active in *Salmonella typhimurium* TA1538 and TA98 strains in the presence or absence of S9 metabolic system (Gentile et al., 1982; Moriya et al., 1983; Hour et al., 1998). Whereas the insecticide did not induce reverse mutations in *Escherichia coli* (Simmon, 1979), it has been claimed as a relatively weaker mutagen with the repair defective Ames *Escherichia coli* K-12 test (Saxena et al., 1997). Similarly, positive results have been found after exposure in *Vibrio fischeri* regardless the absence or presence of S9 metabolic system (Canna-Michaelidou & Nicolaou, 1996). When DNA damage and repair assays were performed, carbofuran was also negative in both *Escherichia coli* and *Bacillus subtilis* bacterial systems (SRI, 1979). Similar negative results were also found after carbofuran exposure in *Saccharomyces cerevisiae* mitotic recombination assay (Simmon, 1979).

The mammalian *in vitro* gene mutation assay systems generated results consistent with the microbial gene mutation assays, although they were generally more responsive. When a mammalian cell system was employed for mutagenic screening, carbofuran was found to be positive in V79 cells (Wojciechowski et al., 1982). Similar results were reported for the cell mutation assay in mouse lymphoma L5178 cells (Kirby, 1983a, b). Unscheduled DNA synthesis was monitored in human fibroblasts and primary rat hepatocytes following treatment with the insecticide with and without S9 fraction. Both negative and positive results were obtained for the same endpoint in human primary fibroblasts regardless of the presence or absence of a rat liver metabolic activation system (Simmon, 1979; Gentile et al., 1982) but negative results were obtained in primary rat hepatocyte cultures (SRI, 1979). Single-strand breaks detected by alkaline comet assay were induced in *in vitro* human peripheral lymphocytes (Das et al., 2003; Naravaneni & Jamil, 2005). The induction of DNA fragmentation on human skin fibroblasts have been found to be enhanced after *in vitro* carbofuran treatment (Blevins et al., 1977b).

As opposed to mutation assays that detect specific gene defects, the chromosomal assays evaluate the structure of the whole chromosome. Five studies of carbofuran have evaluated the induction of sister chromatid exchanges in mammalian cell cultures. In one of the first studies, carbofuran was negative in Chinese hamster ovary cells regardless of the presence or absence of S9 fraction (Thilagar, 1983b). However, other authors reported positive results for the same cell system (Gentile et al., 1982; Thilagar, 1983c; Lin et al., 2007; Soloneski et al., 2008) as well as human lymphocytes (Georgian et al., 1985). Similarly, the effects on chromosomal structure following exposure to carbofuran were investigated in Chinese hamster ovary and primary human lymphocytes cells. While carbofuran did not induce *in vitro* chromosome damage in Chinese hamster ovary cells with or without metabolic system activation (Thilagar, 1983a), positive results were reported to occur not only in the same cellular system (Lin et al., 2007) but also in human lymphocytes *in vitro* (Pilinskaia & Stepanova, 1984; Das et al., 2003). However, inconclusive response for this endpoint has been also reported to occur in the latter system after carbofuran exposure (Naravaneni & Jamil, 2005). Positive results have been also reported for the ability of carbofuran to induce micronuclei in both Chinese hamster ovary cells and human lymphocytes *in vitro* with and without S9 metabolic fraction (Soloneski et al., 2008; Mladinic et al., 2009).

Several assays have been developed to assess the ability of carbofuran to cause cytotoxic effects on different cellular systems. Negative response was observed in both *Escherichia coli* and *Bacillus subtilis* bacterial systems (Simmon, 1979). When the analysis of cell-cycle

End-point/Test System	Concentration[a]	Results	References
In vitro assays			
Ames test			
Salmonella typhimurium, S9 +/-	100 – 10 000 µg/plate	+/-	Blevins et al., 1977a; Waters et al., 1982; Haworth & Lawlor, 1983; Yoon et al., 2001
Salmonella lactam assay, S9 +/-	1 – 10 000 µg/plate	-	Hour et al., 1998
	0.1 – 100 µg/plate	+	Gentile et al., 1982; Moriya et al., 1983; Hour et al., 1998
Pol A reverse mutation			
Escherichia coli (WP$_2$), S9 +/-	1 – 5 000 µg/plate	-	Simmon, 1979
Escherichia coli (K-12)	1 – 5 000 µg/plate	+	Saxena et al., 1997
Mutatox test			
Vibrio fischeri (M169), S9 +/-	175 µg/plate	+	Canna-Michaelidou & Nicolaou, 1996
DNA damage and repair			
Escherichia coli (W3110-p3478)	0 – 5 mg/6-mm disk	-	SRI, 1979
Bacillus subtilis (H17-M45)	0 – 5 mg/6-mm disk	-	SRI, 1979
Mitotic recombination			
Saccharomyces cerevisiae (D3), S9 +/-	1 – 50 mg/ml	-	Simmon, 1979
Gene mutation assay			
V79 cells	NA	+	Wojciechowski et al., 1982
Cell mutation tk locus			
Mouse lymphoma L5178 Y cells, S9 +/-	16 – 1 780 µg/ml	+/-	Kirby, 1983a, b
UDS	0.1 – 1 000 µg/ml	-	Simmon, 1979
Human fibroblasts (WI-38), S9 +/-	0.1 – 1 000 µg/ml	+	Gentile et al., 1982
Human lung fibroblasts	0 – 100 µg/ml	-	SRI, 1979
Primary rat hepatocytes			
Alkaline comet assay			
HL	NA	+	Naravaneni & Jamil, 2005
	0.5 – 4.0 µM	+	Das et al., 2003
DNA fragmentation analysis			
Human skin fibroblasts	NA	+	Blevins et al., 1977b
SCE assay			
CHO cells, S9 +/-	12.5 – 312.5 µg/ml	-	Thilagar, 1983b
CHO cells, S9 +/-	12.5 – 2 500 µg/ml	+	Thilagar, 1983c
CHO-K1 cells	5 – 100 µg/ml	+	Gentile et al., 1982; Soloneski et al., 2008
CHO-W8 cells	0.04 – 0.32 µg/ml	+	Lin et al., 2007
HL	NA	+	Georgian et al., 1985

Chromosomal aberrations

CHO cells, S9 +/-	50 – 2 500 µg/ml	-	Thilagar, 1983a
CHO-W8 cells	0.04 – 0.32 µg/ml	+	Lin et al., 2007
HL	NA	+/-	Naravaneni & Jamil, 2005
HL	100 – 300 µg/ml	+	Pilinskaia & Stepanova, 1984
HL[b]	NA	+	Das et al., 2003

Micronuclei assay

CHO-K1 cells	10 – 100 µg/ml	+	Soloneski et al., 2008
HL, S9 +/-	0.008 µg/ml	+	Mladinic et al., 2009

Growth inhibition

Escherichia coli	1 – 500 mg/ml	-	Simmon, 1979
Bacillus subtilis	1 – 500 mg/ml	-	Simmon, 1979

Alteration in CCP

CHO-W8 cells	0.04 – 0.32 µg/ml	-	Lin et al., 2007
CHO-K1 cells	50 – 100 µg/ml	+	Soloneski et al., 2008
CHL cells	30 µM	-	Yoon et al., 2001

Brain tubulin assembly assay

Porcine cells	100 – 2 000 µmol/l	+	Stehrer-Schmid & Wolf, 1995

Cell viability

CHL cells	30 µM	-	Yoon et al., 2001
CHO-K1 cells	50 – 100 µg/ml	+	Soloneski et al., 2008

Apoptosis

CHL cells	30 µM	-	Yoon et al., 2001
Mouse brain microvascular endothelial cells	3 – 30 µM	-	Jung et al., 2003
Rat cortical cells	500 µM	+	Kim et al., 2004

In vivo assays

Reverse mutation

Zea mays	NA	-	Gentile et al., 1982

Sex-linked recessive lethal test

Drosophila melanogaster	0 – 10 ppm	-	DeGraff, 1983; Gee, 1983

Dominant –lethal mutagenicity

Mice	0.025 – 0.5 mg/Kg/day	-	FMC, 1971

UDS

Rat hepatocytes	5 – 10 ppm	-	Valencia, 1981; 1983

Alkaline comet assay

Mouse peripheral lymphocytes	0.1 – 0.4 mg/Kg bw	-	Zhou et al., 2005
HL*	NA	+	Castillo-Cadena et al., 2006

SCE			
Mouse peripheral lymphocytes	NA	+	Gentile et al., 1982
Rat	NA	+	Aly, 1998
Chromosomal aberrations			
Allium cepa	20 – 80 ppm	+	Saxena et al., 2010
Allium sativum	20 – 80 ppm	+	Saxena et al., 2010
Drosophila melanogaster	NA	-	Woodruff et al., 1983
Mouse bone marrow cells	3.8 – 1.9 (for 4 days) mg/Kg bw	+	Chauhan et al., 2000
Mouse bone marrow cells	0.1 – 1.0 mg/Kg bw	-	Pilinskaia & Stepanova, 1984
Rat bone marrow cells	0.6 – 10 mg/Kg bw	-	Putman, 1983b, a
HL[b]	NA	+	Zeljezic et al., 2009
Micronuclei			
Mouse peripheral lymphocytes	0.1 – 0.4 mg/Kg bw	-	Zhou et al., 2005
Mouse bone marrow cells	5.7 – 1.9 (for 4 days)	-	Chauhan et al., 2000
Alteration in CCP			
Allium cepa	20 – 80 ppm	+	Saxena et al., 2010

UDS, unscheduled DNA synthesis; SCE, sister chromatid exchange; HL, human lymphocytes; CCP, cell-cycle proliferation; NA, data not available.

Table 1. Evaluation of carbofuran-induced genotoxicity and cytotoxicity on different target systems. [a], expressed as reported by authors; [b], exposed to pesticide mixture containing carbofuran; [*], from agricultural workers occupationally exposed to carbofuran.

progression on mammalian cells was studied, carbofuran gave negative results in Chinese hamster ovary cells and lung fibroblasts (Yoon et al., 2001; Lin et al., 2007). On the other hand, Soloneski and co-workers (2008) reported a delay in the cell-cycle progression of Chinese hamster ovary cells after the insecticide treatment. Carbofuran was tested *in vitro* in the porcine brain tubulin assembly assay for detecting whether the chemical can be considered as a microtubule poison and an aneuploidy agent. A dose-dependent reduction in the degree of polymerization of tubulins was reported in porcine cells after *in vitro* treatment (Stehrer-Schmid & Wolf, 1995). Controversial results were reported for the cell viability assay in mammalian cells, e.g., Chinese hamster lung and ovary cells after the exposure (Yoon et al., 2001; Soloneski et al., 2008). Finally, whereas carbofuran-induced apoptosis has been reported in rat cortical cells (Kim et al., 2004), negative results have been also observed in mouse brain microvascular entothelial cells and Chinese lung fibroblasts (Yoon et al., 2001; Jung et al., 2003). Similar end-points for both genotoxicity and cytotoxicity were also applied in *in vivo* systems. Carbofuran has been reported as a non inducer agent of mutations in plants cells, at least in *Zea mays* (Gentile et al., 1982), in the *Drosophila melanogaster* sex-linked recessive lethal test (DeGraff, 1983; Gee, 1983), and in the mice dominant-lethal mutagenicity test (FMC, 1971). Negative results have been obtained for the induction of unscheduled DNA synthesis in primary rat hepatocytes (Valencia, 1981, 1983). Controversial observations have been reported for the induction of DNA single-strand breaks assayed by the alkaline comet assay. Positive results were reported in circulating erythrocytes from occupationally exposed workers (Castillo-Cadena et al., 2006) whereas no induction was observed in mouse peripheral lymphocytes exposed *in vivo* (Zhou et al., 2005). It should be noted that the former positive results could not be totally committed to carbofuran but to other pesticides, since the cohort of donors included in the study was exposed to a panel of other pesticides. Several reports were able to revealed that carbofuran increased the frequency of sister chromatid exchanges in mammalian cells from mouse and rats exposed *in vivo* (Gentile et al., 1982; Aly, 1998), and chromosomal aberrations in plants from *Allium* (Saxena et al., 2010), and mammals including occupationally exposed workers (Putman, 1983a, b; Pilinskaia & Stepanova, 1984; Chauhan et al., 2000; Zeljezic et al., 2009) but not in insects (Woodruff et al., 1983) as well as in rodent cells (Putman, 1983a, b; Pilinskaia & Stepanova, 1984). When the micronuclei induction end-point was employed in mouse, no induction was found either in bone marrow cells (Chauhan et al., 2000) or circulating lymphocytes (Zhou et al., 2005). Finally, alterations in the progression of the cell-cycle were reported to occur after carbofuran exposure in plants when the *Allium cepa* model was employed (Saxena et al., 2010).

4. Pirimicarb. Genotoxicity and cytotoxicity profiles

Pirimicarb (2-dimethylamino-5,6-dimethylpyrimidin-4-yl dimethylcarbamate, CASRN: 23103-98-2) is a dimethylcarbamate insecticide member with both contact and systemic activity. Similar to carbofuran, pirimicarb is a derivative of carbamic acid being its chemical structure formula shown in Fig. 2.

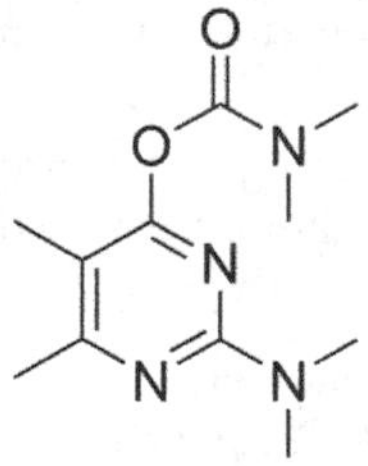

Fig. 2. Chemical structure of pirimicarb. Source: INCHEM (www.inchem.org)

Based on its acute toxicity, pirimicarb has been classified as a moderately hazardous compound (class II) by WHO (http://www.who.int/ipcs/publications/pesticides hazard/en/) and slightly to moderately toxic (category II-III) by EPA (1974a). Among carbamate pesticides, pirimicarb is registered as a fast-acting selective aphicide mostly used in a broad range of crops, including cereals, sugar beet, potatoes, fruit, and vegetables, and is relatively non-toxic to beneficial predators, parasites, and bees (WHO-FAO, 2004, 2009). It acts by contact, translaminar, vapor, and systemic action. Its mode of action is inhibiting acetylcholinesterase activity (WHO-FAO, 2004, 2009).

Available information on the genotoxic and cytotoxic properties of pirimicarb is limited and inconsistent. Only few data are available in the literature (WHO-FAO, 2004, 2009). Genotoxicity and cytotoxicity studies have been conducted with this carbamate using several end-points on different cellular systems. A summary of the results reported so far is presented in Table 2.

Pirimicarb has been generally recognized as non-genotoxic in bacteria, yeast and fungi as well as in mammalian cells (EPA, 1974b). It has been reported to be non-mutagenic in *Salmonella typhimurium* when the Ames reversion mutagenicity test for the TA1535, TA1538, TA98, and TA100 strains after S9 metabolic activation has been used (Trueman, 1980; Callander, 1995). Furthermore, similar situation was observed in both *Escherichia coli* and *Aspergillus nidulans* when the reverse mutation assay or recessive lethal gene mutation test

End-point/System	Concentration[a]	Results	References
In vitro assays			
Ames test			
Salmonella typhimurium, S9 +/-	2 500 µg/plate	-	Trueman, 1980
	5 000 µg/plate	-	Callander, 1995
Pol A reverse mutation			
Escherichia coli	5 000 µg/plate	-	Callander, 1995
Recessive lethal gene mutation			
Aspergillus nidulans	NA	-	Käfer et al., 1982
Cell mutation tk locus			
Mouse lymphoma L5178 Y cells	1 400 mg/ml - S9	-	Clay, 1996
	100 mg/ml + S9	+	Clay, 1996
Alkaline comet			
HL	50 - 500 µg/ml	+	Ündeger & Basaran, 2005
SCE assay			
CHO-K1 cells	100 - 200 µg/ml	+	Soloneski & Larramendy, 2010
Chromosomal aberrations			
CHO-K1 cells	10 - 300 µg/ml	+	Soloneski & Larramendy, 2010
HL, S9 +/-	500 µg/ml	-	Wildgoose et al., 1987
Alteration in CCP			
CHO-K1 cells	100 - 300 µg/ml	+	Soloneski & Larramendy, 2010
In vivo assays			
Eye mosaic system w/w+			
Drosophila melanogaster	NA	+	Aguirrezabalaga et al., 1994
Dominant lethal mutation			
Mice	20 mg/Kg bw	-	McGregor, 1974
UDS			
Rat liver cells	200 mg/Kg bw	-	Kennelly, 1990
Chromosomal aberrations			
Rat bone marrow cells	50/-100 mg/Kg bw	-	Anderson et al., 1980
HL*	NA	+	Pilinskaia, 1982
Micronuclei assay			
Cnesterodon decemmaculatus	50 - 157 mg/L	+	Vera Candioti et al., 2010b
Rhinella arenarum	80 - 250 mg/L	+	Vera Candioti et al., 2010a
Rat bone marrow cells	69.3 mg/Kg bw	-	Jones & Howard, 1989

Table 2. Evaluation of Pirimicarb-induced genotoxicity and cytotoxicity on different target systems. [a], expressed as reported by authors; UDS, unscheduled DNA synthesis; SCE, sister chromatid exchange; HL, human lymphocytes; CCP, cell-cycle proliferation; NA, data not available.

were respectively applied (Käfer et al., 1982; Callander, 1995). Negative and positive results were obtained for the induction of mutagenicity in mouse lymphoma L5178Y cells regardless of the presence or absence of a rat liver metabolic activation system (Clay, 1996). Furthermore, the induction of DNA single strand breaks, estimated by the alkaline comet assay, was evaluated revealing positive results in human lymphocytes exposed *in vitro* to pirimicarb (Ündeger & Basaran, 2005). Similar positive results were found when the sister chromatid exchange assay was performed in Chinese hamster ovary cells (Soloneski & Larramendy, 2010). Although a significant increase in chromosomal aberrations has been reported in Chinese hamster ovary cells after pirimicarb exposure (Soloneski & Larramendy, 2010), Wildgoose and coworkers (1987) observed negative results in human lymphocytes with or without S9 metabolic activation. Finally, the induction of alterations in the cell-cycle progression in Chinese hamster ovary cells was reported to occur after *in vitro* exposure to pirimicarb (Soloneski & Larramendy, 2010).

In *in vivo* genotoxic and cytotoxic studies, pirimicarb was able to induce different types of lesions. It has been reported the ability of the insecticide to give positive results by using the eye mosaic system *white/white+* (*w/w+*) somatic mutation and recombination test (SMART) when *Drosophila melanogaster* was employed as experimental model (Aguirrezabalaga et al., 1994). However, McGregor and co-workers (1974) reported negative results in mice when the dominant lethal mutation assay was performed. Similar negative results were found by Kenelly (1990) using the unscheduled DNA synthesis in rat liver cells. At the chromosomal level, pirimicarb did not induce chromosomal alterations in bone marrow cells of Wistar male rats after oral administration (Anderson et al., 1980). Contrarily, Pilinskaia (1982) observed a significant increase of chromosomal aberrations in the peripheral blood lymphocytes from occupational workers after pirimicarb exposure. Finally, when the micronuclei induction end-point was employed, positive results were reported in erythrocytes of the fish *Cnesterodon decemmaculatus* and *Rhinella arenarum* tadpoles by Vera Candioti and collaborators (Vera Candioti et al., 2010a, b). Lastly, when a mammal model was employed for the micronuclei detection, Jones and Howard (1989) found negative results in rat bone marrow cells.

5. Comparison of the genotoxicity and cytotoxicity of carbofuran and pirimicarb and some Argentinean technical formulations

One of the goals of our research group is to compare the genotoxic and cytotoxic effects exerted by the pesticide active ingredients (Pestanal®, Riedel-de Haën, Germany) and their technical formulations commonly used in Argentina on vertebrate cells both *in vitro* and *in vivo*.

The evaluation was performed using end-points for genotoxicity [Sister Chromatid Exchange, Chromosome Aberration, and Micronuclei frequencies] and cytotoxicity [Mitotic Index, Cell Viability, Proliferative Rate Index, Erythroblasts/Erythrocytes Ratio, 3(4,5-Dimethylthiazol-2-yl)-2,5-diphenyltetrazolium bromide (MTT) and Neutral Red assays] (Soloneski et al., 2008; Soloneski & Larramendy, 2010; Vera Candioti et al., 2010a, b).

We comparatively evaluated the genotoxic and cytotoxic *in vitro* effects on mammalian Chinese hamster ovary cells induced by the pure insecticide carbofuran and its commercial formulation Furadan® (47% carbofuran, FMC Argentina S.A., Buenos Aires, Argentina). Similarly, the genotoxic and cytotoxic *in vitro-in vivo* effects induced by the pure insecticide pirimicarb and its commercial formulation Aficida® (50% pirimicarb, Syngenta Agro S.A.,

Buenos Aires, Argentina) on mammalian Chinese hamster ovary cells as well as circulating erythrocytes of the fish *Cnesterodon decemmaculatus* and the amphibian *Rhinella arenarum* tadpoles were also estimated.

A summary of the results obtained is presented in Fig. 3. The figure clearly reveals that all compounds assayed were able to inflict damage at chromosomal and cellular level regardless of the cellular system used as target.

We observed that carbofuran/Furadan® and pirimicarb/Aficida® caused SCEs on mammalian cells indicating that they have a clastogenic activity (Fig. 3A). It has been suggested that at the chromosomal level, the induction of SCEs is a reliable indicator for the screening of clastogens, since the bioassay is more sensitive than the analysis of clastogen-induced chromosomal aberrations (Palitti et al., 1982). The results also demonstrate the ability of pirimicarb/Aficida® to induce DNA damage quali- and quantitative analyzed by the frequency of chromosomal aberrations (Fig. 3B). Furthermore, a putative clastogenic/aneugenic activity exerted *in vitro* by carbofuran/Furadan® and *in vivo* by pirimicarb/Aficida® was also demonstrated by the ability of the pesticide-induced micronuclei (Fig. 3C). The analysis of the proliferative replication (Fig. 3D) and the mitotic indexes (Fig. 3E) demonstrated that both carbofuran/Furadan® and pirimicarb/Aficida® were able to delay the cell-cycle progression as well as to exert a marked reduction of the cellular mitotic activity on mammalian cells *in vitro*. Besides, carbofuran/Furadan® and pirimicarb/Aficida® were able to induced a clear cellular cytotoxicity. This deleterious effect was estimated by a loss of lysosomal activity (indicated by a decrease in the uptake of neutral red), as well as alteration in energy metabolism (measured by mitochondrial succinic dehydrogenase activity in the MTT assay), as clearly revealed in insecticides-treated Chinese hamster ovary cells (Fig. 3F).

Overall, the results revealed, depending upon the endpoint employed, that the damage induced by the commercial formulations of both insecticides is, in general, greater than that produced by the pure pesticides (Fig. 3). Unfortunately, the identity of the components present in the excipient formulations was not made available by the manufacturers. These final remarks are in accord with previous observations not only reported by us but also by other research groups indicating the presence of xenobiotics within the composition of the commercial formulations with genotoxic and cytotoxic effects (David, 1982; Lin & Garry, 2000; Soloneski et al., 2001, 2002, 2003, 2008; González et al., 2007a, b, 2009; Elsik et al., 2008; Molinari et al., 2009; Soloneski & Larramendy, 2010). These observations highlight that, in agriculture, agrochemicals are generally not used as a single active ingredient but as part of a complex commercial formulations. Thus, both the workers as well as non-target organisms are exposed to the simultaneous action of the active ingredient and a variety of other chemical/s contained in the formulated product. Hence, risk assessment must also consider additional geno-cytotoxic effects caused by the excipient/s.

Finally, the results highlight that a whole knowledge of the toxic effect/s of the active ingredient of a pesticide is not enough in biomonitoring studies as well as that agrochemical/s toxic effect/s should be evaluated according to the commercial formulation available in market. Furthermore, the deleterious effect/s of the excipient/s present within the commercial formulation should be neither discarded nor underestimated. The importance of further studies on this type of pesticide in order to achieve a complete knowledge on its genetic toxicology seems to be, then, more than evident.

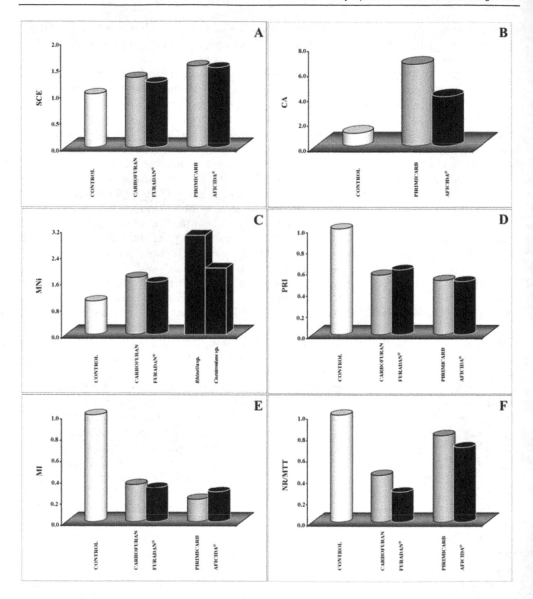

Fig. 3. Comparative genotoxicity and cytotoxicity effects induced by carbofuran and pirimicarb pure herbicides Pestanal® (grey) and their technical formulations Furadan® and Aficida® (black) commonly used in Argentina on mammalian Chinese hamster ovary cells (cylinders) and *in vivo* piscine and amphibian erythrocytes (bars). Results are expressed as fold-time values over control data. Evaluation was performed using end-points for genotoxicity [Sister Chromatid Exchanges (A), Chromosome Aberrations (B), Micronuclei (C)] and cytotoxicity [Proliferative Rate Index (D), Mitotic Index (E), 3(4,5-Dimethylthiazol-2-yl)-2,5-diphenyltetrazolium bromide (MTT) and Neutral Red (NR) (F)].

6. Acknowledgements

This study was supported by grants from the National University of La Plata (Grant Numbers 11/N564, 11/N619) and the National Council for Scientific and Technological Research (CONICET, PIP N° 0106) from Argentina.

7. References

Aguirrezabalaga, I., Santamaría, I., Comendador, M.A., 1994. The w/w⁺ SMART is a useful tool for the evaluation of pesticides. *Mutagenesis* 9, 341-346.

Aly, M.S., 1998. Chromosomal damage induced in adult mice by carbofuran. *Egyptian German Society of Zoology* 26C, 1-10.

Anderson, D., Richardson, C.R., Howard, C.A., Bradbrook, C., Salt, M.J., 1980. Pirimicarb: a cytogenetic study in the rat. World Health Organization.

Blevins, R.D., Lee, M., Regan, J.D., 1977a. Mutagenicity screening of five methyl carbamate insecticides and their nitroso derivatives using mutants of *Salmonella typhimurium* LT2. *Mutation Research* 56, 1-6.

Blevins, R.D., Lijinksy, W., Regan, J.D., 1977b. Nitrosated methylcarbamate insecticides: effect on the DNA of human cells. *Mutation Research* 44, 1-7.

Bolognesi, C., 2003. Genotoxicity of pesticides: a review of human biomonitoring studies. *Mutation Research* 543, 251–272.

Bolognesi, C., Carrasquilla, G., Volpi, S., Solomon, K.R., Marshall, E.J., 2009. Biomonitoring of genotoxic risk in agricultural workers from five colombian regions: association to occupational exposure to glyphosate. *Journal of Toxicological and Environmental Health. Part A* 72, 986-997.

Bolognesi, C., Creus, A., Ostrosky-Wegman, P., Marcos, R., 2011. Micronuclei and pesticide exposure. *Mutagenesis* 26, 19-26.

Bolognesi, C., Parrini, M., Reggiardo, G., Merlo, F., Bonassi, S., 1993. Biomonitoring of workers exposed to pesticides. *International Archives of Occupational and Environmental Health* 65, S185-S187.

Callander, R.D., 1995. Pirimicarb: an evaluation of the mutagenic potential using *S. typhimurium* and *E. coli*. Central Toxicology Laboratory Report No. CTL/P/4798 GLP, Unpublished.

Canna-Michaelidou, S., Nicolaou, A.S., 1996. Evaluation of the genotoxicity potential (by Mutatox test) of ten pesticides found as water pollutants in Cyprus. *The Science of the Total Environmental Research* 193, 27-35.

Cantelli-Forti, G., Paolini, M., Hrelia, P., 1993. Multiple end point procedure to evaluate risk from pesticides. *Environmental Health Perspectives* 101, 15-20.

Castillo-Cadena, J., Tenorio-Vieyra, L.E., Quintana-Carabia, A.I., García-Fabila, M.M., Ramírez-San Juan, E., Madrigal-Bujaidar, E., 2006. Determination of DNA damage in floriculturists exposed to mixtures of pesticides. *Journal of Biomedicine and Biotechnology* 2006, 1-12.

Clark, H.A., Snedeker, S.M., 2005. Critical evaluation of the cancer risk of dibromochloropropane (DBCP). *Journal of Environmental Science and Health - Part C Environmental Carcinogenesis and Ecotoxicology Reviews* 23, 215-260.

Clay, P., 1996. Pirimicarb: L5178Y TK+/- Mouse Lymphoma Mutation Assay. Central Toxicology Laboratory. Report No: CTL/P/5080 GLP, Unpublished.

Cole, D.C., Carpio, F., Julian, J., Léon, N., 1998. Assessment of peripheral nerve function in an Ecuadorian rural population exposed to pesticides. *Journal of Toxicology and Environmental Health. Part A* 55, 77-91.

Chauhan, L.K., Pant, N., Gupta, S.K., Srivastava, S.P., 2000. Induction of chromosome aberrations, micronucleus formation and sperm abnormalities in mouse following carbofuran exposure. *Mutation Research* 465, 123-129.

Das, A.C., Chakravarty, A., Sukul, P., Mukherjee, D., 2003. Influence and persistence of phorate and carbofuran insecticides on microorganisms in rice field. *Chemosphere* 53, 1033-1037.

David, D., 1982. Influence of technical and commercial decamethrin, a new synthetic pyrethroid, on the gonadic germ population in quail embryos. *Archives d'Anatomie, d'Histologie et d'Embryologie* Vol. 65, 99-110.

DeGraff, W.G., 1983. Mutagenicity evaluation of MC 10242 for the sex-linked recessive lethal test in *Drosophila melanogaster*. Study No. A 83-1060. Unpublished report prepared by Litton Bionetics, Inc. Kensington, MD, USA. Submitted to WHO by FMC Corp., Philadelphia, PA, USA.

Dorough, H.W., Casida, J.E., 1964. Insecticide metabolism, nature of certain carbamate metabolites of insecticide Sevin. *Journal of Agricultural and Food Chemistry* 12, 294-304.

Elsik, C.M., Stridde, H.M., Tann, R.S., 2008. Glyphosate adjuvant formulation with glycerin. ASTM Special Technical Publication, pp. 53-58.

EPA, 1974a. Compendium of Registered Pesticides. US Government Printing Office, Washington, DC.

EPA, 1974b. Pesticide Fact Sheet: Pirimicarb. US Government Printing Office, Washington, DC.

EPA, 2004. ECOTOX: Ecotoxicology Database. USEPA/ORD/NHEERL, Mid-Continent Ecology Division. URL:http://www.epa.gov/ecotox/ecotox_home.htm.

EPA, 2006. Carbofuran I.R.E.D. Facts. U.S. Environmental Protection Agency, Office of Pesticide Programs, U.S. Government Printing Office: Washington, DC.

EPA, 2007. Revised *N*-methyl carbamate cumulative risk assessment. U.S. Environmental Protection Agency. Office of Pesticide Programs.

FMC, 1971. Dominant-lethal mutagenicity study [carbofuran]. NCT. 438.99. DPR Vol. 254-029, Record 939879.

Gee, J., 1983. *Drosophila* sex-linked recessive lethal assay of 2,3dihydro-2,2-dimethyl-7-benzofuranyl N-methyl carbamate (carbofuran, T-2047). University of Wisconsin. Lab. Project 103c. DPR Vol. 254-109, Record 47742.

Gentile, J.M., Gentile, G.J., Bultman, J., Sechriest, R., Wagner, E.D., Plewa, M.J., 1982. An evaluation of the genotoxic properties of insecticides following plant and animal activation. *Mutation Research* 101, 19-29.

Georgian, L.I., Moraru, I., Draghicescu, T., Tarnavschi, R., 1985. The effect of low concentrations of carbofuran, cholin salt of maleic hydrazine, propham and chlorpropham on sister-chromatid exchange (SCE) frequency in human lymphocytes *in vitro. Mutation Research* 147, 296-301.

González, N.V., Soloneski, S., Larramendy, M., 2009. Dicamba-induced genotoxicity oh Chinese hamster ovary (CHO) cells is prevented by vitamin E. *Journal of Hazardous Materials* 163, 337-343.

González, N.V., Soloneski, S., Larramendy, M.L., 2007a. The chlorophenoxy herbicide dicamba and its commercial formulation banvel induce genotoxicity in Chinese hamster ovary cells. *Mutation Research* 634, 60-68.

González, N.V., Soloneski, S., Larramendy, M.L., 2007b. Genotoxicity analysis of the phenoxy herbicide dicamba in mammalian cells *in vitro. Toxicology In Vitro* 20, 1481-1487.

Gupta, R.C., 1994. Carbofuran toxicity. *Journal of Toxicology and Environmental Health. Part A* 43, 383-418.

Haworth, S.R., Lawlor, T.E., 1983. *Salmonella*/mammalian microsome plate incorporation mutagenicity assay (Ames test). Carbofuran technical. FMC Study No. A 83-868 (MBA Study No. 1921.501). Unpublished report prepared by Microbiological Associates, MD, USA.Submitted to WHO by FMC Corp., Philadelphia, PA, USA.

Hodgson, E., Levi, P.E., 1996. Pesticides: an important but underused model for the environmental health sciences. *Environmental Health Perspectives* 104, 97-106.

Hour, T.C., Chen, L., Lin, J.K., 1998. Comparative investigation on the mutagenicities of organophosphate, phthalimide, pyrethroid and carbamate insecticides by the Ames and lactam tests. *Mutagenesis* 13, 157-166.

HSDB, 2011. Carbofuran. National Library of Medicine.

IARC, 1976. Some carbamates, thiocarbamates and carbazides. International Agency for Research on Cancer, Lyon.

IARC, 1987. Genetic and related effects: an updating of selected IARC monographs from volumes 1 to 42. International Agency for Research on Cancer, Lyon.

IARC, 2003. Monographs on the evaluation of carcinogenic risk to human. Vols. 5 - 53. International Agency for Research on Cancer, Lyon.

Jones, K., Howard, C.A., 1989. Pirimicarb (technical): an evaluation in the mouse micronucleus test. Unpublished report No. CTL/P/2641 from Central Toxicology Laboratory, Zeneca. Submitted to WHO by Syngenta Crop Protection AG. Conducte according to OECD 474 (1983). GLP compliant.

Jung, Y.S., Kim, C.S., Park, H.S., Sohn, S., Lee, B.H., Moon, C.K., Lee, S.H., Baik, E.J., Moon, C.H., 2003. N-nitroso carbofuran induces apoptosis in mouse brain microvascular endothelial cells (bEnd.3). *Journal of Pharmacology Sciences* 93, 489-495.

Käfer, E., Scott, B.R., Dorn, G.L., Stafford, R., 1982. *Aspergillus nidulans*: systems and results of tests for chemical induction of mitotic segregation and mutation. I. Diploid and duplication assay systems. A report of the U.S. EPA Gene-Tox Program. *Mutation Research* 98, 1-48.

Kamel, F., Boyes, W.K., Gladen, B.C., Rowland, A.S., Alavanja, M.C., Blair, A., Sandler, D.P., 2000. Retinal degeneration in licensed pesticide applicators. *American Journal of Industrial Medicine* 37, 618-628.

Karczmar, A., 1998. Anticholinesterases: dramatic aspects of their use and misuse. *Neurochemistry International* 32, 401-411.

Kennelly, J.C., 1990. Pirimicarb: Assessment for the induction of unscheduled DNA synthesis in rat hepatocytes *in vivo*. Unpublished report No. CTL/P/2824 from Central Toxicology Laboratory, Zeneca. Submitted to WHO by Syngenta Crop Protection AG. Conducte according to OECD 486 (1983). GLP compliant.

Kim, S.J., Kim, J.E., Ko, B.H., Moon, I.S., 2004. Carbofuran induces apoptosis of rat cortical neurons and down-regulates surface alpha7 subunit of acetylcholine receptors. *Molecules and Cells* 17, 242-247.

Kirby, P.E., 1983a. L5178Y TK+/- mouse lymphoma mutagenesis assay. FMC Study No. A 83-962 and A 83-988 (MBA Study No. T1982.701). Unpublished report prepared by Microbiological Associates, Bethesda, MD, USA. Submitted to WHO by FMC Corp., Philadelphia, PA, USA.

Kirby, P.E., 1983b. L5178Y TK+/- mouse lymphoma mutagenesis assay. FMC Study No. A 83-1064 (MBA Study No. T2124.201). Unpublished report prepared by Microbiological Associates, Bethesda, MD, USA. Submitted to WHO by FMC Corp., Philadelphia, PA, USA.

Lin, C.M., Wei, L.Y., Wang, T.C., 2007. The delayed genotoxic effect of N-nitroso N-propoxur insecticide in mammalian cells. *Food and Chemical Toxicology* 45, 928-934.

Lin, N., Garry, V.F., 2000. *In vitro* studies of cellular and molecular developmental toxicity of adjuvants, herbicides, and fungicides commonly used in Red River Valley, Minnesota. *Journal of Toxicology and Environmental Health* 60, 423-439.

McGregor, D.B., 1974. Dominant lethal study in mice of ICI PP062. Zeneca unpublished report No. CTL/C/256 from Inveresk Research International. Submitted to WHO by Syngenta Crop Protection AG. Conducte according to OECD 478 (1983). GLP compliant.

Metcalf, R., Fukuto, R., Collins, C., Borck, K., Abd El-Aziz, A., Munoz, R., Cassil, C., 1968. Metabolism of 2,2-dimethyl-2,3-dihydrobenzofuran-7-N-methylcarbamate (Furadan) in plants, insects, and mamals. *Journal of Agricultural and Food Chemistry* 16, 300-311.

Mladinic, M., Perkovic, P., Zeljezic, D., 2009. Characterization of chromatin instabilities induced by glyphosate, terbuthylazine and carbofuran using cytome FISH assay. *Toxicology Letters* 189, 130-137.

Molinari, G., Soloneski, S., Reigosa, M.A., Larramendy, M.L., 2009. *In vitro* genotoxic and citotoxic effects of ivermectin and its formulation ivomec® on Chinese hamster ovary (CHO$_{K1}$) cells. *Journal of Hazardous Materials* 165, 1074-1082.

Moriya, M., Ohta, T., Watanabe, K., Miyazawa, T., Kato, K., Shirasu, Y., 1983. Further mutagenicity studies on pesticides in bacterial reversion assay systems. *Mutation Research* 116, 185-216.

Naravaneni, R., Jamil, K., 2005. Cytogenetic biomarkers of carbofuran toxicity utilizing human lymphocyte cultures *in vitro*. *Drug and Chemical Toxicology* 28, 359-372.

Palitti, F., Tanzarella, C., Cozzi, R., Ricordy, R., Vitagliano, E., Fiore, M., 1982. Comparison of the frequencies of SCEs induced by chemical mutagens in bone-marrow, spleen and spermatogonial cells of mice. *Mutation Research* 103, 191-105.

Pavanello, S., Clonfero, E., 2000. Biomarkers of genotoxic risk and metabolic polymorphisms. *Indicatori biologici di rischio genotossico e polimorfismi metabolici* 91, 431-469.

Peter, J.V., Cherian, A.M., 2000. Organic insecticides. *Anaesthesia and Intensive Care* 28, 11-21.

Pilinskaia, M.A., 1982. Cytogenetic effect of the pesticide pirimor in a lymphocyte culture of human peripheral blood *in vivo* and *in vitro*. *Tsitologiia Genetika* 16, 38-42.

Pilinskaia, M.A., Stepanova, L.S., 1984. Effect of the biotransformation of the insecticide furadan on *in vivo* and *in vitro* manifestations of its cytogenetic activity. *Tsitologiia Genetika* 18, 17-20.

Putman, D.L., 1983a. Activity of FMC 10242 (T1982) in the *in vivo* cytogenetics assay in sprague-dawley rats. Unpublished report No: A83-972 prepared by Microbiological Associates, Bethesda, MD, USA. Submitted to WHO by FMC Corp., Philadelphia, PA, USA.

Putman, D.L., 1983b. Activity of FMC 10242 in the subchronic *in vivo* cytogenetics assay in male rats. Study No. A 83-1065 (MBA No. T2124.102). Unpublished report prepared by Microbiological Associates, Bethesda, MD, USA. Submitted to WHO by FMC Corp., Philadelphia, PA, USA USA.

Saxena, P.N., Gupta, S.K., Murthy, R.C., 2010. Carbofuran induced cytogenetic effects in root meristem cells of *Allium cepa* and *Allium sativum*: A spectroscopic approach for chromosome damage. *Pesticide Biochemistry and Physiology* 96, 93-100.

Saxena, S., Ashok, B.T., Musarrat, J., 1997. Mutagenic and genotoxic activities of four pesticides: captan, foltaf, phosphamidon and furadan. *Biochemistry and Molecular Biology International* 41, 1125-1136.

Shirasu, Y., 1975. Significance of mutagenicity testing on pesticides. *Environmental and Quality Safety* 4, 226-231.

Simmon, U.F., 1979. *In vitro* microbiological mutagenicity and unscheduled DNA synthesis studies of eighteen pesticides. Study No. 68-01-2458 (FMC Study No. A81-509, EPA 600/1-79-041). Report prepared by SRI International, Menlo Park, California, for Health Effects Research Laboratory Office of Research and Development. US Environmental Protection Agency, Research Triangle Park, NC, USA. Submitted to WHO by FMC Corp., Philadelphia, PA.

Soloneski, S., A, R.M., Larramendy, M.L., 2003. Effect of the dithiocarbamate pesticide zineb and its commercial formulation, the azzurro. V. Abnormalities induced in the spindle apparatus of transformed and non-transformed mammalian cell lines. *Mutation Research* 536, 121-129.

Soloneski, S., González, M., Piaggio, E., Apezteguía, M., Reigosa, M.A., Larramendy, M.L., 2001. Effect of dithiocarbamate pesticide zineb and its commercial formulation azzurro. I. Genotoxic evaluation on cultured human lymphocytes exposed *in vitro*. *Mutagenesis* 16, 487-493.

Soloneski, S., González, M., Piaggio, E., Reigosa, M.A., Larramendy, M.L., 2002. Effect of dithiocarbamate pesticide zineb and its commercial formulation azzurro. III. Genotoxic evaluation on Chinese hamster ovary (CHO) cells. *Mutation Research* 514, 201-212.

Soloneski, S., Larramendy, M.L., 2010. Sister chromatid exchanges and chromosomal aberrations in chinese hamster ovary (CHO-K1) cells treated with the insecticide pirimicarb. *Journal of Hazardous Materials* 174, 410-415.

Soloneski, S., Reigosa, M.A., Molinari, G., González, N.V., Larramendy, M.L., 2008. Genotoxic and cytotoxic effects of carbofuran and furadan® on Chinese hamster ovary (CHO$_{K1}$) cells. *Mutation Research* 656, 68-73.

SRI, 1979. *In vitro* microbiological mutagenicity and unscheduled DNA synthesis studies of eighteen pesticides. Contract 68-0102458. DPR Vol. 254-110, Record 47751 *(E. coli* & *B. subtilis)*, 47752 *(S. cerevisiae)*, 47753 (WI-38).

Stehrer-Schmid, P., Wolf, H., 1995. Effects of benzofuran and seven benzofuran derivatives including four carbamate insecticides in the *in vitro* porcine brain tubulin assembly assay and description of a new approach for the evaluation of the test data. *Mutation Research/Reviews in Genetic Toxicology* 339, 61-72.

Thilagar, A., 1983a. Chromosome aberrations in Chinese hamster ovary (CHO) cells. FMC Study No. A 83-1096. Unpublished report prepared by Microbiological Associates, Bethesda, MD, USA. Submitted to WHO by FMC Corp., Philadelphia, PA, USA.

Thilagar, A., 1983b. Sister chromatid exchange assay in Chinese hamster ovary (CHO) cells. FMC Study No. A 83-1095 (MBA No.1982.334001). Unpublished report prepared by Microbiological Associates, Bethesda, MD, USA. Submitted to WHO by FMC Corp., Philadelphia, PA, USA.

Thilagar, A., 1983c. Sister chromatid exchange assay in Chinese hamster ovary (CHO) cells. FMC Study No. A 83-1097 (MBA No. 2124.334001). Unpublished report prepared by Microbiological Associates, Bethesda, MD, USA. Submitted to WHO by FMC Corp., Philadelphia, PA, USA.

Thilagar, A., 1983d. Unscheduled DNA synthesis in rat primary hepatocytes. FMC Study No. A 83-969 (MBA No. T1982.380) Unpublished report prepared by Microbiological Associates, Bethesda, MD, USA. Submitted to WHO by FMC Corp., Philadelphia, PA, USA.

Trueman, R.W., 1980. An examination of pirimicarb for potential mutagenicity using the *Salmonella*/microsome reverse mutation assay. Unpublished report No: CTL/P/540 form Central Toxicology Laboratory, Zeneca. Submitted to WHO by Syngenta Crop Protection AG. Conducted to a protocol that was consistent with the OECD guideline 471 (1983). GLP compliant.

Ündeger, Ü., Basaran, N., 2005. Effects of pesticides on human peripheral lymphocytes *in vitro*: induction of DNA damage. *Archives in Toxicology* 79, 169-176.

Valencia, R., 1981. Mutagenesis screening of pesticides using *Drosophila*. FMC Study No. A 83-1042 (EPA 600/1-81-017). Report prepared by Warf Institute Inc., Madison, WI, for Health Effects Research Laboratory. Office of Research and Development, US Environmental Protection Agency, Research Triangle Park, NC, USA. Submitted to WHO by FMC Corp., Philadelphia, PA, USA.

Valencia, R., 1983. *Drosophila* sex-linked recessive lethal assay of 2,3-dihydro-2,2-dimethyl-7-benzofuranyl- *N*-methyl carbamate (carbofuran). FMC Study No. A 83-1019 (MBA No. T1047.160). Unpublished report prepared by Microbiological Associates, Bethesda, MD and the University of Wisconsin, Zoology Department, USA. Submitted to WHO by FMC Corp., Philadelphia, PA, USA.

Vera Candioti, J., Natale, G.S., Soloneski, S., Ronco, A.E., Larramendy, M.L., 2010a. Sublethal and lethal effects on *Rhinella arenarum* (Anura, Bufonidae) tadpoles exerted by the pirimicarb-containing technical formulation insecticide Aficida. *Chemosphere* 78, 249-255.

Vera Candioti, J., Soloneski, S., Larramendy, M.L., 2010b. Genotoxic and cytotoxic effects of the formulated insecticide aficida on *Cnesterodon decemmaculatus* (Jenyns, 1842) (Pisces: Poeciliidae). *Mutation Research* 703, 180-186.

Waters, M.D., Sandhu, S.S., Simon, V.F., 1982. Study of pesticide genotoxicity. *Basic Life Sciences* 21, 275-326.

WHO-FAO, 2004. Pesticides residues in food-2004. FAO Plant Production and Protection paper World Health Organization and Food and Agriculture Organization of the United Nations, Rome, pp. 154-161.

WHO-FAO, 2009. Pesticides residues in food-2009. FAO Plant Production and Protection paper World Health Organization and Food and Agriculture Organization of the United Nations, Rome, pp. 1-426.

WHO, 1988. The WHO recommended classification of pesticides by hazard and guidelines to the classification 1988-1989. World Health Organization, Geneva.

WHO, 1990. Public health impacts of pesticides used in agriculture (WHO in collaboration with the United Nations Environment Programme, Geneva, 1990). World Health Organization, Geneva.

WHO, 2000-2002. International programme of chemical safety. World Health Organization, Geneva.

WHO, 2009. The WHO recommended classification of pesticides by hazard. World Health Organization, Geneva.

Wildgoose, J., Howard, C.A., Richardson, C.R., Randall, V., 1987. Pirimicarb: A cytogenetic study in human lymphocytes *in vitro*. Central Toxicology Laboratory. Report No. CTL/P/1655, GLP, Unpublished.

Wojciechowski, J.P., Kaur, P., Sabharwal, P.S., 1982. Induction of ouabain resistance in V-79 cells by four carbamate pesticides. *Environmental Research* 29, 48-53.

Woodruff, R.C., Phillips, J.P., Irwin, D., 1983. Pesticide-induced complete and partial chromosome loss in screens with repair-defective females of *Drosophila melanogaster*. *Environmental Mutagenesis* 5, 835-846.

Yoon, J.Y., Oh, S.H., Yoo, S.M., Lee, S.J., Lee, H.S., Choi, S.J., Moon, C.K., Lee, B.H., 2001. *N*-nitrosocarbofuran, but not carbofuran, induces apoptosis and cell cycle arrest in CHL cells. *Toxicology* 169, 153-161.

Zeljezic, D., Vrdoljak, A.L., Lucas, J.N., Lasan, R., Fucic, A., Kopjar, N., Katic, J., Mladinic, M., Radic, B., 2009. Effect of occupational exposure to multiple pesticides on translocation yield and chromosomal aberrations in lymphocytes of plant workers. *Environmental Sciences Technololgy* 43, 6370-6377.

Zhou, P., Liu, B., Lu, Y., 2005. DNA damaging effects of carbofuran and its main metabolites on mice by micronucleus test and single cell gel electrophoresis. *Science in China. Series C, Life Sciences* 48, 40-47.

12

The Role of *Anopheles gambiae* P450 Cytochrome in Insecticide Resistance and Infection

Rute Félix and Henrique Silveira
Centro de Malária e Outras Doenças Tropicais, UEI Parasitologia Médica, Instituto de Higiene e Medicina Tropical, Universidade Nova de Lisboa
Portugal

1. Introduction

Anopheles gambiae is the major vector of malaria transmission in sub-Saharan Africa where the disease is responsible for the highest morbidity and mortality worldwide. Malaria, nowadays, is still a major burden causing the death of nearly one million people each year, mostly children under the age of five, and affecting those living in the poorest countries (World Health Organization [WHO], 2010).

Currently, the major obstacles to malaria eradication are the absence of a protective vaccine, the spread of parasite resistance to anti-malarial drugs and the mosquito resistance to insecticides. Controlling mosquito vectors is fundamental to reduce mosquito-borne diseases. In fact, it has been one of the most used and effective method to prevent malaria, namely trough insecticides spraying and impregnated bed nets. These methods are highly dependent on a single class of insecticides, the pyrethroids, which are the most frequently used compounds for indoor residual spraying, and the only insecticide class used for insecticide treated nets (WHO, 2010). The extensive use of a single class of insecticides further increases the risk of mosquitoes developing resistance, which could rapidly lead to a major public health problem mainly in sub-Saharan countries where insecticidal vector control is being used widely (WHO, 2010). Strategies to control malaria are still not enough to totally eliminate malaria transmission, having yet to overcome several difficulties as the development of parasite drug resistance and mosquito-vector insecticide resistance (Yassine & Osta, 2010). Unfortunately the emergence of mosquito populations capable of withstanding insecticide exposure is threatening the efficiency of these control measures.

2. Insecticide resistance

Resistance has been defined as 'the inherited ability of a strain of some organisms to survive doses of a toxicant that would kill the majority of individuals in a normal population of the same species' (Scott, 1999). The evolution of insecticide-resistant mosquito strains is an increasing problem and one of the major obstacles for the control of medical and agricultural arthropod pests. Therefore, a better understanding of its genetic and biological basis is critical. Insecticide resistance can also lead to outbreaks of human diseases when vectors cannot be controlled. Hence, the elucidation of resistance mechanisms is extremely important for the development of tools to monitor resistance in populations, thereby contributing to mosquito control programs. Although the mechanisms by which insecticides become less effective are similar across all vector taxa, each resistance problem is potentially unique and may involve a complex pattern of resistance *foci* (Brogdon & McAllister, 1998). The main forms of resistance mechanisms can be divided in two groups: target site resistance, which occurs when the insecticide no longer binds to its target, and metabolic resistance, which occurs when enhanced levels of modified activities of detoxification enzymes prevent the insecticide from reaching its site of action. Alone or in combination these mechanisms confer resistance, sometimes at high levels, to all classes of insecticides.

2.1 Target site resistance

Target site resistance is based on alterations of amino acids in the site of action where the insecticide is supposed to bind, causing the insecticide to be less effective or ineffective at all. Knock down resistance (*Kdr*) occurs due to a single or multiple substitutions in the sodium channel (Martinez-Torres et al., 1998; Ranson et al., 2000a); and alteration in acetylcholinesterase results in decreased sensitivity to insecticides (Mutero et al., 1994). Insecticide resistance has been reported from many insects including *A. gambiae* that showed the presence of insensitive acetylcholinesterase in two different populations that were resistant to carbosulfan, a carbamate insecticide (N'Guessan et al., 2003). Mutations at a single codon in the *Rdl* (resistance to dieldrin) gene have been documented in all dieldrin-resistant insects, and confer both insensitivity to the insecticide and a decrease rate of desensitisation (ffrench-Constant et al., 1998). However, in *A. gambiae* this type of resistance mechanism has not been described so far. Those are examples of target site resistance that is not the object of the present review.

2.2 Metabolic resistance

Metabolic resistance usually involves over-expression of enzymes capable of detoxifying insecticides or modifications in the amino acid sequences that cause alterations in the levels and activity of detoxifying proteins. There are three major enzyme families involved in this type of resistance, glutathione-S-transferases (GST), carboxylesterases and P450 cytochromes. Carboxylesterases are mainly involved in organophosphate and carbamate and to a lesser extent in pyrethroid resistance, while P450 cytochromes are mainly involved in the metabolism of pyrethoids and to a lesser extent, detoxification of organophosphates and carbamates (Hemingway & Ranson, 2000). Glutathione S-transferases are involved in the detoxification of a wide range of xenobiotics, including the organochloride insecticide DDT (Enayati et al., 2005). In *A. gambiae* metabolic resistance to insecticides can be conferred by elevation in the activity of these three classes of detoxifying enzymes.

The over-expression of carboxylesterases as an evolutionary response to organophosphorus and carbamate insecticide selection pressure has been reported in several insects, including mosquitoes (Newcomb et al., 1997; Vulule et al., 1999; Zhu et al., 1999). Organophosphorus and carbamate inhibit B esterases by rapid esterification of the serine residue in the active site, usually followed by a slow hydrolysis of the new ester bond. Therefore, these insecticides can be considered as inhibitors of esterases, because they are poor substrates which have a high affinity for these enzymes (Hemingway & Karunaratne, 1998). Carboxylesterases in large amounts causes resistance as the insecticides are rapidly sequestered, even before reaching the target-site acetylcholinesterase (Hemingway &

Karunaratne, 1998). There are many reports of over expression of carboxylesterases in insecticide resistant mosquitoes including *A. gambiae*, where enhanced production of carboxylesterases was observed in permethrin-resistant mosquitoes (Vulule et al., 1999). Glutathione S-transferases are a major class of detoxification enzymes that possess a wide range of substrates specificities (Enayati et al., 2005). Elevated GST activity has been implicated in resistance to several classes of insecticides (Ranson et al., 2001). Higher enzyme activity is usually due to an increase in the amount of one or more enzymes, either as a result of gene amplification or more commonly through increases in transcriptional rate, rather than qualitative changes in individual enzymes (Hemingway et al., 2004). The primary function of GSTs is the detoxification of both endogenous and xenobiotic compounds either directly or by catalysing the secondary metabolism of a vast array of compounds oxidised by P450 cytochromes (Wilce & Parker, 1994). GST enzymes metabolise insecticides by facilitating their reductive dehydrochlorination or by conjugation reactions with reduced glutathione to produce water soluble metabolites that are more readily excreted (Wilce & Parker, 1994). They also contribute to the removal of toxic oxygen free radical species produced through the action of pesticides (Enayati et al., 2005). In *A. gambiae* elevated GST levels were shown to be associated with DDT resistance (Ranson et al., 2001). Furthermore genetic mapping of the major *loci* conferring DDT resistance in *A. gambiae* implicate both *cis-* and *trans*-acting factors in the overexpression of GSTs (Ranson et al., 2000b). GSTs in *A. gambiae* were over expressed in a DDT-resistant strain, but only one *GSTE2-2* was able to metabolise DDT (Ortelli et al., 2003).

P450 cytochromes are a complex family of enzymes that are involved in the metabolism of xenobiotics and have a role in the endogenous metabolism. P450 cytochromes mediated resistance is probably the most frequent type of insecticide resistance. They are involved in the metabolism of virtually all insecticides, leading to activation of the molecule in the case of organophosphorus insecticides, or more generally to detoxification (Scott & Wen, 2001). In most cases where a link between insecticide resistance and elevated P450 activity has been shown, the P450 cytochrome belongs to the *CYP6* family (Nikou et al., 2003; Djouaka et al., 2008; Müller et al., 2007; McLaughlin et al., 2008). Although being difficult the identification of the specific P450 cytochrome associated with resistance, several P450 cytochromes were already isolated from insecticide resistant strains (Dunkov, et al., 1997; Kasai & Scott, 2000; Sabourault et al., 2001).

3. Insect P450 cytochromes

P450 Cytochromes are hemoproteins which act as terminal oxidases in monooxygenase systems. P450 cytochromes, whose name originated on its characteristic absorbance peak at 450 nm that appears when these enzymes are reduced and saturated with carbon-monoxide, constitute one of the oldest and largest super families of enzymes being found in almost all living organisms. In the literature, P450 enzymes are known by several names: cytochromes P450 monooxigenases, mixed functions oxidases, microsomal oxidases and heme thiolate proteins.

Insect P450s play a critical role in the metabolism of a wide variety of endogenous and exogenous compounds such as steroids, fatty acids and a wide range of xenobiotics and have also been implicated in vital processes like growth, development, feeding, reproduction, insecticide resistance and tolerance to plant toxins (Feyereisen, 1999; Scott et al., 1998; Scott, 1999). P450 cytochromes are also intimately involved in the synthesis and degradation of insect hormones and pheromones, including 20-hydroxyecdysone and juvenile hormone (Feyereisen, 1999).

3.1 Nomenclature

To distinguish one of these cytochromes among all the P450s, a standardized nomenclature system was implemented (Nebert et al., 1991; Nelson et al., 1996). Each P450 is named with CYP, followed by an Arabical number for the gene family, a letter for the sub-family and another Arabical number for the gene. Cytochromes P450s with share more then 40% of the amino acids are usually grouped into the same family and members with >55% of the amino acids identical are normally grouped in the same sub-family. However, there are exceptions to these rules (Nelson et al., 1996). As it is based on amino acid similarities, no information regarding the function of each P450 should be assumed from its name.

3.2 Structure

P450s can be divided into classes depending on how electrons from NAD(P)H are delivered to the catalytic site. Class I P450s are found in eukaryotes and are associated with mitochondrial membranes. This class of enzymes requires both a FAD-containing reductase and an iron sulphur redoxin, and catalyzes several steps in the biosynthesis of steroid. Class II enzymes are the most common in eukaryotes and are found in the endoplasmic reticulum. These enzymes only require an FAD/FMN-containing P450 reductase for transfer of electrons. Their functions are extremely diverse and, in eukaryotes, include aspects of the biosynthesis and catabolism of signalling molecules and steroid hormones (Feyereisen, 1999). Class III enzymes are self-sufficient and require no electron donor. They are involved in the synthesis of signalling molecules. Finally, class IV enzymes receive electrons directly from NAD(P)H. Class I and II P450s from all organisms participate in the detoxification or sometimes the activation of xenobiotics and class III and IV enzymes are considered remains of the ancestral forms of P450s involved in detoxification of damaging activated oxygen species (Werck-Reichhart & Feyereisen, 2000).

Most P450s are approximately 500 amino acids long. The core of these proteins is formed by a four-helix bundle, two sets of β sheets, two helices and a coil called the "meander". A characteristic consensus sequence known as the P450 "signature" FXXGXXXCXG, located on the C-terminus of the heme binding region, contains a conserved cysteine that serves as a fifth ligand to the heme iron. There are two other conserved motifs specific of the P450 proteins. One is the DGXXT domain, which corresponds to the proton transfer groove on the distal site of the heme. Another is the EXXR domain, which is probably needed to stabilize the core structure located on the proximal side of heme (Werck-Reichhart & Feyereisen, 2000).

3.3 Microssomal / mitochondrial

In insects both mitochondrial and microssomal P450 systems have been described. The majority of P450 in insects are microssomal, located in the endoplasmic reticulum, and require the flavoprotein NADPH cytochrome P450 reductase as the main electron donor; however cytochrome b_5 is sometimes needed, depending of the substrate and of the P450 cytochrome involved. Mitochondrial P450 are also present, but, differently from microssomal P450, require ferridoxin and a NADPH ferridoxin reductase as electron donor (Scott & Wen, 2001).

3.4 Characterization / function

Cytochromes P450 enzymes catalyse thousands of different reactions, which are based on the activation of molecular oxygen, with insertion of one of its atoms into the substrate, and reduction of the other to form water (Guengerich, 1991). P450s use electrons from NAD(P)H to catalyse the activation of molecular oxygen, leading to the regiospecific and stereospecific oxidative attack of structurally diverse chemicals (Werck-Reichhart & Feyereisen, 2000).

The interaction that occurs between P450 cytochromes and the NADPH–cytochrome P450 reductase is better expressed as a cyclic reaction (Guenguerich, 1991) as it is depicted in Figure 1.

The cycle is initiated by the binding of the substrate to the ferric form of the enzyme to form an enzyme-substrate complex, followed by a reduction of the ferric complex by an electron transferred from NADPH via NADPH-cytochrome P450 reductase. Next, the binding of molecular oxygen to the reduced complex forms an enzyme-oxygen-substrate complex followed by the transference of a second electron from NADPH via NADPH-cytochrome P450 reductase or from cytochrome b_5. A second proton is added, which results in the breaking of the oxygen-oxygen bond, releasing one atom of oxygen as water. The oxygen atom remaining is transferred to the substrate, originating an oxidized product, which is released, and a ferric form of the enzyme is once more generated. Then the cycle is re-initiated (Guenguerich, 1991).

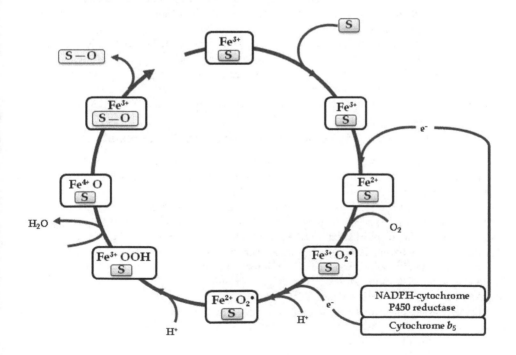

Fig. 1. Catalytic mechanism of P450 enzymes, where S is the substrate.

3.5 Diversity and specificity

The huge diversity of P450 cytochromes is probably due to an extensive process of gene duplication and cases of gene amplification, conversion, genome duplication, gene loss and lateral transfer (Werck-Reichhart & Feyereisen, 2000). Due to their extremely diverse functions, they can be found with different patterns of expression in all types of tissues and in almost all types of organisms. Although being expressed in a wide range of tissues, insect P450s have their highest activity associated with midgut, fat body and malpighian tubules (Feyereisen, 1999; Scott, 1999).

Additionally, P450s metabolise a large number of substrates, probably due to the existence of numerous P450 isoforms and to the broad specificity of some isoforms (Scott & Wen, 2001). Nevertheless the substrate specificity and type of reaction catalysed by each P450 cytochrome is still not well understood.

Their diversity enables individual P450 cytochromes to display different expression patterns related to life stages, tissues, inducers/inhibitors and substrates. There are P450s that are expressed in all life stages (CYP12 genes) while others are only expressed in adults (CYP6Z1) or in larval stages (CYP6Z3) (Nikou et al., 2003). Although being found expressed in almost all types of tissues, there are P450s which are tissue specific, while others are expressed everywhere (Feyereisen, 1999; Scott et al., 1998; Scott & Wen, 2001). Expression of P450 cytochromes may also be sex specific, as some P450s showed higher levels of expression in males compared with females (Muller et al., 2007; Nikou et al., 2003).

A large variation in substrate specificity can also be found among different P450s, some being capable of metabolising several substrates while others have only one known substrate (Scott, 1999; Scott et al., 1998). There can be also some overlapping substrate specificity among P450 cytochromes, so that one compound could be metabolised by several enzymes. The production of one or several metabolites from a single substrate also differs depending on the P450s. P450s show a vast variation in response to inducers and inhibitors, each P450 can be induced/inhibited by one or several compounds. Some P450s can also remain unaltered while others are induced or repressed (Scott et al., 1998).

4. *Anopheles gambiae* P450 cytochromes and insecticide resistance

The *A. gambiae* genome has 111 annotated P450 cytochromes (Ranson et al., 2002). The great interest in these cytochromes derives from their role in the oxidative metabolism of insecticides, but only in few cases a definitive link between an increased expression of a specific P450 cytochrome and increased insecticide metabolism has been established.

Increasing reports of specific *A. gambiae* P450 cytochromes being involved in insecticide resistance have been published in the past. The involvement of P450s in pyrethroid resistance started to be demonstrated in *A. gambiae* from Kenyan villages, in synergistic studies using specific P450 cytochrome inhibitors and also given the detection of increased heme levels in resistant mosquitoes (Vulule et al., 1999).

In 2003, Nikou et al., verified that a P450 cytochrome (CYP6Z1) was over-expressed in a pyrethroid-resistant strain of *A. gambiae,* and the development of her work pointed to an implication of the involvement of this P450 in conferring pyrethroid resistance to this mosquito (Nikou et al., 2003).

Later, a microarray *chip* was constructed containing fragments from 230 genes associated with detoxification (David et al., 2005) to further study the metabolic based insecticide resistance in *A. gambiae*. From this work resulted the identification of, among other genes, several P450 cytochromes that were highly expressed in the *A. gambiae* permethrin or DDT-resistant strains (David et al., 2005). Of notice is the P450 cytochrome *CYP325A3*, which belongs to a class that was not associated with insecticide resistance before and which was highly over-expressed in an *A. gambiae* permethrin resistant strain. Additionally, *CYP325A3* was later reported as constitutively over-expressed in a Nigerian pyrethroid resistant strain of *A. gambiae* (Awolola et al., 2009).

In 2007, studies regarding a recently colonised strain of *A. gambiae* from Ghana identified genes whose expression levels were associated with pyrethroid resistance. Among these were three P450 cytochromes (*CYP6M2, CYP6Z2* and *CYP6Z3*) (Muller et al., 2007). These results, together with their location within a cluster of P450 cytochromes in the right arm of chromosome 3 (3R), which is in close association with a pyrethroid resistance QTL (Ranson et al., 2004), strongly support their involvement in insecticide resistance. A subsequent study showed that *CYP6Z2* displays broad substrate specificity, which may be associated with xenobiotics metabolism and detoxification (Mclaughlin et al., 2008). Despite, *CYP6Z2* being able to bind to permethrin and cypermethrin, *CYP6Z2* does not metabolise neither one of these insecticides (Mclaughlin et al., 2008).

In 2008, Djouaka et al. also identified several P450 cytochromes over-expressed in one or more pyrethroid resistant populations of *A. gambiae*. Among these were *CYP6P3* and once again *CYP6M2*. Both genes showed high levels of over-expression in all the resistant populations, but the first was the gene that showed greatest differences. In the same year, *CYP6P3* was also identified as being up-regulated in another highly permethrin resistant *A. gambiae* population (Müller et al., 2008).

Recent studies on *A. gambiae* recombinant proteins *CYP6M2* (Stevenson et al., 2011) and *CYP6P3* (Müller et al., 2008) demonstrated that these enzymes could metabolise pyrethroids. Thus, the up regulation of these P450 cytochromes in pyrethroid resistant populations, strongly supports a key role for these genes to confer pyrethroid resistance in *A. gambiae*.

Highly expressed P450s have been also reported in DDT resistant strains of *A. gambiae* (David et al., 2005). *CYP6Z1* and *CYP12F1* were strongly over-expressed together with other genes, suggesting that multiple genes could contribute to the DDT resistance phenotype. The slightly over-expression of the electron donor cytochrome P450 reductase in the DDT resistant strain further supported a P450-based resistance mechanism in *A. gambiae* (David et al., 2005).

As the above P450 cytochromes, *CYP314A1* was also found to be over-expressed in a DDT resistant strain of *A. gambiae* from Kenia (Vontas et al., 2005), suggesting a possible involvement in the insecticide resistance phenotype. Both *CYP6Z1* and *CYP6Z2* were over-expressed in DDT resistant strains of *A. gambiae* (David et al. 2005). Although being very similar, these two cytochromes have predicted substrate cavities dramatically different and *CYP6Z1* was predicted to be the only one capable of metabolizing DDT. Chiu et al. (2008) through biochemical characterisations supported these predictions and identified *CYP6Z1* as the only P450 cytochrome capable of metabolising DDT, demonstrating its potential as a target to reduce *A. gambiae* resistance to DDT (Chiu et al., 2008).

Another evidence of the involvement of P450s in insecticide resistance is the fact that silencing the main electron donor of P450 cytochromes, the cytochrome P450 reductase, by RNAi, greatly increased the susceptibility of *A. gambiae* to permethrin, emphasising the important chemoprotective role of P450 cytochromes in this process (Lycett et al., 2006).

Nevertheless, although P450s have been clearly associated with insecticide resistance, the identification of specific P450 cytochromes responsible for insecticide resistance is still extremely difficult.

5. *Anopheles gambiae* P450 cytochromes and malaria infection

P450 cytochromes have also been implicated in other vital processes as in *A. gambiae* response to bacterial challenge and to parasite invasion, but the real importance and function of these cytochromes in this process is still not well understood.

A genome expression analysis of *A. gambiae* was made to identify which genes responded to injury, bacterial challenge and malaria infection (Dimopoulos et al., 2002). This study identified three P450 cytochromes, one associated with injury, microbial challenge and oxidative stress; the second associated with the response to septic injury which is similar to a bacterial infection *in vivo*; and the third associated with the response to malaria infection and the presence of lipopolysaccharide (Dimopoulos et al., 2002).

The involvement of P450 cytochromes in response to microbial challenge was established when two P450 cytochromes (*CYP4C27* and *CYP306A1*) were differently expressed in the presence of Gram negative (*Salmonella thyphimurium*) or Gram positive (*Staphylococcus aureus*) bacteria (Aguilar et al., 2005). This involvement was even more evident when a study, trying to implicate the mosquito midgut microbiota in the defense against malaria parasites, showed that there were ten P450s differently expressed in response to *Escherichia coli* and *S. aureus* in the *A. gambiae* midgut twelve hours after an uninfected blood meal (Dong et al., 2009). Between the P450 cytochromes differently expressed there were *CYP4H17*, *CYP6M3*, *CYP6AG1*, *CYP9J5*, two of them were mitochondrial cytochromes, *CYP49A1* and *CYP12F4* (Dong et al., 2009).

Regarding the relation between P450 cytochromes and the response to malaria infection, it was partly unveiled for the first time in a study about the midgut epithelial responses to *Plasmodium* invasion (Vlachou et al., 2005). The study revealed that P450 cytochromes were differentially expressed during different phases of the midgut invasion (before invasion, during invasion and after invasion) as well as when they compared *Plasmodium* wild-type infection with *Plasmodium* that were unable to invade the epithelium. (Vlachou et al., 2005). P450s that stood out in this study were *CYP305A1*, *CYP304B1*, *CYP6Z1* and *CYP6M4* (Vlachou et al., 2005). The role of P450 cytochromes in the *A. gambiae* response to malaria infection has been reinforced in the last years. Comparing the *A. gambiae* response to two different *Plasmodium* parasites -*P. berghei* and *Plasmodium falciparum* - showed that the mosquito induced slightly different immune responses to each parasite, and that the mosquito was capable of sensing infected blood constituents and mount an immune response, even in the absence of invading ookinetes (Dong et al., 2006). Although there were different responses between the three experimental groups, in all of them there were P450s differentially expressed in the midgut (*CYP6AG1*, *CYP6M4*, *CYP6M1*, *CYP9J5* and *CYP12F3*) and in the fat body (*CYP6AG1* and *CYP4G17*), reinforcing, their involvement in response to malaria infection.

Further evidence of the link between P450 cytochromes and the mosquito's response to malaria infection came from different studies. First, the effect on gene regulation of the presence of chloroquine in an uninfected blood meal and in a *Plasmodium* infected blood meal was investigated (Abrantes et al., 2008). This work showed that chloroquine affects the abundance of transcripts which encode proteins involved in a variety of processes, including P450 cytochromes that were differently expressed in the *P. berghei* infected blood meal (*CYP9L1, CYP304B1* and *CYP305A1*). A second study focused on the role of *A. gambiae* detoxification enzymes, from the three major families involved in detoxification, GSTs, carboxylesterases and P450 cytochromes, in the response to *Plasmodium* infection (Félix et al., 2010). In this study the impact of *P. berghei* infection was analyzed at two time points: one day following the blood meal, during which parasites invade the midgut epithelium, and eleven days after the blood meal when sporozoites were starting to be released to the hemolymph; in two different tissues, midgut and fat body. At day one after the *Plasmodium* infected blood meal they found 17 P450 cytochromes down-regulated and 5 P450 cytochromes up-regulated, including *CYP9L1, CYP304B1, CYP325H1, CYP6M2* and *CYP6Z2* in the midgut, and 5 P450 up-regulated and 1 down-regulated in the fat body, including *CYP12F2, CYP6M2, CYP6M3* and *CYP4G17*. At eleven days after an infected blood meal they found 2 P450 cytochromes up-regulated and 3 down-regulated in the midgut and 1 P450 cytochrome up-regulated and 1 down-regulated in the fat body. The high number of P450 cytochromes differently expressed by the presence of *P. berghei* parasites in different phases of infection and in different tissues suggests that P450 cytochromes are deeply involved in the mosquito response to *Plasmodium* infection, having an important role in different development stages of the parasite and covering different tissues of the mosquito. More specifically, these P450 cytochromes might have a direct role in *Plasmodium* response during the parasite invasion of the midgut epithelium as this is the moment and tissue where more P450 were differentially expressed. The over expression of these P450 cytochromes could be part of a mosquito response mechanism to parasite invasion occurring in the midgut. One possibility is that P450s are involved in the cytoskeleton rearrangement (Vlachou et al., 2005; Vlachou & Kafatos, 2005), or alternatively P450s could be involved in the production of nitric oxide and other reactive oxygen radicals that are induced by *Plasmodium* invasion of the midgut epithelium (Han et al., 2000; Luckhart et al., 1998). The blood meal *per si* generates metabolic changes that are also expected to increase the oxidative stress in the mosquito midgut, which is augmented by the presence of *Plasmodium* parasites (Molina-Cruz et al., 2008). Moreover, other parasite killing mechanisms also induce oxidative stress inside the host which, although helping to eliminate the parasite, are also toxic to the host cell. The high level of oxidative stress inside the host cell could trigger cellular and molecular regulation of these P450 cytochromes, at this time point, being responsible for host detoxification and parasite elimination.

Mosquito hemocytes mediate important cellular immune responses including phagocythosis, encapsulation and secrete immune factors such as antimicrobial peptides and mediate melanization. Recently, studies were made to characterize the role of *A. gambiae* hemocytes in mosquito immunity, consisting in a genome-wide transcriptomic analysis of adult female hemocytes following infection by bacteria and *Plasmodium* parasites (Baton et al., 2009). This work showed that *CYP325H1* and *CYP6M1* were differently expressed in the presence of *Micrococcus luteus*, a Gram-positive bacteria (Baton et al., 2009), reinforcing the role of P450 cytochromes in response to microbial challenge. This work also showed that *CYP325H1* was differently expressed 24 hours after the infected blood meal, during *P. berghei* ookinete invasion of the midgut epithelium. Moreover, *CYP6AG1* and *CYP6M3* were also differentially expressed 19 days after the infected blood meal, during *P. berghei* sporozoite migration through the hemolymph (Baton et al., 2009), suggesting that P450 cytochromes have a role in the response to malaria infection by hemocytes. Another study aiming to analyze the transcriptional profile of circulating *A. gambiae* hemocytes during *P.*

berghei infection showed that *CYP6Z1*, *CYP6M2*, *CYP6M3* and *CYP12F2* were differently expressed at 24-28 hours after an infective blood meal (Pinto et al., 2009), valuing the importance of P450 cytochromes on the hemocyte response to malaria parasite invasion.

6. Conclusion

The role of P450 cytochromes during *Plasmodium* invasion is still poorly understood, but it may play out to be of utmost importance to combat malaria transmission. Here, we intend to bring an update review on the connection between P450 cytochromes and the *A. gambiae* response to malaria infection, identifying several P450 cytochromes that probably are, directly or indirectly, involved in the response to *Plasmodium* invasion. We have also reviewed the implication of P450 cytochromes in *A. gambiae* insecticide resistance. However, uncovering the objective role of these cytochromes in insecticide resistance, that is naming specific cytochromes and describing in detail the processes in which those specific P450s are involved is still extremely difficult.

The consistent detection of differential expression of P450 cytochromes, in studies about either insecticide resistance or the response to malaria infection, suggests that the role of these P450s could be similar in these two processes. Nevertheless, the real importance and function of P450 cytochromes in these processes is still not well understood neither the possibility of interplay between infection and insecticide resistance. One of the P450 cytochromes with expression altered in response to insecticides and *Plasmodium* infection was *CYP6M2* that, was highly over-expressed in a pyrethroid-resistant strain of *A. gambiae* mosquitoes (Muller et al., 2007) and also highly over-expressed in response to *Plasmodium* infection in both the midgut and the fat body 1 day after an infected blood meal (Félix et al., 2010). These results suggest that the role of *CYP6M2* might be the same in response to insecticides and infection, or that these two processes might share the activation mechanism of *CYP6M2* expression. *CYP6M2* could also function as an endogenous mediator, acting as the first response to different challenges, which would explain being increased by parasite infection and insecticide exposure. Similar to *CYP6M2* is *CYP6Z1*, yet another P450 cytochrome that was over-expressed in insecticides-resistant strains of *A. gambiae* (David et al., 2005; Nikou et al., 2003) and was also over-expressed in response to *Plasmodium* infection (Vlachou et al., 2005). The increase in the expression of this P450 could function as an immediate response to an exogenous challenge or *A. gambiae* could have the same mechanism of response, including over-expression of specific P450 cytochromes, to parasite infection and insecticide exposure. *CYP6Z2* was highly over-expressed in a pyrethroid-resistant strain (Müller et al., 2007), but opposite to *CYP6M2*, was down-regulated in the midgut of *A. gambiae* at day 1 and day 11 after an infected blood meal (Félix et al., 2010). These results suggest a different role for *CYP6Z2* in response to the insecticide and to parasite infection, however, we have to take into account that, although being able to bind to permethrin and cypermethrin, *CYP6Z2* does not metabolise these compounds (Mclaughlin et al., 2008). So the over-expression of *CYP6Z2* in a pyrethroid-resistant strain might be associated with different processes other than insecticide resistance.

A more complete knowledge about the factors involved in P450 cytochromes response to malaria infection and insecticide resistance is extremely needed for the implementation of efficient malaria and vector control programmes, including strategies able to adapt to different types of resistance. Although the interaction of insecticides with P450 enzymes has been studied, many of its aspects still remains poorly understood. Grasping the underlying processes in this interaction might help mitigate the problem of insecticide resistance, and therefore contribute to the control of malaria and other human diseases.

7. Acknowledgments

We would like to thank to Fundação da Ciência e Tecnologia (FCT) for for the financial support provided by a PhD fellowship grant (SFRH/BD/28024/2006) and research funds from projects POCTI/SAU-IMI/59489/2004 and PTDC/SAUMII/102596/2008.

8. References

Abrantes, P.; Dimopoulos, G.; Grosso, A.R.; do Rosário, V.E. & Silveira, H. (2008). Chloroquine mediated modulation of *Anopheles gambiae* gene expression. *PLoS One*, 3(7), e2587, ISSN 1932-6203

Aguilar, R.; Dong, Y.; Warr, E. & Dimopoulos, G. (2005). *Anopheles* infection responses; laboratory models versus field malaria transmission systems. *Acta Tropica*, 95(3), 285-291, ISSN 0001-706X

Awolola,T.S.; Oduola, O.A.; Strode, C.; Koekemoer, L.L.; Brooke, B. & Ranson, H. (2009). Evidence of multiple pyrethroid resistance mechanisms in the malaria vector *Anopheles gambiae* sensu stricto from Nigeria. *Transactions of the Royal Society of Tropical Medicine and Hygiene*, 103, 1139-1145, ISSN 0035-9203

Baton, L.A.; Robertson, A.; Warr, E.; Strand, M.R. & Dimopoulos, G. (2009). Genome-wide transcriptomic profiling of *Anopheles gambiae* hemocytes reveals pathogen-specific signatures upon bacterial challenge and *Plasmodium berghei* infection. *BMC Genomics*, 5, 10:257, ISSN 1471-2164

Brogdon, W.G. & McAllister, J.C. (1998). Insecticide resistance and vector control. *Emerging Infectious Diseases*, 4(4), 605-613, ISSN: 1080-6059

Chiu, T.L.; Wen, Z.; Rupasinghe, S.G. & Schuler, M.A. (2008). Comparative molecular modeling of *Anopheles gambiae* CYP6Z1, a mosquito P450 capable of metabolizing DDT. *Proceedings of the National Academy of Sciences*, 105(26), 8855-8860, ISSN 0027-8424

David, J.P.; Strode, C.; Vontas, J.; Nikou, D.; Vaughan, A.; Pignatelli, P.M.; Louis, C.; Hemingway, J. & Ranson, H. (2005). The *Anopheles gambiae* detoxification chip: a highly specific microarray to study metabolic-based insecticide resistance in malaria vectors. *Proceedings of the National Academy of Sciences*, 102(11), 4080-4084, ISSN 0027-8424

Dimopoulos, G.; Christophides, G.K.; Meister, S.; Schultz, J.; White, K.P.; Barillas-Mury, C. & Kafatos, F.C. (2002). Genome expression analysis of *Anopheles gambiae*: responses to injury, bacterial challenge, and malaria infection. *Proceedings of the National Academy of Sciences*, 99(13), 8814-8819, ISSN 0027-8424

Djouaka, R.F.; Bakare, A.A.; Coulibaly, O.N.; Akogbeto, M.C.; Ranson, H.; Hemingway, J. & Strode, C. (2008). Expression of the cytochrome P450s, CYP6P3 and CYP6M2 are significantly elevated in multiple pyrethroid resistant populations of *Anopheles gambiae s.s.* from Southern Benin and Nigeria. *BMC genomics*, 9, 538, ISSN 1471-2164

Dong, Y.; Aguilar, R.; Xi, Z.; Warr, E.; Mongin, E. & Dimopoulos, G. (2006). *Anopheles gambiae* immune responses to human and rodent *Plasmodium* parasite species. *PLoS Pathogens*, 2(6), e52, ISSN 1553-7366

Dong, Y.; Manfredini, F. & Dimopoulos, G. (2009). Implication of the mosquito midgut microbiota in the defense against malaria parasites. *PLoS Pathogens*, 5(5), e1000423, ISSN 1553-7366

Dunkov, B.C.; Guzov, V.M.; Mocelin, G.; Shotkoski, F.; Brun, A.; Amichot, M.; Ffrench-Constant, R.H. & Feyereisen, R. (1997). The *Drosophila* Cytochrome P450 Gene Cyp6a2: Structure, Localization, Heterologous Expression, and Induction by Phenobarbital. DNA and Cell Biology, 16(11), 1345-1356, ISSN 1044-5498

Enayati, A.A.; Ranson, H. & Hemingway, J. (2005). Insect glutathione transferases and insecticide resistance. *Insect Molecular Biology*, 14(1), 3-8, ISSN 0962-1075

Félix, R.C.; Müller, P.; Ribeiro, V.; Ranson, H. & Silveira, H. (2010). *Plasmodium* infection alters *Anopheles gambiae* detoxification gene expression. *BMC genomics*, 11, 312, ISSN 1471-2164

Feyereisen, R. (1999). Insect P450 Enzymes. *Annual Review of Entomology*, 44, 507-533, ISSN 0066-4170

ffrench-Constant, R.H.; Pittendrigh, B.; Vaughan, A. & Anthony, N. (1998). Why are there so few resistance-associated mutations in insecticide target genes? *Phil.Trans. R. Soc. Lond. B*, 353, 1685-1693, ISSN 1471-2970

Guengerich, F.P. (1991). Reactions and Significance of Cytochrome P-450 Enzymes. *The Journal of Biological Chemistry*, 266(16), 10019-10022, ISSN 0021–9258

Han, Y.S.; Thompson, J.; Kafatos, F.C. & Barillas-Mury, C. (2000). Molecular interactions between *Anopheles stephensi* midgut cells and *Plasmodium berghei*: the time bomb theory of ookinete invasion of mosquitoes. *Embo Journal*, 19(22), 6030-6040, ISSN 0261-4189

Hemingway, J. & Karunaratne, S.H. (1998). Mosquito carboxylesterases: a review of the molecular biology and biochemistry of a major insecticide resistance mechanism. *Medical and veterinary Entomology*, 12(1), 1-12, ISSN 1365-2915

Hemingway, J. & Ranson, H. (2000). Insecticide resistance in insect vectors of human disease. *Annual Review of Entomology*. 41, 371-391, ISSN 0066.4170

Hemingway, J.; Hawkes, N.J.; McCarrol, L. & Ranson, H. (2004). The molecular basis of insecticide resistance in mosquitoes. *Insect Biochemistry and Molecular Biology*, 34, 653-655, ISSN 0965-1748

Kumar, S.; Christophides, G.K.; Cantera, R.; Charles, B.; Han, Y.S.; Meister, S.; Dimopoulos, G.; Kafatos, F.C. & Barillas-Mury, C. (2003). The role of reactive oxygen species on *Plasmodium* melanotic encapsulation in *Anopheles gambiae*. *Proceedings of the National Academy of Sciences*, 100(24), 14139-14144, ISSN 0027-8424

Kumar, S.; Gupta, L.; Han, Y.S. & Barillas-Mury, C. (2004). Inducible peroxidases mediate nitration of *anopheles* midgut cells undergoing apoptosis in response to *Plasmodium* invasion. *Journal of Biological Chemistry*, 279(51), 53475-53482, ISSN 0021-9258

Kasai, S. & Scott, J.G. (2000). Overexpression of Cytochrome P450 CYP6D1 Is Associated with Monooxygenase-Mediated Pyrethroid Resistance in House Flies from Georgia. *Pesticide Biochemistry and Physiology*, 68, 34-41, ISSN 0048-3575

Luckhart, S.; Vodovotz, Y.; Cui, L. & Rosenberg, R. (1998). The mosquito *Anopheles stephensi* limits malaria parasite development with inducible synthesis of nitric oxide. *Proceedings of the National Academy of Sciences*, 95(10), 5700-5705, ISSN 0027-8424

Lycett, G.J.; McLaughlin, L.A.; Ranson, H.; Hemingway, J.; Kafatos, F.C.; Loukeris, T.G. & Paine, M.J. (2006). Anopheles gambiae P450 reductase is highly expressed in oenocytes and in vivo knockdown increases permethrin susceptibility. *Insect Molecular Biology*, 15(3), 321-327, ISSN 0962-1075

Martinez-Torres, D.; Chandre, F.; Williamson, M.S.; Darriet, F.; Bergé, J.B.; Devonshire, A.L.; Guillet, P.; Pasteur, N. & Pauron, D. (1998). Molecular characterization of pyrethroid knockdown resistance (*kdr*) in the major malaria vector *Anopheles gambiae s.s. Insect Molecular Biology*, 7(2), 179-184, ISSN 0962-1075

McLaughlin,L.A.; Niazi, U.; Bibby, J.; David, J.P.; Vontas, J.; Hemingway, J.; Ranson, H.; Sutcliffe, M.J. & Paine, M.J. (2008). Characterization of inhibitors and substrates of *Anopheles gambiae* CYP6Z2. *Insect Molecular Biology*, 17(2), 125-135, ISSN 0962-1075

Molina-Cruz, A.; DeJong, R.J.; Charles, B.; Gupta, L.; Kumar, S., Jaramillo-Gutierrez, G. & Barillas-Mury, C. (2008) Reactive Oxygen Species Modulate *Anopheles gambiae* Immunity against Bacteria and *Plasmodium*. *The Journal of Biological Chemistry*, 283(6), 3217-3223, ISSN 0021-9258

Müller, P.; Donnelly, M.J. & Ranson, H. (2007). Transcription profiling of a recently colonised pyrethroid resistant *Anopheles gambiae* strain from Ghana. *BMC genomics*, 8, 36, ISSN 1471-2164

Müller,P.; Warr, E.; Stevenson, B.J.; Pignatelli, P.M.; Morgan, J.C.; Steven, A.; Yawson, A.E.; Mitchell, S.N.; Ranson, H.; Hemingway, J.; Paine. M.J. & Donnelly, M.J. (2008). Field-caught permethrin-resistant *Anopheles gambiae* overexpress CYP6P3, a P450 that metabolises pyrethroids. *PLoS Genetics*, 4(11), e1000286, ISSN 1553-7390

Mutero, A.; Pralavorio, M.; Bride, J.M. & Fournier, D. (1994). Resistance-associated point mutations in insecticide-insensitive acetylcholinesterase. *Proceedings of the National Academy of Sciences*, 91(13), 5922-5526, ISSN 0027-8424

Nebert, D.W.; Nelson, D.R.; Coon, M.J.; Estabrook, R.W.; Feyereisen, R.; Fujii-Kuriyama, Y.; Gonzalez, F.J.; Guengerich, F.P.; Gunsalus, I.C.; Johnson, E.F.; Loper, J.C.; Sato, R.; Waterman, M.R. & Waxman, D.J. (1991). The P450 Superfamily: Update on New Sequences, Gene Mapping, and Recommended Nomenclature. *DNA and Cell Biology*, 10(1), 1-14, ISSN 1044-5498

Nelson, D.R.; Koymans, L.; Kamataki, T.; Stegeman, J.J; Feyereisen, R.; Waxman, D.J.; Waterman, M.R.; Gotoh, O.; Coon, M.J.; Estabrook, R.W.; Gunsalus, I.C. & Nebert, D.W. (1996). P450 superfamily: update on new sequences, gene mapping, accession numbers and nomenclature. *Pharmacogenetics*, 6(1), 1-42, ISSN 0960-314X

Newcomb, R.D.; Campbell, P.M.; Ollis, D.L.; Cheah, E.; Russell, R.J. & Oakeshott, J.G. (1997). A single amino acid substitution converts a carboxylesterase to an organophosphorus hydrolase and confers insecticide resistance on a blowfly. *Proceedings of the National Academy of Sciences*, 94(14), 7464-7468, ISSN 0027-8424

N'Guessan, R.; Darriet, F.; Guillet, P.; Carnevale, P.; Traore-Lamizana, M.; Corbel, V.; Koffi, A.A. & Chandre, F. (2003). Resistance to carbosulfan in *Anopheles gambiae* from Ivory Coast, based on reduced sensitivity of acetylcholinesterase. *Medical and veterinary Entomology*, 17(1), 19-25, ISSN 1365-2915

Nikou, D.; Ranson, H. & Hemingway, J. (2003). An adult-specific CYP6 P450 gene is overexpressed in a oyrethroid-resistant strain of the malaria vector, *Anopheles gambiae*. *Gene*, 318, 91-102, ISSN 0378-1119

Ortelli, F.; Rossiter, L.C.; Vontas, J.; Ranson, H. & Hemingway, J. (2003). Heterologous expression of four glutathione transferase genes genetically linked to a major insecticide-resistance locus from the malaria vector *Anopheles gambiae*. *Biochemical Journal*, 373, 957-963, ISSN 0264-6021

Peterson, T.M.; Gow, A.J. & Luckhart, S. (2007). Nitric oxide metabolites induced in *Anopheles stephensi* control malaria parasite infection. *Free Radical Biology & Medicine*, 42(1), 132-142, ISSN 0891-5849

Pinto, S.B.; Lombardo, F.; Koutsos, A.C.; Waterhouse, R.M.; McKay, K.; Chunju, A.; Ramakrishnan, C.; Kafatos, F.C. & Michel, K. (2009). Discovery of *Plasmodium* modulators by genome-wide analysis of circulating hemocytes in *Anopheles gambiae*. *Proceedings of the National Academy of Sciences*, 106(50), 21270-21275, ISSN 0027-8424

Ranson, H.; Jensen, B.; Vulule, J.M.; Wang, X.; Hemingway, J. & Collins, F.H. (2000a). Identification of a point mutation in the voltage-gated sodium channel gene of Kenyan Anopheles gambiae associated with resistance to DDT and pyrethroids. Insect Molecular Biology, 9 (5), 491-497, ISSN 0962-1075

Ranson, H.; Jensen, B.; Wang, X.; Prapanthadara, L.; Hemingway, J. & Collins, F.H. (2000b). Genetic mapping of two loci affecting DDT resistance in the malaria vector *Anopheles gambiae*. *Insect Molecular Biology*, 9(5), 499-507, ISSN 0962-1075

Ranson, H.; Rossiter, L.; Ortelli, F.; Jensen, B.; Wang, X.; Roth, C.W.; Collins, F.H. & Hemingway, J. (2001). Identification of a novel class of insect glutathione S-transferases involved in resistance to DDT in the malaria vector *Anopheles gambiae*. *Biochemical Journal*, 359, 295-304, ISSN 0264-6021

Ranson, H.; Claudianos, C.; Ortelli, F.; Abgrall, C.; Hemingway, J.; Sharakhova, M.V.; Unger, M.F.; Collins, F.H. & Feyereisen, R. (2002). Evolution of supergene families associated with insecticide resistance. *Science*, 298, 179-181, ISSN 0036-8075

Ranson, H.; Paton, M.G.; Jensen, B.; McCarroll, L.; Vaughan, A.; Hogan, J.R.; Hemingway, J. & Collins, F.H. (2004). Genetic mapping of genes conferring permethrin resistance in the malaria vector, *Anopheles gambiae*. *Insect Molecular Biology*, 13 (4), 379-386, ISSN 0962-1075

Sabourault, C.; Guzov, V.M.; Koener, J.F.; Claudianos, C.; Plapp. F.W.Jr. & Feyereisen, R. (2001). Overproduction of a P450 that metabolizes diazinon is linked to a loss-of-function in the chromosome 2 ali-esterase (MdalphaE7) gene in resistant house flies. *Insect Molecular Biology*, 10 (6), 609-618, ISSN 0962-1075

Scott, J.G.; Liu, N. & Wen, Z. (1998). Insect cytochromes P450: diversity, insecticide resistance and tolerance to plant toxins. *Comparative Biochemistry and Phisiology Part C*, 121(1-3), 147-155, ISSN 1532-0456

Scott, J.G. (1999). Cytochromes P450 and insecticide resistance. *Insect Biochemistry and Molecular Biology*, 29(9), 757-777, ISSN 0965-1748

Scott, J.G. & Wen, Z. (2001) Cytochromes P450 of insects: the tip of the iceberg. *Pest Management Science*, 57(10), 958-967, ISSN 1526-4998

Stevenson, B.J.; Bibby, J.; Pignatelli, P.; Muangnoicharoen, S.; O'Neill, P.M.; Lian, L.Y.; Müller, P.; Nikou, D.; Steven, A.; Hemingway, J.; Sutcliffe, M.J. & Paine, M.J. (2011). Cytochrome P450 6M2 from the malaria vector *Anopheles gambiae* metabolizes pyrethroids: Sequential metabolism of deltamethrin revealed. *Insect Biochemistry and Molecular Biology*, in press, ISSN 0965-1748

Vlachou, D. & Kafatos, F.C. (2005). The complex interplay between mosquito positive and negative regulators of *Plasmodium* development. *Current Opinion in Microbiology*, 8(4), 415-421, ISSN 1369-5274

Vlachou, D.; Schlegelmilch, T.; Christophides, G.K. & Kafatos, F.C. (2005). Functioal Genomic Analysis of Midgut Epithelial Responses in *Anopheles* during *Plasmodium* Invasion. *Current Biology*, 15(13), 1185-1195, ISSN 0960-9822

Vontas, J., Blass, C. Koutsos, A.C., David, J.-P., Kafatos, F.C., Louis, C., Hemingway, J., Christophides, G.K. & Ranson, H. (2005). Gene expression in insecticide resistant and susceptible *Anopheles gambiae* strains constitutively or after insecticide exposure. *Insect Molecular Biology*, 14(5), 509-521, ISSN 0962-1075

Vulule, J.M., Beach, R.F., Atieli, F.K., McAllister, J.C., Brogdon, W.G., Roberts, J.M., Mwangi, R.W. & Hawley, W.A. (1999). Elevated oxidase and esterase levels associated with permethrin tolerance in *Anopheles gambiae* from Kenyan villages using permethrin-impregnated nets. *Medical and veterinary Entomology*, 13(3), 239-244, ISSN 1365-2915

Werck-Reichhart, D. & Feyereisen, R. (2000). Cytochromes P450: a success history. *Genome Biology*, 1(6), 3003, ISSN 1759-6653

Wilce, M.C. & Parker, M.W. (1994). Structure and function of glutathione S-transferases. *Biochim Biophys Acta*, 1205, 1-18, ISSN 0006-3002

World Health Organization (2010). World Malaria Report 2010, ISBN 978 92 4   156410. Geneve.

Yassine H. & Osta M.A. (2010). *Anopheles gambiae* innate immunity. *Cellular Microbiology*, 12(1), 1-9, ISSN 1462-5814

Zhu, Y.C.; Dowdy, A.K. & Baker, J.E. (1999). Differential mRNA expression levels and gene sequences of a putative carboxylesterase-like enzyme from two strains of the parasitoid *Anisopteromalus calandrae* (Hymenoptera: Pteromalidae). *Insect Molecular Biology*, 29(5), 417-421, ISSN 0962-1075

Permissions

The contributors of this book come from diverse backgrounds, making this book a truly international effort. This book will bring forth new frontiers with its revolutionizing research information and detailed analysis of the nascent developments around the world.

We would like to thank Farzana Perveen, for lending her expertise to make the book truly unique. She has played a crucial role in the development of this book. Without her invaluable contribution this book wouldn't have been possible. She has made vital efforts to compile up to date information on the varied aspects of this subject to make this book a valuable addition to the collection of many professionals and students.

This book was conceptualized with the vision of imparting up-to-date information and advanced data in this field. To ensure the same, a matchless editorial board was set up. Every individual on the board went through rigorous rounds of assessment to prove their worth. After which they invested a large part of their time researching and compiling the most relevant data for our readers. Conferences and sessions were held from time to time between the editorial board and the contributing authors to present the data in the most comprehensible form. The editorial team has worked tirelessly to provide valuable and valid information to help people across the globe.

Every chapter published in this book has been scrutinized by our experts. Their significance has been extensively debated. The topics covered herein carry significant findings which will fuel the growth of the discipline. They may even be implemented as practical applications or may be referred to as a beginning point for another development. Chapters in this book were first published by InTech; hereby published with permission under the Creative Commons Attribution License or equivalent.

The editorial board has been involved in producing this book since its inception. They have spent rigorous hours researching and exploring the diverse topics which have resulted in the successful publishing of this book. They have passed on their knowledge of decades through this book. To expedite this challenging task, the publisher supported the team at every step. A small team of assistant editors was also appointed to further simplify the editing procedure and attain best results for the readers.

Our editorial team has been hand-picked from every corner of the world. Their multi-ethnicity adds dynamic inputs to the discussions which result in innovative outcomes. These outcomes are then further discussed with the researchers and contributors who give their valuable feedback and opinion regarding the same. The feedback is then collaborated with the researches and they are edited in a comprehensive manner to aid the understanding of the subject.

Apart from the editorial board, the designing team has also invested a significant amount of their time in understanding the subject and creating the most relevant covers. They scrutinized every image to scout for the most suitable representation of the subject and create an appropriate cover for the book.

The publishing team has been involved in this book since its early stages. They were actively engaged in every process, be it collecting the data, connecting with the contributors or procuring relevant information. The team has been an ardent support to the editorial, designing and production team. Their endless efforts to recruit the best for this project, has resulted in the accomplishment of this book. They are a veteran in the field of academics and their pool of knowledge is as vast as their experience in printing. Their expertise and guidance has proved useful at every step. Their uncompromising quality standards have made this book an exceptional effort. Their encouragement from time to time has been an inspiration for everyone.

The publisher and the editorial board hope that this book will prove to be a valuable piece of knowledge for researchers, students, practitioners and scholars across the globe.

List of Contributors

Silvia I. Rondon
Oregon State University, Hermiston Agricultural Research and Extension Center, Hermiston, OR, USA

Stuart R. Reitz
United States Department of Agriculture, Agricultural Research Service, Center for Medical, Agricultural and Veterinary Entomology, Tallahassee, FL,

Joe Funderburk
North Florida Research and Education Center, University of Florida, Quincy, FL, USA

Martina Díaz and Carmen Rossini
Laboratorio de Ecología Química, Facultad de Química, Universidad de la República, Uruguay

Stefano Civolani
Department of Biology and Evolution, University of Ferrara, Ferrara, Agricultural Foundation "Fratelli Navarra", Malborghetto di Boara (Ferrara), Italy

Mohamed Braham and Lobna Hajji
Centre régional de recherche en Horticulture et Agriculture Biologique; Laboratoire d'Entomologie – Ecologie; Chott-Mariem, Tunisia

Carlos García Salazar
Agriculture & Agribusiness Institute, Michigan State University Extension, West Olive, MI

Anamaría Gómez Rodas and John C. Wise
Department of Entomology, Trevor Nichols Research Center, Fennville, MI, USA

K.M. Tubajika
1United States Department of Agriculture (USDA), Animal and Plant Inspection Service (APHIS), Plant Protection and Quarantine (PPQ), Center for Plant Health Science and Technology (CPHST), Raleigh, North Carolina

G.J. Puterka
United States Department of Agriculture (USDA),
Agricultural Research Service (ARS), Stillwater, Oklahoma

N.C. Toscano3
Department of Entomology, University of California, Riverside, California

J. Chen and E.L. Civerolo
United States Department of Agriculture (USDA), Agricultural Research Service (ARS), San Joaquin Valley Agricultural Sciences Center (SJVSC), Parlier, California, USA

Allan T. Showler
USDA-ARS, Kika de la Garza Subtropical Agricultural Research Center, Weslaco, Texas, U.S.A.

C. N. Kurugundla
Water Affairs, Private Bag 002, Maun

P. M. Kgori
Animal Health, Box 14, Maun

N. Moleele
Biokavango Project, Okavango Research Institute, Maun, Botswana

K. R. Dabiré,A. Diabaté, M. Namountougou and J-B. Ouédraogo
IRSS/Centre Muraz, BP 390 Bobo-Dioulasso 3Liverpool School of Tropical Medicine, Burkina Faso

L. Djogbenou
2IRSP/ Ouidah, Bénin

C. Wondji
Liverpool School of Tropical Medicine, UK

F. Chandre
LIN/Montpellier UMR MIGEVEC, France

F. Simard
IRD/IRSS, Bobo-Dioulasso UMR MIGEVEC, Burkina Faso

T. Martin
CIRAD, UR HORTSYS, Montpellier, France

M. Weill
ISEM, CNRS/Université Montpellier 2, Montpellier, France

T. Baldet
IRD/CIRAD/CREC, Cotonou, Bénin

Sonia Soloneski and Marcelo L. Larramendy
Faculty of Natural Sciences and Museum, National University of La Plata, Argentina

Rute Félix and Henrique Silveira
Centro de Malária e Outras Doenças Tropicais, UEI Parasitologia Médica, Instituto de Higiene e Medicina Tropical, Universidade Nova de Lisboa, Portugal